# Lecture Notes in Physics

For information about Vols. 1–151, please contact your bookseller or Springer-Verlag.

# Lecture Notes in Physics

Edited by H. Araki, Kyoto, J. Ehlers, München, K. Hepp, Zürich
R. Kippenhahn, München, H. A. Weidenmüller, Heidelberg
and J. Zittartz, Köln
Managing Editor: W. Beiglböck

## 232

# New Aspects of Galaxy Photometry

Proceedings of the Specialized Meeting of the
Eighth IAU European Regional Astronomy Meeting
Toulouse, September 17–21, 1984

Edited by J.-L. Nieto

Springer-Verlag Berlin Heidelberg GmbH

Jean-Luc Nieto
Observatoire du Pic du Midi et de Toulouse
14, Avenue Edouard Belin, F-31400 Toulouse

ISBN 978-3-540-15657-4       ISBN 978-3-540-39566-9 (eBook)
DOI 10.1007/978-3-540-39566-9

Originally published by Springer-Verlag Berlin Heidelberg New York Tokyo in 1985

## Introductory Remarks

The colloquium "New Aspects of Galaxy Photometry" was one of the three specialized colloquia of the Eighth IAU regional meeting held at Toulouse, September 17-21, 1984. Initially limited to four afternoon sessions, it had to be extended to other sessions, held in the mornings, and to a poster session, because of the large number of communication proposals.

This topic, in a meeting devoted to Extragalactic Astronomy, was decided on since it seemed obvious that several new aspects, related to technical improvements of all kinds, are beginning to completely change our approach to the understanding of the morphology, dynamics, and evolution of galaxies, for which one field remains essential : galaxy photometry. The most evident of these aspects is certainly the recent extensive (not to say exclusive) use of CCD technology and its consequences, namely, fast acquisition of data, high accuracy, extension of the optical wavelength range to the near infrared, new sets of (wide-band and narrow-band) filters, etc... Several other aspects are also very important such as the progress of more classical but still efficient, acquisition techniques : photography or electronography. Both classical and CCD techniques are more and more completed by sophisticated image treatments relying upon either a high degree of mastery of the properties of photographic emulsions or the continual improvement of computer facilities, with color displays, automatic image (coordinated) reduction procedures, and computer-linked microdensitometers. Another important recent insight from computer science in galactic studies is the use of intensive and sophisticated numerical codes in N-body simulations. New aspects of a different nature are the insights from data obtained at other wavelength ranges such as radio, infrared, X-ray, UV, coming from satellites or ground-based technologically advanced equipment. Last, but certainly not least, the launching of the Space Telescope in the near future is not to be neglected in this (non-exhaustive) list.

The afternoon sessions, mainly made up of invited lectures (titles in capital letters in the Table of Contents), followed in general the outline of the above lines : the first two afternoons were devoted to these technological improvements, while the other two were centered on astrophysical applications, namely, dynamics and evolution of galaxies. The morning and poster sessions consisted of more specific results on galaxies, although, for the sake of space, some papers more related to activity of galaxies and cosmology had to be discarded. They were presented in parallel sessions during the general IAU meeting and are not included here. In addition, a summary of this specialized meeting was presented by M. Capaccioli in a plenary session of the general meeting.

These Proceedings follow quite closely the outline of the meeting with a section on each afternoon session, one for the morning sessions and one for the poster session. A panel discussion with A. Bosma, M. Capaccioli, A. Fabian, J.-L. Nieto, F. Schweizer, and R. Terlevich (chairman) ended the meeting. It is not reported here, nor is the review of the posters presented by D. Carter, and one or two papers withdrawn by their authors.

Let me conclude here with my deep thanks to J.-P. Zahn, chairman of the Scientific Organizing Committee of the Eighth IAU regional meeting, who made this specialized meeting possible, as well as to the members of the Scientific Organizing Committee of this specialized meeting for their useful contribution. The perfect organization provided by the Local Organizing Committee, notably J.-C. Augistrou and M. Mauruc is deeply acknowledged. I would like to express also my gratitude to J. Jobard, who typed (or retyped) several papers and helped me very much in the preparation of these Proceedings. My final thanks to G. Paturel for his talented drawings which illustrate this volume here and there.

Jean-Luc NIETO

# TABLE OF CONTENTS

## SCIENTIFIC ORGANIZING COMMITTEE

M. CAPACCIOLI, University of Padova

E. KHACHIKIAN, Byurakan Observatory

P. van der KRUIT, University of Groningen

D. MALIN, Anglo-Australian Observatory

J.-L. NIETO, Observatoire du Pic-du-Midi et de Toulouse (Chairman)

S. OKAMURA, Tokyo Astronomical Observatory

G. de VAUCOULEURS, University of Texas

AFTERNOON SESSION I :


TECHNOLOGICAL
IMPROVEMENTS

NEW TRENDS IN CCD PHOTOMETRY OF GALAXIES

B. Fort

Observatoires du Pic du Midi et de Toulouse

14, avenue Edouard Belin

31400 Toulouse

Abstract : This paper summarizes some possibilities of CCDs for photometry of galaxies. After a quick-look at CCD development, and at some basic information for data reduction, we mention the relevance of CCD/focal reducer techniques.

Introduction

This paper is mainly addressed to the new-comers to CCD observations of galaxies. Its purpose is to give an idea on CCD photometry, restricting the discussion to extended objects. The approach will be centered on the instrument and the data reduction procedure, but we shall try to illustrate this paper with some observational examples. Most of them come from the author's own experience, and therefore cannot cover all the works in this field of interest. However, we shall try to give some important results coming from CCD users, which seem to emerge, at least as viewed from Europe. In this respect, concerning the CCD device itself, the future scientific CCD has been forecast by J. Janesick and collaborators (1984 a,b).

The future CCD

With the continuous and impressive progress in CCD development, mainly in the thinning technique for backside illumination, it will be possible to have a very high quantum efficiency from 1 Å to 10 000 Å.
Figure 1 shows the expected quantum efficiency for a thin backside illuminated TI 3PCCD. The solid line corresponds to results already obtained in laboratory, and the dotted line what is expected very soon (Janesick 1984 a, b). In the UV range, a phosphorus coating can bring the quantum efficiency of frontside thick CCD to about 15 %. With thin backside illuminated silicium native layer and UV flooding technique it can even increase up to 60 % (Hilvack et al. 83). Optical coating can also minimize Fresnel losses and bring the quantum efficiency very close to one. (Janesick et al. 1984b)

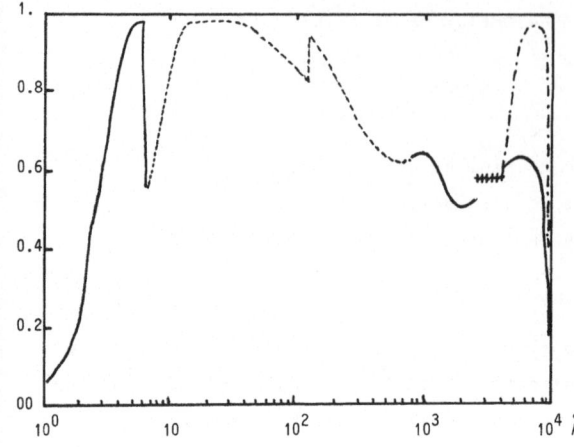

Fig. 1 : Spectral response of a backside illuminated and UV flood CCD. Solid line and crosses correspond to measurements (Janesick et al 1984a, Hilvak et al 1983); the dotted line and the dashed line correspond respectively to the expected values for extreme UV and visible range with special antireflection coating (Janesick et al 1984b).

| Parameter | Manufacturer | | | | |
|---|---|---|---|---|---|
| | TI | TI | RCA | GEC | TH-CSF* |
| Format | 800x800 | 800x800 | 512x320 | 576x385 | 576x384 |
| Technology | 3PCCD | VPCCD | 3PCCD | 3PCCD | 4PCCD |
| Pixel size micron | 15 | 15 | 30 | 22 | 23 |
| Array size mm | 12x12 | 12x12 | 15.4x9.6 | 12.7x8.5 | 13.2x8.8 |
| Full well | 50 000 | 50 000 | 350 000 | 500 000 | 200 000 |
| Spectral range | 1-10000 | 1-100 | 3000-10000 | 4500-10000 | 5000-11000 |
| Charge Transfer inefficiency x $10^{-6}$ | < 30 | > 100 | < 100 | < 100 | < 10 |
| Readout noise | 5 - 9 | 25-150 | 35-80 | 3 - 10 | 20-30 |

Table 1

* from Mellier and al 1985

Finally mosaicked CCDs are now under development in many places, and readout noise as low as 3 elec.x pixel$^{-1}$ will be obtained in slow scan mode. We can deduce that we are now confronted with the genitor of a future outstanding image detector. Even more, the photoelectric performances of the present state of the art seem so good, that we should learn to use it with its ultimate capabilities. Table 1 gives a summary of the photometric characteristics of some CCDs in use for astronomy.

There is no doubt that within a few years new chips coming from these manufacturers or others could be extensively used with better performances. (Philips 600 x 600, Tek 1931 x 1931, GEC 1500 x 1500).

Basics in CCD two-dimensional photometry

It is not possible to give in a few pages a complete review of the procedure to follow in order to reach the ultimate photometric accuracy obtainable with a CCD. The deviation from linearity of a good CCD camera can be lower than $5 \times 10^{-4}$ (Vigroux 1985) (except at extremely low flux levels where bad transfer occurs). This figure should allow an absolute photometry around 0.1%, probably very close to that of photoelectric measurements with a photomultiplier or photodiode (Walker 1984). Let us just see how such a goal can be reached with CCD images.

The simplest way to describe the response curve of a linear CCD is the equation (1) :

$$S_{ij} = \eta_{ij} \, \phi_{*ij} \, \tau + \alpha_{ij} + \beta_{ij} \, \tau + \Delta\gamma_{ij} + \Delta_{ij}(\phi\tau) \tag{1}$$

where :

$S_{ij}$ = signal on the pixel index i, j

$\phi_{*ij}$ = flux of the celestial object

$\eta_{ij}$ = quantum efficiency pattern

$\tau$ = exposure time

$\alpha_{ij}$ = electronic offset (signal level for $\tau = 0$)

$\beta_{ij}$ = dark current per pixel i, j x seconde$^{-1}$

$\Delta\gamma_{ij}$ = readout noise

$\Delta_{(\phi\tau)ij}$ = photoelectron shot noise.

We suppose that the detector is cooled enough so that the dark current is negligible during the exposure time $\tau$.

From the measurement $S_{ij}$, it is possible to derive an approximation to the flux $\phi$, using equation (2)

$$\phi_{ij} \sim \frac{S_{ij} - \alpha_{ij}}{\eta_{ij} \, \tau} . \tag{2}$$

The accuracy should depend only on the readout noise and photoelectron shot noise if the two quantities $\alpha_{ij}$ and $\eta_{ij}$ are perfectly determined.

a) Determination of $\alpha_{ij}$

The offset level $\alpha_{ij}$ is due to electronic bias voltage, and could be determined with a very short exposure in the dark. Figure 2 represents such an offset in the center of a RCA chip (CEA/INAG CCD Camera, Bouerre et al 1981). It displays a relatively flat granular pattern corresponding to the Gaussian readout noise. It might be possible to use the mean value of a large number of offsets in order to derive a good determination of $\alpha_{ij}$ (residual error of readout noise becoming negligible).

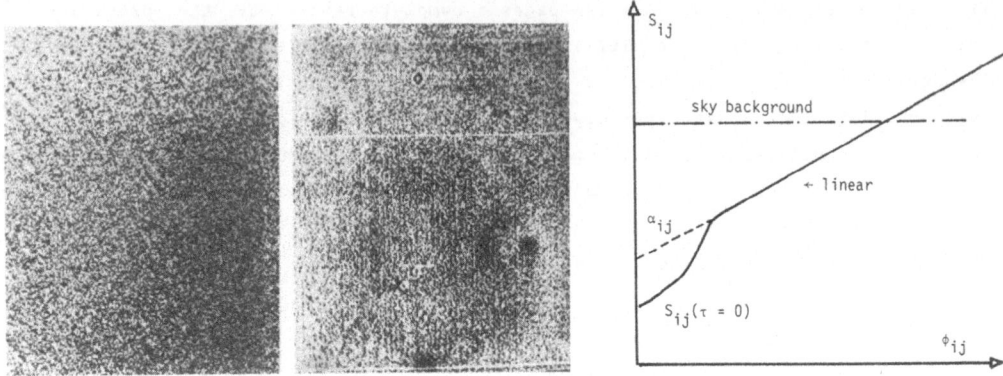

Fig. 2 : $S_{ij}$ ($\tau$=o)          Fig. 3 : $\alpha_{ij}$          Fig. 4 : Response curve Sij

However, for some chips as VPTI, 3PGEC or 3PRCA, a non-linearity with a threshold effect can accur at very low flux levels, so that :

$$S_{ij}(\tau = 0) \neq \alpha_{ij} \, .$$

The true offset $\alpha_{ij}$ can differ from the electronic offset $S_{ij}$ ($\tau = 0$) from several hundreds of electrons and vary from column to column, line to line or pixel to pixel. It is then necessary to calculate the "true offset" corresponding to the linear part of the response curve. This can be done by using a stabilized light as a Tritium source (stability 1 ppm per hour), which projects a relatively flat field on the CCD. The linear response of each pixel on the CCD between $\phi_{ijmin}$ and the full well capacity $\phi_{ijmax}$, is extrapolated for $\tau = 0$ (fig. 4).
In many observations of galaxies, the sky background brings the CCD up to the linear part of the response curve S. It is then possible to calculate a very good "true offset" from the regression curve using measurements between $S_{sky}$ and $S_{max}$ (fig. 5).

b) Quantum efficiency $\eta_{ij}$
The second major parameter to derive in order to obtain good photometry is the quantum efficiency map $\eta_{ij}$. This is obtained with the classical measurement of a

flat-field (FF) through the optical system (telescope + filter). For relatively narrow-band filters (200 Å), the best procedure is to use a CCD image of a clear sky at sunset or sunrise. Another possibility is dome illumination, which, from the only author's experience, does not seem to give stable results.

After removing possible star tracks on FF exposures, a quantum efficiency map $\eta_{ij}$ within $10^{-3}$ accuracy can be derived.

The data reduction procedure described above (a,b) does not generally allow a photometric accuracy better than 0.03 mag which is a limit obtained now by CCD users in a routine way.

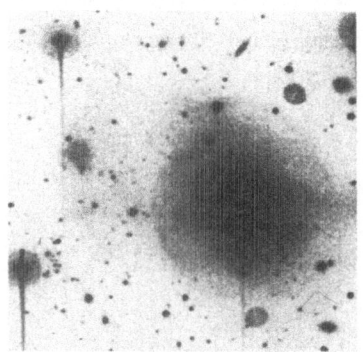

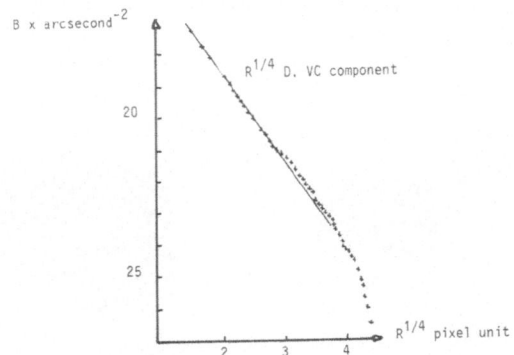

Fig.5 : Faint shells of NGC 2865 Blue Filter 4000 Å - Δλ = 400 Å - τ = 20mn - F/2 CFHT 3.6 m (Fort et al 1984)

Fig.6 : Profile of NGC 3923. Blue Filter 4500 Å - Δλ = 400 Å - τ = 20 mn - F/2 - CFHT 3.6 m (Fort et al 1984)

Such an accuracy around 1% is good enough for many astronomical programs, and can be favorably compared with that of other images detectors. To illustrate this point, we give on figure 5 an image of the elliptical galaxy NGC 2865. It displays several shells which can be directly seen on the direct image, without any image enhancement techniques (Fort et al. 1984). On figure 6, we give also the typical CCD profile of an elliptical galaxy NGC 3923. In this last exemple, the sky background subtraction is very important and can be a problem with the small size of the CCD. We shall not discuss this point, the filter choice, nor the residual fringe pattern coming from the emission lines of the sky. All these aspects are developped in these proceedings (Vigroux 1985). These two results given on figure 5 and 6 illustrate how efficient the CCD can be for the detection of very faint structures ($m_V \leqslant 27$).

## Possibility of better photometry

However, we must go ahead in improving the performances. Therefore, we have to ask ourselves why such a detector with such a good linearity and large dynamics does not commonly allow a photometry better than 1%.

One answer seems to be found in a residual error coming from the FF correction $\eta_{ij}$. In general the FF exposure cannot be obtained with a light having the color of the celestial object. The effective wavelength is therefore different for the FF and the object.

For rather large filters such as B, V, R, the result is generally an additional shot noise around 1% which corresponds to a small quantum efficiency variation from pixel to pixel.

Let us illustrate this effect with an example coming from a Cine-CCD experiment (Fort et al. 1984).

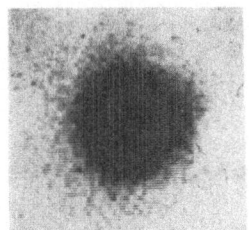 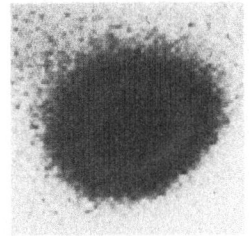 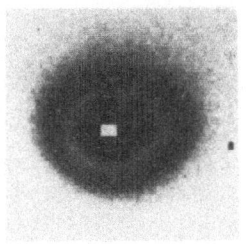

Fig. 7a            Fig. 7b            Fig. 7c

A star was imaged on a small part of a CCD and very short exposures were taken with a different position of the star on the CCD (fig 7a). This random position was obtained by not guiding the telescope. It also comes from the random shift of the star due to mean wavefront atmospheric tilts.

When we add such 100 elementary pictures after FF correction we can see the FF noïse in the wing of the star halo (fig. 7b).

If we use a recentering procedure before adding the 100 CCD images (fig. 7c), we have a residual FF noise pattern less important, due to smoothing of the CCD response on several pixels for the same point of the object.

One of the conclusions in the experiment described above is the possibility of increasing the photometric accuracy above 1% using composite images. There is another and __elegant way__ to generalize this concept with the scanning CCD procedure (Mc Graw 1982, Mackay 1983, Boronson et al 1983 and some others).

The technique consists in letting the galaxy move along the columns of the CCD synchroneously with a vertical transfert clocking. In such a mode, the data for each point of the galaxy have been sampled by 512 different pixels on the chip. Thus pixel to pixel FF variations are greatly reduced.

It is interesting to illustrate that in another way with one galactic shell of NGC 3923. We have supposed that such a faint structure is identical along any section perpendicular to an isodensity curve. If we add all these sections across the shell, we can derive a profile with a significant signal far below the lecture noise of the CCD, down to one coding step or less. In this case it is necessary to account for the non linearity at very low flux levels introduced by the A/D convector which rounds the signal to an integer number (Fort et al. 1984). Compositing all these sections yields a smoothing of the residual FF and readout error and noise.

Fig. 8 : Profile of an outer faint shell of NGC 3923 as compared to the r.m.s. noise (readout + photoelectron + FF noise). RCA/CEA/INAG CCD camera (Bouerre et al 1981) Blue Filter 4500 Å $\Delta\lambda$ = 400 Å - $\tau$ = 30 minutes F/2 CFHT 3.6 m.

An empirical rule is that we do not generally increase the signal to noise ratio with a pixel charge above 20000 electrons if we do not use a scanning CCD. This latter technique makes possible to overcome the accuracy of photoelectric measurements (0.005 mag) (Walker 1984) in the particular case of a crowded field (thanks to the accurate determination of the sky background).

Photometry of galaxies with a focal reducer

New trends for the time being seem to be the development of fast dioptric or cathadioptric focal reducers for galaxy photometry. Such optical devices turn out to be a perfect match for the small sizes of available fields of view and for obtaining good S/B ($\sim$ 100) ratios in a short time (Enard 1984, Fort et al 1984). As an example, we give in fig. 9 a deep image of the interacting galaxies NGC 6240 which clearly displays an outer arm extension.

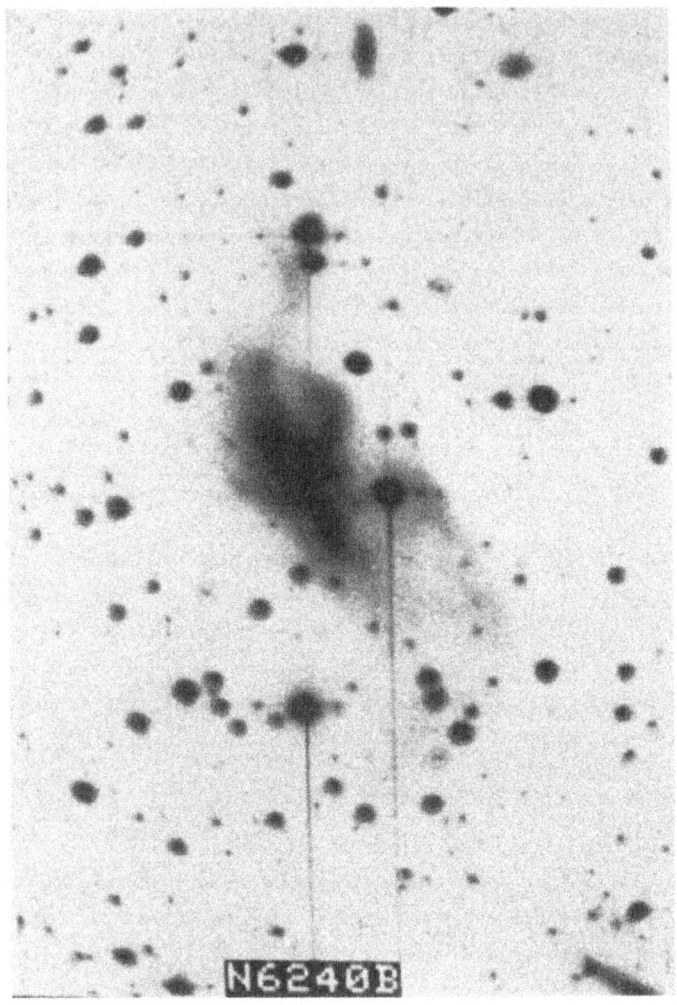

Fig 9 : The interacting galaxies NGC 6240 (see the text).

However, such systems can introduce optical defects like ghost images and scattering halos coming from brighter objects in the field (optical coating effect).

They can also be very powerful tools for multiobject spectroscopy. For instance, focal reducers can accomodate metallic masks which are punched out in real time by the computer driving the CCD camera (Fort et al 1984). This kind of device will probably be intensively developped on large telescopes for stellar spectroscopy in galaxies, or spectroscopy of clusters (Butcher 1982, Dressler and Gunn 1983, Fort et al 1984).

## Conclusion

As very well demonstrated by numerous papers, CCDs are outstanding image detectors that we have not yet learned to use at its best of its possibilities. They will probably allow photometric measurements with an accuracy as low as 0.005 mag and overcome the classical photoelectric photometers, thanks to their two-dimensional aspects. The present limitation comes from a rather poor transfer efficiency and a small format. This second point can be soon overcome by the use of mosaicked CCD. It seems that focal reducer techniques should be pushed with a high priority on large telescopes, mainly for their multiaperture spectroscopy capabilities.

## Bibliography

Boronson T, Shectman S. : 1983 Astron. J., Vol 88, p. 1707

Bouerre A., Cretolle, J., Fort B., Jouan R., Gorrisse M., Leconte A., Rio Y., Vigroux L. : 1981, Solid Stage Imagers for astronomy SPIE 290.

Butcher H., : 1982, Instrumentation in astronomy IV SPIE vol. 331.

Dressler A. and Gunn J. : 1983, Ap. J., 270, 7-19.

Enard D. : 1984, VLT, instrumentation and programs, IAU colloquium n° 79, ESO Garching, Ulrich MH and Kjär editors.

Fort B., Lelièvre G., Picat J.P., Rio Y., Vigroux L. : 1984 VLT, instrumentation and programs, IAU colloquium n° 79, ESO Garching Ulrich MH and Kjär editors.

Fort B., Picat J.P., Cailloux M., Mauron N., Dreux M., Fauconnier T., : 1984, A & A 135, 356-360.

Carter D., Fort B., Meatheringham S.J., Prieur J.L., Vigroux L., 1985 surface photometry of shell galaxies. to be submitted.

Henry R., Mac Alpine G., Kirshner R. : 1984 Ap. J. 278, 619-629.

Hilvack R., Henry J.P., Pilcher C., : 1983, Instrumentation in Astronomy V, SPIE vol 445 p. 122.

Janesick J., Elliott T., Marsh H., Collins S., Mc Carthy J., Blooke M., 1984 a.
    The potential of CCDs for UV and X-Ray plasma diagnostics. Review of
    scientific instruments. J.P.L. preprint.

Janesick J., Elliott T., Collins S., Marsh H., Blooke M., Freeman J. 1984 b, The
    futur scientific CCD, SPIE, J.P.L. preprint.

Mc Graw J., Stokman H.S., Angel J.R., : 1982, Instrumentation in Astronomy IV.
    SPIE vol 331, p. 212.

Mackay C.D., : 1982, Instrumentation in Astronomy IV, SPIE, vol. 331, p. 146.

Mellier Y., Dupin J.P., Cailloux M., Fort B., Lours C., Tilloles P., Picat J.P.,
    1985, Photometric evaluation of the Thomson 31133 CCD for astronomical use
    (to be submitted to A & A)

Vigroux L., : 1985,"New aspect of galaxy photometry", Eighth European regional UAI
    Meeting, J.-L. Nieto editor, to be published in Springer Verlag, Lectures in
    Physics series, 1985.

Walker D.R., : 1984, MNRAS vol 209 p. 83.

## Discussion

R. Terlevich : Can the photometric accuracy be improved by using intermediate
band (200 Å) instead of very broad ones and this reducing the colour gradient
change across the filter for Flat Field.

B. Fort : It seems so, but I did not check carefully this point because I
used mainly broad filters except a 210 Å green filter ($B_1$ $B_2$ Z Vigroux system)

S. Djorgovski : For the data dominated by the sky signal (e.g., faint
galaxian envelopes), the best flat fielding is achieved with sky median frames, or
their combinations with projector lamp flat fields.

R. Michard : Why to put many optics in front of your CCD for galaxy
photometry ?

B. Fort : The first idea was to have a fast instrument with a field of 6
arcminutes. I agree that focal reducer techniques seem more interesting for faint
spectroscopy, or Perot-Fabry observations. Direct imaging should definitively take
place on prime focus with the futur large size mosaicked CCDs.

# A NEW FILTER SYSTEM FOR CCD SURFACE PHOTOMETRY

Laurent Vigroux[*]

Service d'Astrophysique CEN Saclay

91191 Gif-sur-Yvette, Cedex, France

[*] Part of this work is based on data obtained (partly) at the European Southern Observatory and (partly) at the Canada France Hawaii Telescope

Classical surface photometry made with photographic plate has been restricted to wide band filter, such as the UBV or the    Gunn system. In contrast the very high sensitivity of CCD allows the use of narrow filters centered on emission or absorption features which are characteristic of the physical state of the observed object.

Emission line intensity maps have already been given for several classes of objects, such as planetary nebulae or ionized gas envelopes around quasars. Moreover CCD's have even also been used as detectors with a tunable Fabry Perot interferometer, which is the ultimate best technique of narrow band photometry.

Absorption line measurements in galaxies have perhaps seemed less attractive, and all the published work of CCD surface photometry has been done in classical broad band systems. In the first part of this paper we shall make a list of absorption features which can be used for surface photometry, and in the second part we shall present some results that we have obtained in early type galaxies using intermediate band photometry.

## I - ABSORPTION FEATURES USEFUL FOR SURFACE PHOTOMETRY

To be used for surface photometry, an absorption feature must be broad and deep. Obviously the deeper it is, the easier it can be measured. It must be broad for 2 reasons. First, to obtain a reasonable accuracy for absolute photometric measurements, the exposure time must not be too long : 1 hour seems to be a maximum. Secondly, to achieve an accuracy of 1%, one needs at least 10000 photoelectrons of night sky alone. In one hour, with a 4 meters class telescope, this requires at least a 5nm wide filter in the red, and a 20nmm wide filter in the blue. Moreover, to derive useful information in galaxy evolution, one must observe a large sample of galaxies in the same system. Even if the galaxies are selected inside the same cluster, the velocity dispersion of cluster galaxies, which is of the order of 1000 km/s, requires that filters must allow for the Doppler shift, and therefore cover a range of 1.5nm in the blue to 3 nm in the red. These conditions lead to features which are at least 10nm broad.

Most of the standard absorption features which are used in stellar narrow band photometry, such as the DDO system, are too narrow to be used for surface photometry, and one is restricted to a very limited number of features.

- In UV, the 390 nm CN band is the most prominent. It is very sensitive to metallicity.

- In the blue, we can use the blue depression at 400nm. This corresponds to the B1 band of the Geneva photometric system. It is a good metallicity indicator for stars of type F2 III to K5 III (Grenon 1978).

- In the green the Mg band (Mg I + Mg H) at 517.5nm is the most prominent band. It was first included in the 10 color system of Faber (1973), and it is widely used as a metallicity indicator for early type galaxies (eg. Terlevich et al 1980).

- The red domain is perhaps less familiar to astronomers. Fig 1 shows a spectrum of the elliptical galaxy NGC 4472 obtained at the 3.6 meter telescope of ESO in Chili with the Boller and Chivens Cassegrain spectrograph and a Reticon. Unfortunately, all the nicest absorption features are due to our own atmosphere.

Figure 1: Red spectra of the elliptical galaxy NGC 4472

The only band which fulfills our criteria is the TiO band at 712nm. Unfortunately this band is located between 2 $H_2O$ absorption features and its use requires some caution. This band is not very sensitive to metallicity but rather to the number of M stars in a galaxy. This band has been included in the Wing system (1971), and its use for an integrated light has been discussed by Mould (1976), and Mould et al (1980).

This absorption strength must be determined with rerefence to a continuum level. Several filters can be used to this effect :

- U filter : a 350 nm filter can be used. This is out of the range of a thick CCD. However it is in the range of observable wavelengths with a thin backside illuminated CCD such as the RCA chip. However, due to the strong variation of the quantum efficiency of these CCD'S with wavelength in this domain, one must use a filter not wider than 50 nm to ensure that the effective wavelength of the filter + CCD system is not too redshifted.

- Blue continuum. The best compromise seems to be the B2 filter of the Geneva system. It is centered at 450 nm and is 50 nm wide. It can be used to determine the blue depression, B1-B2, and, together with the U and CN 390 filters, the Balmer discontinuity and the CN absorption strength.

- Green continuum can be determined with a normal V filter. However the OI 577.4 atmospheric line gives a very high sky level which decreases the accuracy of the photometry.

- The red continuum can be determined with a filter centered beyond $H_\alpha$ . For a thin CCD, one must choose a filter which completely avoids the 630 nm OI atmospheric line that produces a fringed background pattern. This is not the case in the 2 most used R filters, the Gunn and the Mould ones, which have the OI line in their wings. A 680 nm centered filter, 50 nm wide must be prefered. Such a filter is very similar to the Cousin R filter for which a large amount of stellar photometry is available (Bessel 1980).

- For the red continuum, I filter. Two possibilities exist : either a narrow band filter at 750 nm between the TiO band and the $H_2O$ atmospheric band at 760 nm, or a wider filter at 800 nm. The first one may be prefered for low redshift galaxies since it allows a good determination of the TiO band. However, for large redshifts ($\sim$ 3000 km/s) the redder filter must be used. At longer wavelengths the sky brightness dramatically increases. Moreover, the presence of large atmospheric absorption bands, the intensities of which change on a time scale of fifteen minutes (M.Dennefeld, private communication), prevents to obtain any absolute photometry for exposures longer than half an hour.

Observations made with combination of these absorption features and continuum filters can help to determine the stellar content of galaxies, and we present some of our results in the second part of this paper. However, some care must be taken to obtain meaningful results. First some of these filters require exposures as long as 1 hour even with a 4 meter telescope. Classical stellar photometric methods are no longer valid. In particular, it is hopeless to determine accurately the atmospheric absorption. The best compromise we have found between accuracy and a minimum of telescope time for standard observations is to :

1) observe galaxies with low air mass

2) scale each long exposure of a galaxy with a short ($\sim$ 5min) exposure with the fastest filter, usually the R filter.

3) take from time to time standard stars of different spectral types.

During reasonable nights this procedure leads to an absolute photometric accuracy of the order of O.1 magnitude. But the accuracy from point to point in a galaxy is much better, of the order of 0.02 magnitude. This kind of observation is not able for determining the absolute strenghs of absorption features, but it can be used to determine their spatial variation inside a galaxy. Their absolute strength can only be determined by high resolution spectra.

## II - RESULTS ON EARLY TYPE GALAXIES

We have observed a sample of 20 early type galaxies, mainly in the Virgo cluster, to look for color and metallicity gradients. The galaxy types range from EO to Sa, and all galaxies have about the same absolute magnitude, from - 20 to - 18, to avoid any systematic differences with luminosity. These observations were made in may 1983 behind the Cassegrain focal reducer of the Canada France Hawaii Telescope with the CEA-INAG CCD Camera. This camera was set up with a thin backside illuminated RCA CCD. The pixel size was 0.83 arc second. Due to the large chromatism of the focal reducer we restricted ourselves to 3 filters : B2, a Gunn R filter for the continuum, and a Mg band filter (herein after Z filter) 20 nm wide.

A complete description of the observations and of the data reduction can be found elsewhere (Souviron 1984, Vigroux et al 1984). We shall give here only the main results.

From the B2 and R filters we have computed a continuum intensity, V, at the wavelength of the Z filter. We have checked on field stars, on stars in the open cluster NGC 2682, and on stars in the globular cluster M92, that the V-Z index does not present any variation with the spectral type of the stars. However, we have found a difference of 0.1 magnitude between the absolute value of V-Z of the solar metallicity cluster NGC 2682 and the metal poor globular cluster (M92). Inspite of the fact that we do not have a more precise transformation between V-Z and metallicity it seems reasonable to assume that V-Z is a metallicity indicator.

A comparison of our results for NGC 3379 with published data (Vigroux et Nieto 1984) shows that the accuracy of our photometry is 0.02 magnitude down to 1% of the sky brightness. Several checks, including galaxy simulation, have shown that we can derive color gradients with an accuracy of 0.04 magnitude down to 10% of the sky brightness. For each galaxy of our sample we have computed a color and a metallicity gradient, defined as the variation of the B2-R and V-Z indices, respectively, between the core radius and the $\mu_R = 24/\square$" R isophote.

Figure 2 displays the trend of color gradient with galaxy type. In addition to the de Vaucouleur RC2 type, which ranges from -5 for E galaxies to O for Sa, we have introduced other types ranging from -10 for EO to -5 for E5 to take into account the apparent ellipticity of elliptical galaxies. There is a continuous trend of an increase of the color gradient with galaxy type. 2 galaxies depart from the mean correlation, NGC 4636, an E1 galaxy, and NGC6340, a Sa galaxy. However, NGC 6340 is face-on and its disk is highly reddened by dusty spiral arms. The surprising result is that there is also a trend within a given elliptical galaxy with ellipticity.

Since the ellipticity is only apparent, we have looked for a correlation between the color gradient of ellipticals and a more physical parameter. We have found only one correlation, that with the ratio of maximum rotation velocity to central velocity dispersion, $V_m/\sigma$, as given by Davies et al (1983) (figure 3). The latter ratio is related to the amount of rotational energy as compared to random energy in these galaxies. The larger is the rotational energy, the larger is the color gradient. The case of NGC 4636 is particularly informative. This galaxy does not obey the mean correlation of color gradient with type, but, as it is a fast rotator, it follows well the mean correlation of color gradient with Vm/$\sigma$.

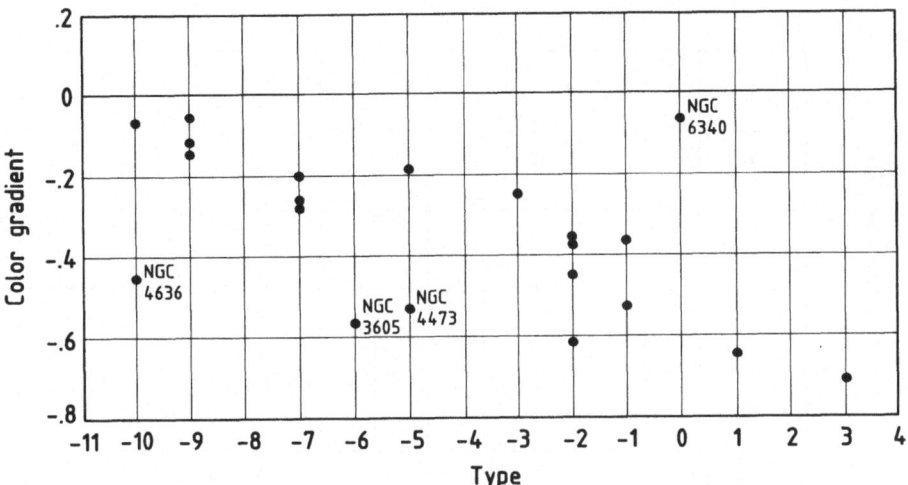

Figure 2 : Color gradient, as defined in the text, in function of the galaxy type.
    - 10 correspond to EO, -5 to E5
    - 4 to - 1 to So and positive number to spiral. The color gradient increase from type -10 to 3.

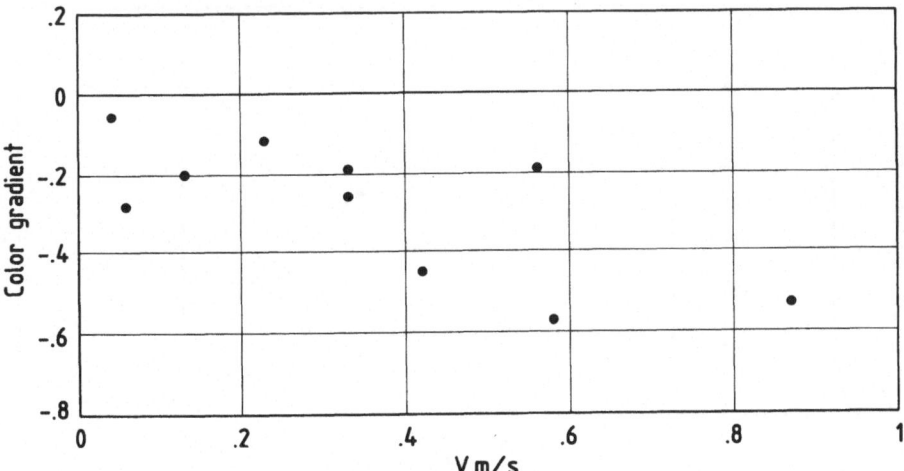

Figure 3 : Color gradient, for elliptical galaxies only, versus their maximum rotation velocity to their central dispersion velocity dispersion ratio

On the contrary the metallicity gradient, which is present in all the observed galaxies, is not correlated with either galaxy type (figure 4) or, for the subsample of ellipticals with the Vm/σ ratio (figure 5).

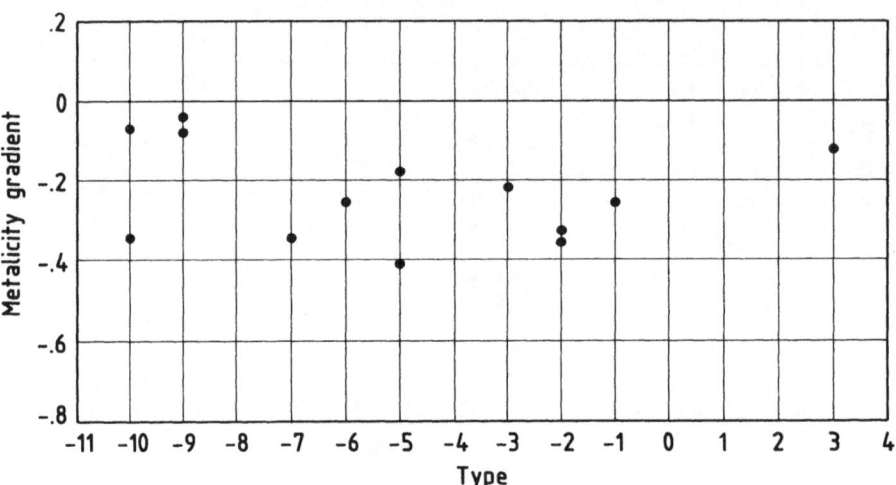

Figure 4 : Metallicity gradient, as defined in the text; versus galaxy types

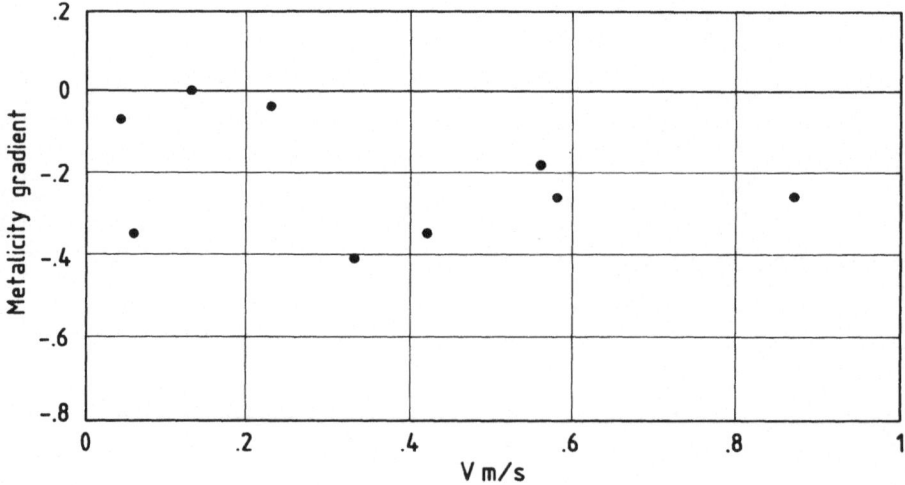

Figure 5 : Metallicity gradient, for elliptical galaxies only, versus their maximum rotation velocity to their central dispersion velocity ratio

Another way to look at these data is to make a color–color diagram for each galaxy. Pixels were selected on the R image according to their R magnitude. The mean B2-R and V-Z indices were obtained for pixels with the same R magnitude and are represented as a point in the B2-R, V-Z plane. Figure 6 displays such a diagram for 3 galaxies, the E3 galaxy NGC 4365 (triangles) the E5 galaxy NGC 4473 (dots) and the edge on Sb galaxy NGC 5908 (crosses). There is a clear shift of the curves from the E to the Sb galaxy. The color gradient increases while the metallicity gradient decreases. Of particular interest is the intermediate position of the E5 galaxy. Most of the SO galaxies in our sample also occupy this intermediate position in their diagram.

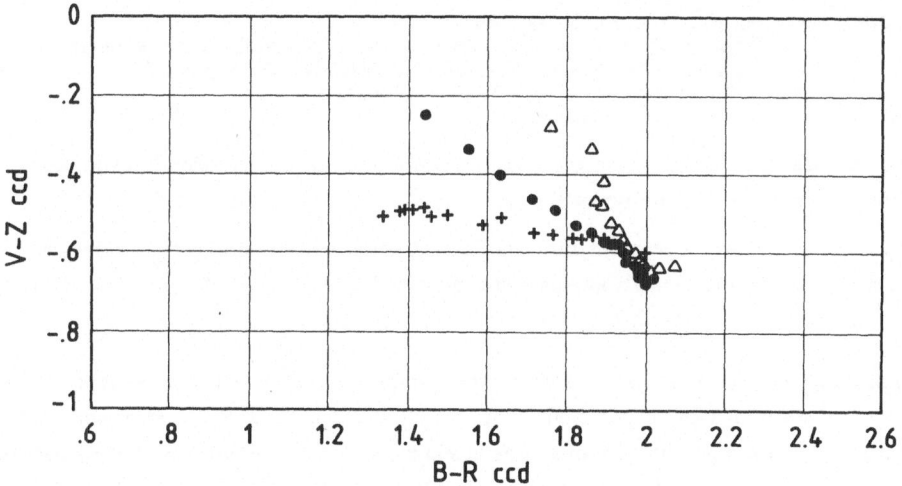

Figure 6 : Metallicity index versus color index for 3 galaxies, NGC 4365 (E3, triangles), NGC 4473 (E5, dots), NGC 5908 (edge on Sb, crosses)

The most straightforward interpretation of these results is that there exists a smooth transition between elliptical and spiral galaxies; the size of the disk component increases along this sequence, which results in an increase of the color gradient. This disk may even be present in elliptical galaxies since we found a correlation between rotation velocity and color gradient. In this picture the rotation velocity appears as the main parameter which governs the transition between galaxy types.

At present this interpretation should be considered as tentative since it is derived from a small sample of galaxies; several observations can strengthen it. In particular one can look for a disk component in rapidly rotating elliptical galaxies by performing very accurate surface photometry. This work is currently in progress by our group.

This discussion is illustrative of what can be done at present with CCD detectors. Because of their high sensitivity and their linearity, they allow to perform very accurate photometry with narrow or intermediate band filters, which yields a better knowledge of the physical parameters of a galaxy. This kind of work has only began 2 or 3 years ago, but it must certainly be considered as a revival of surface photometry.

## REFERENCES

Bessel, M.S., 1979, in problems of calibration of multicolors photometric systems, Dudley observatory report n° 14, A.G.Davis Philip Editor.
Davies, R.L., Efstathiou, G., Fall, S.M., Illingworth, G., and Schechter, P.L., 1983, Ap.J. 266, 41.
Faber, S.M., 1973, Ap.J., 179, 731.
Mould, J.R., 1978, Ap.J. 220, 434.
Mould, J.R., Aaronson, M., 1980, Ap.J., 240, 464.
Souviron, J., 1984, Thèse de 3ème cycle, Université de Paris VII (available on request).
Terlevich et al., Davies, R.L., Faber,S.M., Burstein,D.,1981,MNRAS,196,381.
Vigroux, L., Souviron, J., Lachieze-Rey, M., 1984, in preparation
Vigroux, L., Nieto J.L., 1984, this conference

R. Terlevich: Can you comment on the possible contamination by emission lines in your filter. especially for the latest Hubble types ?

L. Vigroux : We may have a problem with the Gunn R filter which contains $H_\alpha$ . Indeed we found that most of the nuclei of our E galaxies are quite red. This may be due to some ionized gas emission rather than to an intrinsically redder stellar population.

M. Capaccioli : In your talk you alluded to the possibility that E galaxies possess hidden (or barely visible) disks. I wish to point out that in 1976 Barbon, Benacchio and myself made the same suggestion on the ground of similarities in some geometrical properties of the isophotes of E and SO's. Our suggestion got almost ignored; I wish that yours will have better luck.

L. Vigroux : The crucial test for this hypothesis will be a precise surface photometry of the fastest rotator of genuine elliptical galaxies in which, if our hypothesis is correct we may found quite strong disk.

G. Bruzual : How did you test for a correlation between the color gradients that you find in these elliptical galaxies and their absolute luminosity ?

L. Vigroux : We have taken the opposite way. We have selected galaxies in a small range of absolute magnitude, $M_B$ = - 19    -20 to avoid the possible correlations with absolute luminosity.

# ELECTRONOGRAPHY : PERFORMANCES, RESULTS, EXPECTATIONS

J.-P. PICAT

Observatoire du Pic du Midi et de Toulouse

LA 285

31400 Toulouse, France

## Abstract

Properties related to the spatial continuity of electronographic data are shown. Performances are compared to those of sampling detectors such as .CCDs. Applications are proposed where, resolution, large field and low noise are in favour of electronography.

## I - Introduction

This session was devoted to the "New aspects of Galaxy photometry" and besides scientific results, one major conclusion was that no instrument was available to reach a reasonable accuracy in two dimensional photometry before the coming of CCDs. So what can we say about electronography which has not been recognized for at least two decades whereas it was the only way of doing two dimensional photometry with an accuracy comparable to CCDs ?

One can ask whether electronography is an overpassed technique or there is anything which can be done in a better way than with new detectors. When looking at the literature, it seems that some people are still working with electronography and even are still developping new instrumentation (Blecha 1984 ; Scarrott et al. 1983 ; Bely et al 1982).

Several models are in use all around the world based on electrostatic (Lelièvre 1983, Baudrand et al 1982, Kron 1976) and magnetic (Griboval and Jia 1984, Wlérick et al 1982, Mac Mullan et al 1979, Carruthers 1979) focussings. Electrostatic cameras are easier to handle, have lower background noise, allow an electronic magnification or demagnification but suffer from the size of the field which is limited by geometrical aberrations at about 40 mm. On the other hand, magnetic focussing allows very high resolution and low distorsion to be achieved in large fields and a 20 cm field camera could be built without major technical problems.

The French school (Lallemand school) differs from others by the conjunction of two ideas which keep the background noise as low as possible :

- The nuclear plates and the photocathode must be in the same vacuum.

- The photocathodes must be prepared outside the camera and then transferred into the camera in a very clean vacuum.

To allow the recovery of the nuclear plates without destroying the photocathode, a tight valve has been placed between the optics chamber and the plate holder by the Felenbok group in Meudon (Baudrand et al 1982).

It is not easy to compare different detectors since they all have their own application domain. As an illustration, photocounting techniques give the lowest read out noise among all detectors but are valid only for very particular applications. But even if CCD will give comparable results in a near future with the advantage of a much higher quantum efficiency, photocounting techniques will hold in the field of Speckle interferometry or Perot-Fabry experiments.

Electronographic and CCD cameras can be thought of having the same kind of applications and their comparison seems to be more meaningful. Some properties of both techniques are listed in the table below in case of ground observations and commonly available cameras.

II - Ⅰ Ⅰ Ⅰ what could be the advantages of using electronography ?

We will compare in more details two parameters commonly used for qualifying detectors : the DQE and the detectivity.

1) Detective quantum efficiency

Generally the DQE is given for a uniform field but in fact depends very much on the modulation transfert function T(u) as shown by relation (1)

(1) $$\delta(u) = \delta'(o)\, \eta\, T^2(u)$$

where u is the spatial frequency, $\eta$ is the quantum efficiency, $\delta'(o)\,\eta$ the D.Q.E for a uniform field.

This relation clearly shows that results on $\delta(o)$ which are dominated by the quantum efficiency can be completely reversed when looking at high spatial frequencies and that large zero frequency D.Q.E can be drastically reduced by a bad MTF.

At high spatial frequencies, in spite of a lower quantum efficiency, the high resolution of electronographic cameras can preserve a higher D.Q.E than CCDs when using small focal lengths. In addition this allows observations in much larger fields.

2) Detectivity

The ability of a detector for detecting a faint signal against a background can be expressed as the limiting magnitude $m_l$.

An expression is given in (2)

(2) $m_l = C + m_c - 2.5\ \log k - 2.5\ \log \alpha + 2.5\ \log F - 2.5\ \log F(u) + 2.5\ \log (\frac{S}{B})_p$

|  | C.C.D. | E.C. |
|---|---|---|
| quantum efficiency | 40 % - 60 % | 15% |
| wavelength range | 3500 Å - 1 μ | 3000 - 8000 Å |
| pixel size | 15 - 30 μ | few μ continuous medium |
| resolution | 35 - 70 μ | 15 μ |
| number of resolved elements | < 2 $10^5$ | > 2.5 $10^7$ |
| background noise | 10 - 70 e r.m.s | 5 - 7 e r.m.s |
| dynamic range | 1-30 000 ADU 11 magnitudes | 0.002 - 5.5 D 9 magnitudes |
| brightest star obser- vable with sky at S/B ∾ 150 | < 18 - 19 saturation defects | < 20 - 21 no saturation defects |
| response uniformity | pixel to pixel variations | smooth variation on all the photocathode |
| number of exposures | illimited | ∾ 20 for no window models |
| availability of data | on time | delayed a few hours for no window models |
| photometric accuracy | could be better than 1 % | ∾ 1 % |
| photometric stability | 1% | 1% |

where C is a constant, $m_c$ is the sky magnitude, k is a confidence coefficient, α is a measure of the seeing in arcsec, F is the focal length of the telescope, F(u) is a parameter depending on the object profile, $(S/B)_p$ is the best signal to noise ratio reachable on a pixel, u is the size of the integration spot expressed in seeing disk units.

Everything being equal, expression (2) shows that $m_1$ depends on the detector by u (and F(u)) and $(S/B)_p$.
The last term is expressed in relation (3) :

$$(3)\ \left(\frac{B}{S}\right)_p = \sqrt{\frac{1}{N_p} + \beta^2 + \frac{B_L^2 + B_F^2\ t}{N_p^2}}$$

where

$B_L$ is the read out noise, $B_F$ is the background noise, $\beta N_p$ is the flat field correction equivalent noise, $N_p$ is the number of electrons recorded per pixel, t is the exposure time.

This relation shows that at low signal to noise ratio (SNR) - which means $N_p$ small - the read out and background noises are dominant whereas for large values of $N_p$, high SNR are limited by flat field corrections ($\beta$) at levels which are very comparable for electronography and CCDs cameras. This is well confirmed by the photometric accuracy reported in the literature : Sol et al (1984) report $m_v$ = 24.9 ± 0.4 for one hour exposure with the valve electronographic camera at the prime focus of the CFHT at Mauna Kea and Mould et al (1983) report $m_v$ = 25 ± 0.4 for 40 minutes exposure with a CCD TI 800 x 800 at the prime focus of the Mayall Telescope at Kitt Peak.

Using typical values, it is possible to show that up to SNR ∼ 15 the lower quantum efficiency of electronography is overpassed by its lower read out noise and that, compared to CCDs, it must be preferably used in this range.

The first term F(u) is related to the optimization of the integration spot to the object profile. This is always possible with electronography since nuclear plates can be continuously explored. With CCDs this is not always the case and in particular on focal reducer observations where the scale is generally about 1 arsec/pixel. Therefore because of sampling effects, F (u) can vary in the field and the detectivity is no more optimized and uniform.

III - Conclusion

This discussion shows that photometric performances of electronography and CCD cameras are very comparable, each technique having its particular application.

At medium and high SNR, when used in small fields, with an appropriate scale (2 to 3 pixels per arsec), CCDs are much better because of their higher quantum efficiency. The detectivity of both instruments are very comparable but that of CCDs could be improved by using scanning techniques (Boroson et al. 1983) which allow a better flat field correction.

At low SNR (< 15) electronography is better at least until very low read out noise CCDs (a few electrons per pixel) are available in Europe.

In any case, in large fields, the situation favors electronography as the quantity - number of resolved elements x quantum efficiency - is much larger for electronography between 3000 and 8000 Å.

Electronography is particularly well fitted and is used for photometry of galaxies, globular clusters, HII regions, nebulae, nebulosities around quasars because of its large field, high resolution, low noise and no saturation defects.

As a conclusion, where electronographic cameras exist they must be preferentially used against other techniques for observations simultaneously requiring the properties listed above. This has been the choice made at the CFH Telescope in Hawaii (Lelièvre 1983).

As an illustration of high resolution in a large field, figure 1 shows a part of a picture of M33 in the B band obtained with the large field camera of Paris Observatory at the Cassegrain focus of the CFHT (137 μ/arsec ; 9 arcmin field)

As an illustration of the non-pollution of the pictures by saturated objects, figure 2 shows a picture of the nebulosity around R Aquarii in the U band obtained with the valve camera of Paris Observatory at the Prime focus of the CFHT (72 μ/arsec ; 7 arcmin field).

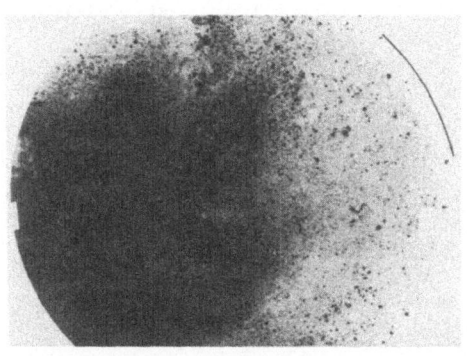 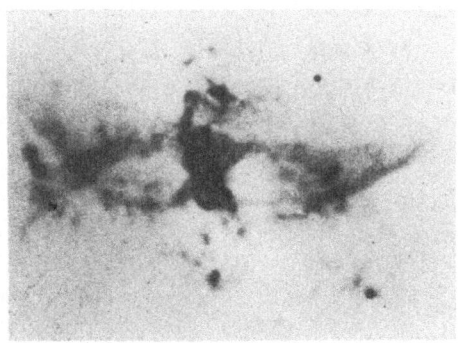

|  fig. 1  |  fig. 2  |
|---|---|

M33 ⊢———⊣ 1arc min     R.Aq. ⊢————————⊣ 1arc min
Courtesy of G. Wlerick

In the future, electronographic cameras could be improved (better photocathodes especially in U and B, larger size, better homogeneity etc...) but it appears that CCDs are much more fashionable and are expected to improve very fast (larger field, better quantum efficiency, lower noise) (Janesick et al. 1984)

References

Baudrand,J., Chevillot,A., Dupin,J.P., Guerin,J., Bellenger,R., Felenbok,P., Picat,J.P., Vanderriest,C. : 1982, J. Optics, 13, 295

Bely, P.Y., Lelièvre, G. : 1982 in Instrumentation for astronomy with large optical telescopes D. Reidel, Ed. Colin M. Humphries, p. 21

Blecha,A. : 1984, Astron. Astrophys. 135, 401.

Boroson,T.A., Thompson,I.B., Schectman,S.A., 1983, Astrophys. J. 88, 1707

Carruthers,G.R. : 1979 in Advances in electronics and electron physics, Academic Press, p. 283

Griboval,P.J. and Jia,X.Z. : 1984 in Astronomy with Schmidt-Type Telescopes, Reidel, Capaccioli Ed. p. 173.

Janesick, J.R., Elliot,T., Collins,S., Marsh,H., Blouke, M., Freeman,J. ; 1984 preprint

Kron,G.E. : 1976, in UAI Colloquium n° 40, Paris p. 2-1

Lelièvre,G. : 1983 in Proceedings of SPIE "Instrumentation in Astronomy", London, p. 151

Mc Mullan,D., Powell,J.R. : 1979, in Advances in electronics and electron physics, Academic Press p. 315

Mould,J.R., Kristian,J., Da Costa, G.S. : 1983, Astrophys. J. 270, 471

Scarrott,S.M., Warren-Smith,R.F., Pallister,W.S., Axon,D.J., Bingham,R.G. : 1983, Mont. Not. R. astr. Soc. 204, 1163

Sol,H., Vanderriest,C., Lelièvre,G., Pedersen,H., Schneider,J. : 1984, Astron. Astrophys., 132, 105

Wlérick,G., Duchesne,M., Servan,B., Gex,F., Munier,J.M. : 1982, in Instrumentation for astronomy with large optical telescopes, R. Reidel, Ed. Colin M. Humphries, p. 291

## Discussion

R. Michard : Could someone comment upon the problem of long term storage of informations taken with digital devices, as compared to photography ?

J.-P. Picat : It is a real problem which will be solved in a near futur by access to optical disks.

D. Axon : Can you comment on the limitation imposed by the emulsion ?

J.-P. Picat : Studies on L4 emulsions have shown that taking into account the microdensitometer it was possible to achieve S/B ∼ 100 with spots of about 30x30 $\mu^2$ for an optimum density 1.5. Generally Kodak emulsions are more uniform than Ilford one.

D. Axon : How linear are the new Kodak emulsions ?

J.-P. Picat : They are comparable to Ilford emulsions of the same speed.

FOOD FOR THE PHOTOMETRISTS - FAINT GALAXIES REVEALED

D.F. Malin
Anglo-Australian Observatory
Epping.  N.S.W.   2121
Australia

The photographic plate was for many years the only useful detector in astronomical photometry and the unique properties of the photographic process ensure that the fruitful union of photography and astronomy will continue well into the electronic age. There is no doubt however that the role of photography is changing and as two-dimensional arrays improve in size and sensitivity the unique properties of the photographic process, which cannot be duplicated by electronic detectors, assume a greater importance.

It is interesting to note in this connection the phenomenal storage capacity of photographic plates and their extreme uniformity over large areas, two apparently unconnected properties, are the very properties which not only justify their continued use but are the most difficult to duplicate electronically.  One has only to use a modern CCD detector at the prime focus of a large telescope to appreciate firstly how much information there is, even in a ˜300 mm$^2$ section of the image plane and secondly how careful one must be to ensure an adequate and appropriate stock of bias frames and flat fields to obtain photometrically or even visually satisfactory results.  Neither of these are significant problems with photographic plates, which, while less sensitive by perhaps an order of magnitude are uniform over areas which are hundreds of times greater that even the largest planned CCD.  (See Cannon, 1984 for a more rigorous comparison).

In this note I intend to emphasize the advantages of the photographic plate in exploring the large-scale structure of objects, especially the fainter features and low surface brightness companions of galaxies and galaxy clusters.

The principal technique for this is photographic amplification (Malin, 1978), a simple, contact copy process which is ideally suited to revealing faint objects, especially on fine grain IIIa-type emulsions.  It has been shown recently that this process can reveal, on a single plate extended objects which are 5.5 magnitudes fainter than the night sky (Malin 1984) and work is now in progress to combine many plates to extend this to fainter limits.  The technique compares well with digital image processing (Couch et al 1984) and is much faster;  a full 356 x 356 mm plate can be copied and examined in less than 60 minutes, compared with many hours for digital scanning.  The techique is for the present entirely qualitative but in principle the results could be calibrated to produce quantitative photometric data.  A fuller discussion, illustrated with many examples has recently appeared (Malin 1984).

Additional examples are described here to illustrate the power of the photographic technique and its ability to extract new information from plates taken for other purposes and to provide new and challenging targets for galaxy photometrists.

NGC 1407 is the dominant E0 galaxy in the extended southern cluster which straddles the Eridanus-Fornax border. Both NGC 1407 and its E1 companion NGC 1400 (galaxy types from de Vaucouleurs et al, 1976-RC2) have extremely smooth radial luminosity profiles with no evidence of tidal interaction between the two. The most remarkable feature of the field however is the large number of low surface brightness galaxies of the kind described by Sandage and Binggeli (1984). These galaxies, arrowed in Figure 1a, are either invisible on the original IIIaJ plate or are just seen as faint, usually spherical blobs without any central condensation. In the few cases where there is a nucleus it appears stellar. These features and their large angular size readily distinguish this class of galaxy from background objects. The distribution of these galaxies is clearly centered on the cluster but without any strong central condensation. The print of Figure 1a was made by superimposing high contrast derivatives from three sky-limited IIIaJ plates taken at the 1.2m UK Schmidt Telescope and the faintest examples shown here have a central surface brightness of about 27.5 mag arc sec$^{-2}$ in the IIIaJ passband (395-~540nm). Figure 1b represents the visual appearance of one of the IIIaJ plates as seen on the usual kind of light table.

It should be emphasized that these galaxies are not rare but the large numbers reported in this example can only be found by the special photographic techniques referred to earlier. The examples reported by Sandage and Binggeli in the Virgo cluster are only the brightest members of a considerable population associated with that cluster.

IC 1459 is the largest angular size of a loose group of galaxies whose most conspicuous member is NGC 7418. The field illustrated, which covers the same angular extent (about 35 x 57 arc min) as Figure 1, also contains several of the distinctive low surface-brightness galaxies which were seen in the earlier illustration; though their number and density is lower in Figure 2, they are mostly of greater angular size. IC 1459 itself is the dominant galaxy in Figure 2 which was made from combining the photographically amplified images from two sky-limited IIIaJ plates from the UK Schmidt Telescope. Although distinct traces of outlying spiral structure are evident on the photograph, this galaxy is classified as E3 in RC2. Apparently associated with IC 1459, and about 7 arc min to the SW is IC 5264, an edge-on disc galaxy which is seen to be crossed by a dust lane typical of a spiral on shorter exposure plates. The evidence of association is slender but compelling; from the western extremity of the edge-on spiral emerges a long thin tail of luminous material whose center of curvature is similar to that of the diffuse spiral arms of IC 1459.

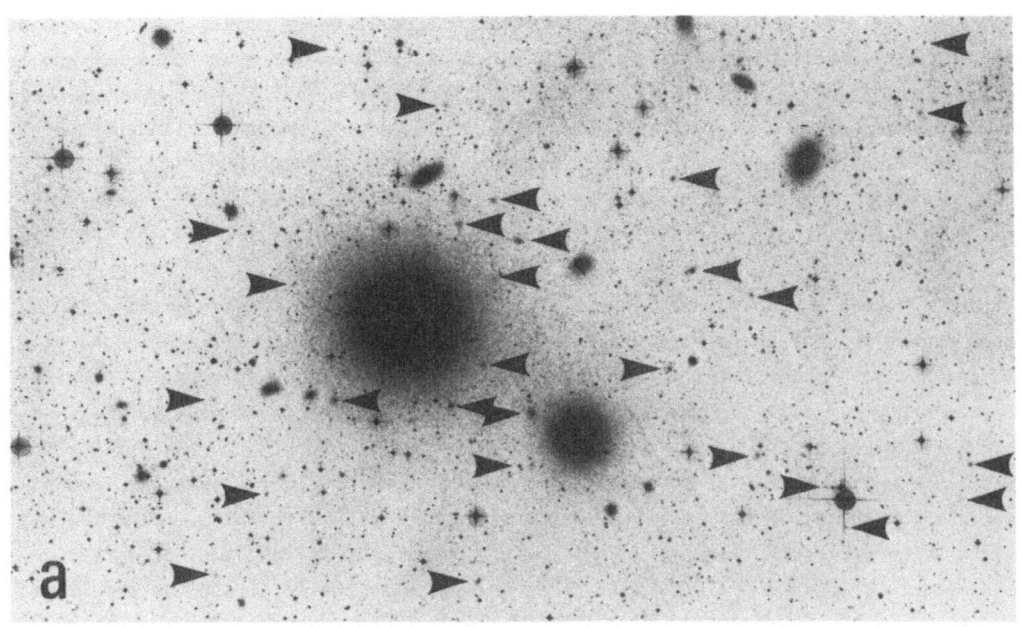

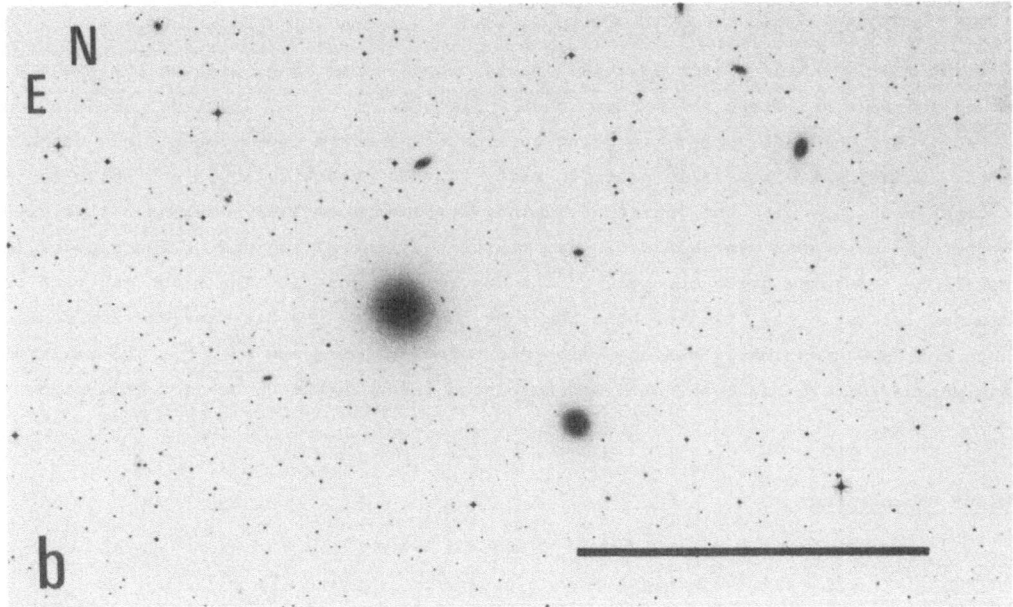

## Figure 1, the field of NGC 1407

Figure 1: The arrows indicate many large, low surface brightness galaxies apparently associated with the NGC 1407 cluster in Eridanus (a, top). These galaxies are mostly invisible on an original deep UKST IIIaJ plate (b, below) and have been revealed here by combining the photographically amplified derivatives from three such plates. Scale bar – 20 arc mins.

About 20 arc min south of these two is NGC 7418A, a large, faint highly disturbed spiral galaxy, almost invisible on the original plate. However there does not appear to be any other object near enough to have caused the evident disruption. The extremely clumpy distribution of stars in the this galaxy, where star formation is apparently vigorously underway contrasts strongly with the smoother, diffuse spiral arms of IC 1459 and suggests that the structures in this latter may be due to interaction with its nearby disc companion.

NGC 4696 is the dominant galaxy of the Centaurus cluster, partly visible in the upper right (NW) corner of Figure 3. This print was made by combining photographically amplified derivatives of three deep IIIaJ plates made by Malcolm G. Smith at the prime focus of the CTIO 4m telescope. As in the previous illustrations, many low surface brightness galaxies are revealed by the photographic technique and on deeper versions of this image the envelopes of NGC 4696 and NGC 4709 appear to overlap. Apart from the smooth large-scale non-uniformities in the field from the outer envelopes of the major galaxies there is some evidence here of patchy nebulosity at the limit of detection of this print. Since these plates were taken in 1975 near the solar minimum we can assume that the night sky was very dark, probably 23 mag arc sec$^{-2}$ which implies that the faintest features seen in this photography surface brightness of 28 mag arc sec $^{-2}$ or even fainter.

The most striking object in this 3-plate composite is long, diffuse but lightly curved jet-like structure which runs roughly diagonally, SE-NW, about 3.3 arc min south of NGC 4709. This jet appears to point towards (or emanate from) the nucleus of NGC 4696, a galaxy which itself is a strong x-ray source (Mushotzky 1980) and which has a distinct dust lane near the centre of luminosity (Shobbrook 1966, Jorgensen et al 1983). The jet can be traced over almost 10 arc min of the central regions of the cluster, but appears to terminate quite abruptly in the centre of Figure 3. The brightest part of this jet is distinctly brighter than the most luminous of the diffuse features mentioned above and has an estimated surface brightness of 27 (B) mag arc sec $^{-2}$. The reality of this unique feature has been confirmed on plates taken on the UK Schmidt Telescope.

Colour Photography

The superimposition technique which was used to combine plates of the same passband to reveal the intriguing objects described above can also be used to combine positive derivatives of plates of different passbands. If the positive from a B plate is combined using blue light with a V band positive in green light and an R positive in red light and these images are recorded in a positive-working colour material, full colour images are obtained. These pictures reveal in one photograph the subtle colour

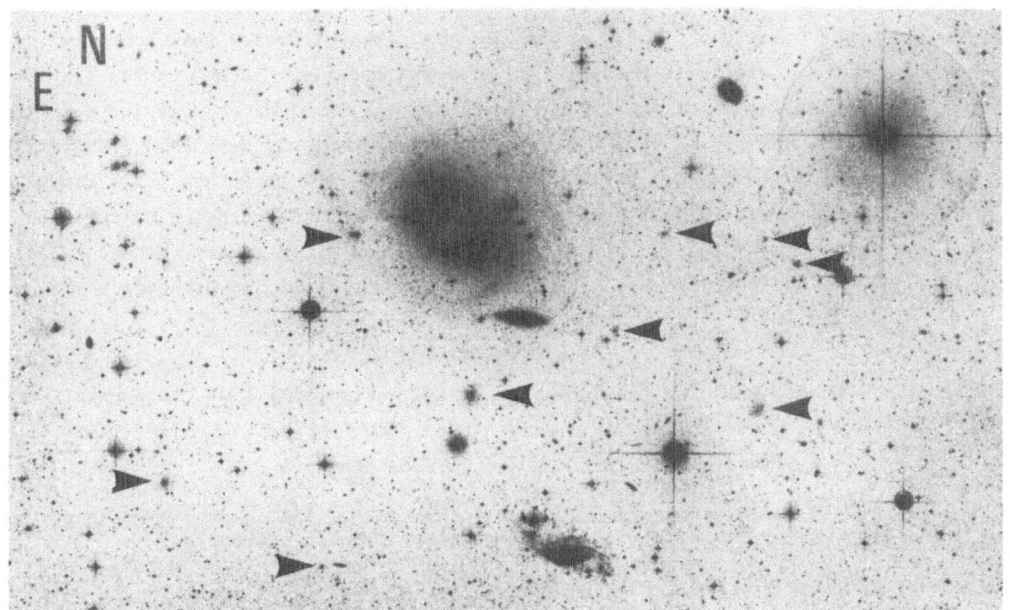

Figure 2, the field of IC 1459
Figure 2:  Low surface brightness galaxies in the field of IC 1459 (upper centre).  This
E3 galaxy shows very faint but extensive spiral structure in this photographically
amplified print.

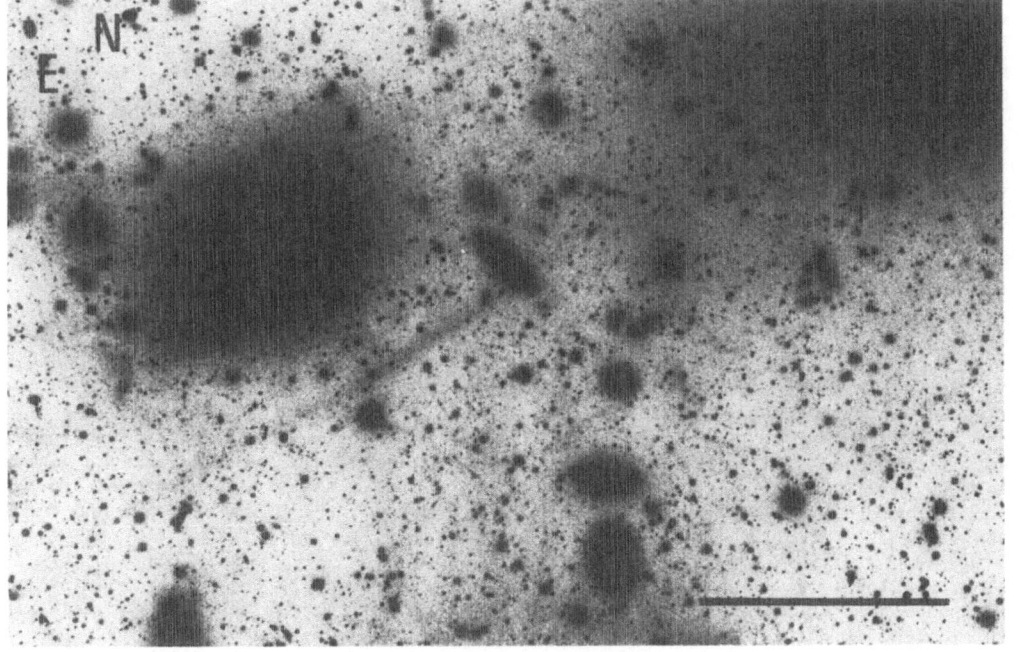

Figure 3, the field of NGC 4696
Figure 3:  A jet-like structure in Centaurus cluster of galaxies revealed by
photographic amplification on plates taken on the CTIO 4m telescope.

gradients which distinguish Population I and II stars in a galaxy or the effect of dust on the colour of starlight. These colour phenomena are of course well-known to astronomers but are difficult to reproduce photographically. Such pictures are a surprisingly useful means of gaining new insights into familiar objects for both the professional astronomers and layman alike and have been widely published. Regrettably it was not possible to reproduce colour pictures in these Proceedings but a recent book (Malin and Murdin, 1984) contains many examples and a full description of the techniques used to produce them. A colour slide showing the Local Group Galaxy, NGC 6822, well resolved into stars of dintinctly different colours was shown and discussed.

### References

Cannon, R.D., 1984. Proc IAU meeting Astronomical Photography 1984, Edinburgh. Sim and Ishida, Eds.

Couch, W.J., et al, 1984. Mon. Not. R astr. Soc. 209 301

de Vaucouleurs, G., de Vaucouleurs, A., and Corwin, H.G., 1976. Second Reference Catalogue of Bright Galaxies Univ. of Texas Press. Austin (RC-2)

Jorgensen, H.E., et al, 1983. Astron Astrophys 122 301

Malin, D.F., 1978. Nature 276 591

Malin, D.F., and Murdin, P.G., 1984. The Colours of the Stars Cambridge University Press, Cambridge UK.

Malin, D.F., 1984 (a), in Astronomy with Schmidt Type Telescopes, IAU Colloquium No78. M. Capaccioli, Ed. Page 57

Mushotzky, R., 1980. X-ray Astronomy p.171. NATO Advanced Study Inst. Series, Giacconi and Setti, Eds.

Sandage, A., Binggeli, B. 1984. Astron J 89 919.

Shobbrook, R.R., 1966. Mon. Not. R. Astr. Soc. 131 351.

### Discussion

F. Schweizer : In your colour combination work, how do you balance the different characteristic curves of different emulsions, or do you use one emulsion with three colour filters?

D. Malin : I use standard B, V and R plate/filter combinations, ie., IIaO/GG385, IIaD/GG495, 098-04/RG610 or 630; these are copied to produce positives and the characteristic curves normalised at that stage, which can also incorporate photographic amplification and unsharp masking techniques. Fuller details can be found in Vistas in Astron. 24 219 (1980) and a complete description of all aspects of this work can be found in the Technical Appendix of our book, Colours of the Stars by Malin and Murdin, CUP, 1984.

R. Terlevich : I would like to point that the resolution in the Local Group Galaxies is independent of the type of detector and depends only on the seeing. The main problem is crowding of the field when trying to reach the main sequence stars.

D. Malin : Of course, but colour differentiation as seen here acts as an extra discriminant which is especially useful in very crowded fields.

COMPUTER-LINKED MICRODENSITOMETERS:  PERFORMANCE, RESULTS

AND EXPECTATIONS

H. T. MacGillivray

Royal Observatory, Blackford Hill, Edinburgh,

EH9 3HJ, Scotland

Summary

The development of fast, computer-controlled microdensitometers has provided a whole new dimension in the high-speed extraction of data from astronomical photographs, enabling the potential of these photographs to be fully exploited by digital means for the first time.  Several such systems are either already in operation throughout the world or will become so in the next few years.

In this paper we summarise the main features of these machines, describe their advantages and disadvantages and discuss their suitability for galaxy photometry. Finally, we describe the main areas where these machines can be of benefit to astronomers, some results which have already been obtained and expectations for the future.

1.    Introduction

The advent of high-speed plate scanning machines represents, in my opinion, one of the most important recent advances in the field of galaxy photometry.  The significance of these machines lies in their ability to produce large quantities of data which are directly amenable to statistical examination.

In this paper, I will not discuss PDS machines although these are generally also computer-linked, and indeed a new generation of higher speed PDS machines is currently under development.  The main discussion herein will be concerned with other types of high-speed microdensitometer which are currently either in operation in different parts of the world or are under development.

But first, some explanation is required as to why these high speed machines are necessary. On any single photographic plate taken with a modern wide-angle Schmidt camera there are ~ $10^9$ pixels (i.e. 1 Giga pixels) of information at 10 micron resolution. Such plates record the images of typically several hundreds of thousands of stars and galaxies down to a limiting magnitude of B ~ 22-23. Each plate represents, in fact, a vast storehouse of information. In order to cover the whole sky in one passband, some 2000 plates are required (~ 2 x $10^{12}$ pixels), and astronomers usually like to work with plates in several passbands (e.g. U, B, V, R and I). Thus, in total there are about ~$10^{13}$ pixels of information for the whole sky down to depths reached by Schmidt telescopes. Added to this must be taken into consideration the significant number of photographs taken with larger-scale telescopes (e.g. of ~ 4m size) and reaching fainter limits. There is, therefore, a vast wealth of information to be extracted, information which the astronomer requires in a form suitable for analysis. The purpose of these machines is to extract that information and make it available in quantised form to astronomers.

We have seen from contributions at this meeting that the use of CCD's is fast producing an alternative information recording medium to photographic plates. However, at present these CCD devices are restricted to exposures of small areas of sky and it will not be for several years yet that they will supergede completely photographic plates. Even so, Capaccioli (these proceedings) has outlined possible dangers in the use of CCD's for galaxy photometry. It is more than likely that high-speed measuring machines will have an important role to play in the field of galaxy photometry for many years to come.

2.  The machines

The main high-speed machines are summarised in the table below. Also indicated is the time taken (or will be taken) for each machine in order to scan a whole Schmidt plate. The figure in brackets is the time that will ultimately be achieved by each machine after all development work is carried out and represents, therefore, the minimum time envisaged.

| Machine | Time taken to scan a Schmidt plate (hours) | |
|---|---|---|
| * COSMOS (UK) | 4 | (3) |
| * APM (UK) | 9 | |
| | | |
| KIDS (Japan) | 16 | (5) |
| ASTROSCAN (Netherlands) | > 20 | |
| * MINNESOTA (USA) | < 10 | |
| USNO (USA) | 8 | |
| | | |
| MAMA (France) | 4-6 | |
| FIRST (ESO) | 8 | (4) |

Machines marked with asterisks are ones which use a flying spot scanning system with single element detection (e.g. photomultipliers), while the others use either line or area illumination with a line or area multi-element detector (CCD or reticon). The machines above the solid line (COSMOS and APM) are at present in production, scanning a whole Schmidt plate in a matter of a few hours. The machines between the solid and broken lines are in production mainly for small areas of plate, while the machines below the broken line are undergoing development and it is hoped that they will be ready in late 1985 or early 1986. At present, the COSMOS machine is scanning at a rate of 3 whole Schmidt plates each day and has the capacity for doubling this figure. Thus, already astronomers have access through these machines to vast quantities of data and much more will become available before long.


3.    Performance of the machines


Although there are many differences between the various machines (mainly in their design characteristics), there are also many similarities, especially in their performance and function. The pixel size used is typically in the range of 7-16 microns, they are (or will be) capable of scanning a Schmidt plate in a few hours and of scanning a few Schmidt plates per day. Some machines do more on-line processing than others, although they all have similar measurement modes (i.e. a 'search' mode for rapid plate scanning and image identification, and a 'fine' mode for more sophisticated types of processing on particular subsets of images), and similar accuracies (i.e. positional accuracy of a few microns, photometric accuracy of 0.1-0.2m for galaxies and major and minor diameters and orientations to a few %). The machines generally produce similar parameters for images and information enabling star/galaxy separation to be carried out.

There are several advantages to the machines. We have already seen that they are very fast. Furthermore, the measurements are carried out accurately, objectively, without effort and with high repeatability and exceedingly large numbers of images can be handled easily. The main disadvantage lies in the fact that they have a limited dynamic range, only being able to reach densities of ~ 2.5. The reason for this lies in the presence of a "halo" surrounding the spot of spot-scanning machines and the presence of scattered light (from the emulsion) in the line or area scanning machines. Certain developments are underway to attempt to minimise these problems. However, the effect is only important when dealing with bright galaxies.

## 4.   High-speed machines and galaxy photometry

What can high-speed machines do in the area of galaxy photometry? For bright galaxies, they are perhaps not ideally suited for the reasons stated above, viz. that with the limited dynamic range they cannot provide any meaningful information in the central regions of bright galaxies. In these situations it is better to use a much slower PDS machine which can reach the high densities required or to use a CCD device on a small telescope.

The main impact of these machines, I believe, is in the photometry of _faint_ galaxies, for which they are well suited and have a photometric accuracy as good as a PDS machine. Because of this, their main use will be in the area of large-scale survey-type work involving large numbers of faint objects on large numbers of plates covering large areas of sky. The very real possibility now presents itself of producing catalogues similar to the Lick galaxy catalogue (Shane and Wirtanen 1967) and the rich cluster catalogue of Abell (1958), but based purely on machine measures. The main advantage of such machine-based catalogues is in their completely objective nature, removing the presence of subjective biases and allowing selection effects to be quantified. These subjective biases have undoubtedly influenced the production of previous catalogues and may be evident in the results obtained therefrom (e.g. MacGillivray and Dodd 1984). Other advantages of a machine-based survey are in the quantitative nature of the catalogues (allowing many parameters to be available for statistical study) and in the accuracy of the parameters recorded.

## 5. Results

First results from machine-based catalogues have been highly encouraging. For example, there is now strong evidence that the two-point correlation function for galaxies shows a break at smaller angular scales, $\sim 3h^{-1}$ Mpc (Shanks et al. 1980), than obtained in the Lick galaxy data, $\sim 9h^{-1}$ Mpc (Groth and Peebles 1977). From analysis of the distribution of very faint galaxies on deep 4m telescope plates (down to B = 24 and R = 22), evidence has been obtained for the possible presence of evolution in the luminosities, colours and clustering of these faint galaxies (see Shanks et al. 1984; Stevenson et al. 1985). If verified from further work then the last result could indicate that clusters of galaxies have grown during relatively recent epochs ($z < 1.0$). However, these are still early days in the analysis of data from high-speed machines, only a small fraction of the sky having been examined in any detail to date. Clearly, continuing these investigations should provide more extremely interesting results and may help with an understanding of the clustering and evolution of galaxies in the Universe. It should be stressed that this type of investigation is only possible in a practical sense through the existence of high speed plate-measuring machines.

## 6. Expectations

What does the future hold for these high-speed machines? Before long large angles of sky will be surveyed to faint limits. A project is already underway with this in mind using COSMOS (e.g. MacGillivray and Dodd 1983, 1984). Such surveys will provide a vast body of information which will be suitable for statistical investigation. Ultimately, this will provide a means for examining (objectively and accurately) the properties and large-scale distribution of galaxies.

In conclusion, high-speed plate scanning machines have an essential part to play in helping with an understanding of the structure and evolution of the Universe on large scales. The future of these machines promises to be extremely fruitful.

## References

Abell, G. O.: 1958, Astrophys. J. Suppl., 3, 211.

Groth, E.J. and Peebles, P.J.E.: 1977, Astrophys. J. Suppl., 217, 385.

MacGillivray, H.T. and Dodd, R.J.: 1983, in "Astronomy with Schmidt-type Telescopes", ed. M. Capaccioli, p.125.

MacGillivray, H.T. and Dodd, R.J.: 1984, in "Clusters and Groups of
   Galaxies", eds. F Mardirossian et al., p.595.

Shane, C.D. and Wirtanen, C.A.: 1967, Pub. Lick Obs. Vol. 22, Part 1.

Shanks, T., Fong, R., Ellis, R.S. and MacGillivray, H.T.: 1980, Mon. Not. R.
   astr. Soc., 192, 209.

Shanks, T., Stevenson, P.R.F., Fong, R. and MacGillivray, H.T.: 1984, Mon.
   Not. R. astr. Soc., 206, 767.

Stevenson, P.R.F., Shanks, T., Fong, R. and MacGillivray, H.T.: 1985, Mon.
   Not. R. astr. Soc., In Press.

DISCUSSION

R Terlevich : If you would have to decide about the future Prime Focus facility on
a 4m telescope, would you decide for a sophisticated photographic camera or an
array of CCDs, bearing in mind your interest in the colours of high redshift
galaxies?

H T MacGillivray : For work on small angles of sky, such as the central regions of
high redshift clusters, I would agree that the use of a CCD would be more
efficient.   However, for work on large angular scales (e.g. studies of the
environment of clusters and the large-scale distribution of galaxies), then at the
present time I think photographic plates are more useful.   For this reason, I do
not think that the ability to do prime focus photography on 4m telescopes should
be excluded.

# RESULTS FROM SURFACE PHOTOMETRY OF EXTENSIVE SAMPLES OF GALAXIES

Sadanori Okamura
Kiso Observatory, Tokyo Astronomical Observatory
Mitake-mura, Kiso-gun, Nagano-ken, 397-01

In spite of increasing use of CCD, wide-angle Schmidt plates will remain important material for years to come to form a photometric data base for large number of galaxies. Huge amount of information contained in a Schmidt plate can be most efficiently extracted using a fast plate scanning machine and a powerful image data processing system. We report the results obtained from a photometric project which we have been carrying out for six years using the 105cm Schmidt telescope at Kiso observatory. These results may demonstrate capabilities of Schmidt plates for extensive photometric survey works. Since details of the project, data reductions and analyses are published elsewhere (Okamura 1984; Watanabe et al. 1982=Paper I, 1984=Paper IV; Watanabe 1983= Paper II; Okamura et al. 1984=Paper III; Kodaira et al. 1983; Ichikawa et al. 1984), only a brief summary is given here.

Two-dimensional V-band luminosity distribution is obtained for 261 galaxies on the basis of the plates taken with Kiso Schmidt telescope. Most of the galaxies are members of either the Virgo Cluster or the Ursa Major Clouds, both lying at nearly equal distance. Four parameters are defined for each galaxy using the generalized radial profile (cf. Papers I and II) within the limiting magnitude of 26 mag arcsec$^{-2}$. They are diameter in logarithmic scale ($\log D_{26}$; in units of 0.1), integrated magnitude ($V_{26}$), mean surface brightness (SB) and mean concentration index [$X_1(P)$]. The last parameter, $X_1(P)$, represents the degree of luminosity concentration towards the galactic center. It is a good measure of both profile shape and intrinsic surface brightness (Paper III).

The $X_1(P)$ versus SB diagram can be used as a tool to investigate galaxy content of distant clusters independent of color indices. It is noted that clusters out to several hundred Megaparsecs could be studied with the data from Wide Field/Planetary Camera of the Space Telescope.

The principal component analysis is performed on the following five samples composed of galaxies for which membership is certain from radial velocity measurement; standard sample (N=201; E, SO, S, Irr), E sample (N=18; E), D sample (N=151; SO, S, Irr), Virgo sample (N=167; E, SO, S, Irr) and UMa sample (N=34; E, SO, S, Irr). It is known from previous studies that both samples of disk galaxies and samples of ellipticals have two-dimensionality. This is

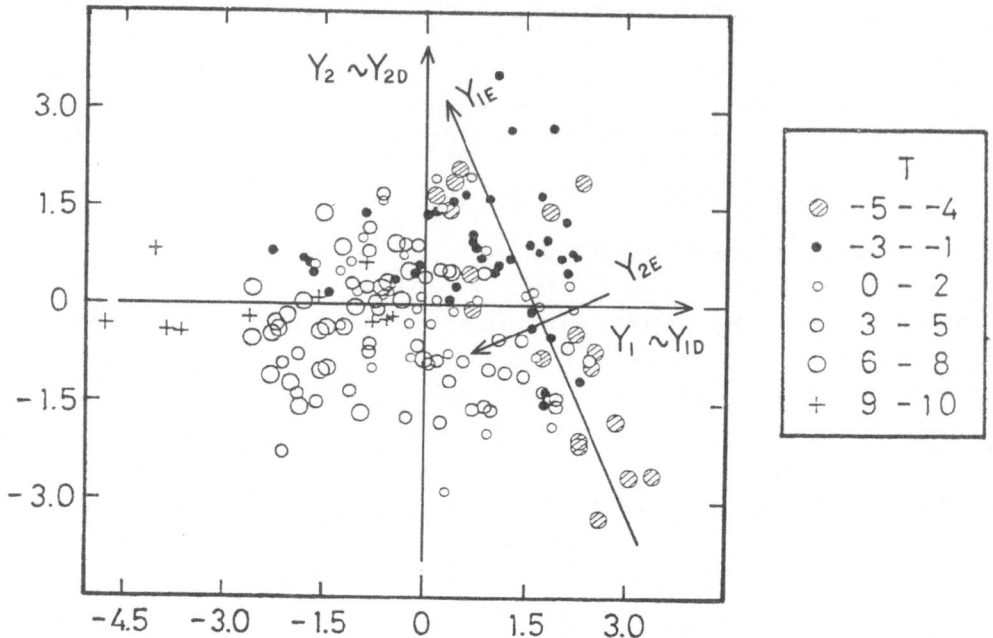

**Figure 1.** Distribution of galaxies in the standard sample projected onto the plane defined by two dominant factors $Y_1$ and $Y_2$. Relations between ($Y_1$, $Y_2$) and two dominant factors for D sample, ($Y_{1D}$, $Y_{2D}$), and E sample, ($Y_{1E}$, $Y_{2E}$), are illustrated.

confirmed for D and E samples. Main result of our principal component analysis is that two-dimensionality is found to be valid for the standard sample comprising galaxies of all morphological types. This means that both ellipticals and disk galaxies are distributed in the same plane in the space of [log$D_{26}$, $V_{26}$, SB, X1(P)]. Figure 1 shows the distribution of galaxies in the standard sample projected onto the plane defined by two dominant factors $Y_1$ and $Y_2$. Relation between ($Y_1$, $Y_2$) and two dominant factors for E sample ($Y_{1E}$, $Y_{2E}$) is also illustrated. It is noted that two dominant factors for D sample, $Y_{1D}$ and $Y_{2D}$, are almost identical with $Y_1$ and $Y_2$. Examinations of the orientation of original parameter axes projected onto the plane defined by two dominant factors reveal that ($Y_1$, $Y_2$) plane is essentially the same plane as those found in previous studies. No significant difference is found in the results of the principal component analysis between Virgo sample and UMa sample (Paper IV).

Projections onto ($Y_1$, $Y_2$) plane of log $D_{26}$ and SB axes are almost orthogonal to each other. Accordingly, we propose to use (log$D_{26}$, SB) instead of ($Y_1$, $Y_2$) as the basis of a quantitative classification and construct the diameter versus surface brightness diagram, which we call DSBD.

Recently, two-dimensional B-band luminosity distribution was obtained for some 60 dwarf ellipticals in the Virgo Cluster on the basis of two large scale plates taken with 2.5m du Pont telescope (Ichikawa et al. 1984). When we plot these dwarf ellipticals in DSBD together with normal ellipticals in a standard sample, we find that dwarf ellipticals appear to form a separate sequence from that of normal ellipticals. This may be a hint of possible structural discontinuity due to different formation histories.

We are going to extend our photometric data base to all northern galaxies in the Revised Shapley-Ames Catalog (Sandage and Tammann 1981) using Kiso Schmidt telescope and a computer system capable of efficient image data processing which will be introduced at Kiso observatory soon.

The author would like to thank all the collaborators in this project and the staff of Kiso observatory for their support. This work was supported in part by the scientific Research Fund of the Ministry of Education, Science and Culture (#59065002).

REFERENCES

Ichikawa, S., Wakamatsu, K., and Okamura, S. 1984, in preparation.

Kodaira, K., Okamura, S., and Watanabe, M. 1983, Astrophys. J. Letters, 274, L49.

Okamura, S. 1984, Proc. ESO Workshop on *The Virgo Cluster of Galaxies* (in press).

Okamura, S., Kodaira, K., and Watanabe, M. 1984, Astrophys. J., 280, 7. (Paper III).

Sandage, A., and Tammann, G. A. 1981, *A Revised Shapley-Ames Catalog of Bright Galaxies* (Carnegie Institution of Washington, Washington, D.C.).

Watanabe, M. 1983, Annals Tokyo Astron. Obs., 2nd Ser., Vol. 19, No. 2, P. 121. (Paper II).

Watanabe, M., Kodaira, K., and Okamura. S. 1982, Astrophys. J. Suppl., 50, 1. (Paper I).

Watanabe, M., Kodaira, K., and Okamura. S. 1984, submitted to Astrophys. J. (Paper IV).

DISCUSSION

F. Simien : How accurate is your determination of integrated magnitudes ? The center of a galaxy is probably overexposed on the Schmidt plate. How do you calibrate ?

S. Okamura : A comparison of our $V_{26}$ with $V_T$ in RC2 for 141 galaxies yields a regression line of $V_T=0.96\ V_{26}+0.38$ with a standard deviation of 0.20 mag. This value, 0.20 mag, may indicate the accuracy of our magnitudes. We exclude the very central region of a galaxy when we compare photographic luminosities with multi-aperture photoelectric magnitudes (cf. Paper II).

M. Capaccioli : How do observational errors affect your conclusions and on which basis did you select $V_{26}$ instead of other levels ?

S. Okamura : Presence of large observational errors, i.e. scatters, would produce spurious components. I do not believe that such large errors are present in our data. More important thing is the selection effect. Our analysis is based upon galaxies with $\log D_{26} \gtrsim 1.2$ and $V_{26} \lesssim 14$. We are interested in the mass distribution in galaxies. Near infrared band would be most appropriate. A large compilation of good plate material in near infrared band would take quite a long time. The V band was chosen as a compromise and efforts were made to reach as faint level as possible. I agree with you on that $V_{26}$ system is rather inconvenient for comparative studies.

R. Terlevich : Those previous studies on the second parameter behaviour of ellipticals and spirals show that it is related to kinematical properties and colour or abundances. Your data refer to the surface photometry of ellipticals and spirals. It is not clear to me that these two different 3-D spaces can be intercompared. I believe your results refer to a different bi-parametric behaviour that is not necessarily related to the $L_B$, $\sigma$, $Mg_2$ described previously. It is certainly very interesting that bi-parametric behaviour occur in two different 3-D spaces. But I do not believe that they are the same. Occurrence of both is important and intriguing.

S. Okamura : Reasons we believe that the two planes are the same are the following. (1) Both our E sample and your samples of ellipticals have two-dimensionality. (2) Orientation of M (magnitude) vector with respect to the factor plane is almost identical among the samples. (3) Ratio of eigenvalues, $\lambda_1/\lambda_2$, is similar between the two studies.

D. Axon : One of the major problems in the principal component analysis is to decide which components are significant. In your case you suggested that the two principal planes in E and D systems are aligned. Most of the power is in the first eigenvalue. Is the dimensionality really 2 only? How did you test whether the two planes are really aligned?

S. Okamura : For E sample, the third factor carries only 3% of the total variance, which is comparable to the scatter due to observational errors. For standard sample, however, the third factor carries 7% of the total variance. Its significance may be marginal. The distribution in factor space of galaxies in standard sample demonstrates that both ellipticals and disk galaxies lie nearly in the same plane.

S. Djorgovski : Several things are necessary for the principal component analysis. First, the data ellipsoid should be a multivariate Gaussian; second, the selection effects can substantially change the shape and orientation of the data ellipsoid; third, one must scale the input quantities very carefully in order to assess the significance of eigenvalues; and finally, the relations between the input quantities and eigenvector components must be linear.

S. Okamura : Thank you for your comments.

# HIGH RESOLUTION IN GALAXY PHOTOMETRY AND IMAGING

J.-L. Nieto (1)   and G. Lelièvre (2)


1 - Observatoire de Toulouse, 14 Avenue Edouard Belin
31400 Toulouse, France
2 - C.F.H. Telescope Corporation, p.o. Box 1597
Kamuela, HI 96743, USA

## I - Introduction

Recent years have witnessed the scope of the use of ground-based high resolution observations to solve some extragalactic astrophysical problems without waiting for the launching of the Space Telescope. It has become already clear to many that this newly accessible resolution range will also be a perfect playground for preparing Space Telescope observations.

Having in mind these two aspects of high resolution in the context of the present conjoncture, we shall present some spectacular examples of the high resolution reached from the ground and discuss a few problems that high resolution observations in galaxy photometry and imaging are able to tackle (Sect. I). This will be illustrated by some recent results (Sect. II). Section III is devoted to a discussion on the several possibilities to still improve the high-resolution obtainable from the ground.

## II.a The high resolution obtainable from the ground

It has been a tradition of high resolution at Pic du Midi Observatory. Spectacular results from solar, planetary and galactic observations have been shown in several instances. Recently, a few results in extragalactic astronomy have started to appear (Nieto and Auriere, 1982 ; Coupinot et al., 1982).

The most spectacular results on the high resolution obtainable at Mauna Kea have been presented in several instances by different observers and are regularly mentioned in CFH information bulletins. Several experiments such as star trails or other VHR observations have reached a resolution of about 0".3 on short exposure images.

- short exposure (duration ranging from a few tenths of a second to a few seconds) images correspond to the accumulation of speckles so that the image is now continuous and its structure is essentially affected by blurring.

- longer exposures are affected by both the variable blurring of the image and the image motion (displacing the centroid of the different short exposure images) during the exposure. These longer exposures depend on only the Fried (1966) parameter, $r_0$, defined as the diameter of a zone where the quadratic mean of the phase displacement of the wavefront is less than 1 square radian. If the diameter of the telescope is $D > 4r_0$, the resolution of a long exposure is limited by the atmosphere at $\approx 1.2 \, \lambda/r_0$, $\lambda$ being the wavelength of the observations.

This value of $r_0$ is between 10 to 20 cm in average observatories (note the influence of the wavelength of the observations since $r_0 \propto \lambda^{6/5}$) but it can become larger than 30 cm at 6 000Å (Racine, 1983; this corresponds to those nights where the FWHM at Mauna Kea is better than 0."5) and even ... 40cm (Thompson and Ryerson, 1984). A 37cm observation for 5 minutes at the Cassegrain focus have been also reported (FWHM = 0."35). At Pic du Midi Observatory, an $r_0$ of 25 cm has been measured at 5000 Å (Coupinot,1984). Unfortunately no histogram of $r_0$ is available for these two sites.

To improve the resolution of the observations, the use of a fast detector certainly allows shorter exposures whose images will be free of any wavefront tilt or image motion. Hecquet and Coupinot (1984) have discussed and observationally confirmed the maximum gain in resolution of short exposures over long exposures being $\approx 2$ for $D/r_0 = 3.4$. It then turns out that too large a telescope is not really requested for improving the resolution in such a way. Apparently, except for exceptionally high values of $r_0$ (which could be the case at Mauna Kea), the above mentioned high values of $r_0$ suggest a 1 to 1.5 meter telescope ideally suited for this purpose.

A further improvement of image quality would be not only to shorten the exposures but also to select the better short exposure images. Again the improvement depends only on $D/r_0$. It is given in table 1 (from Hecquet and Coupinot, 1984).

For a $D/r_0 \sim 8$ or 9 ($r_0 = 40$ cm at the CFH or 25 cm at Pic du Midi), 10 % of the images have a resolution 2.6 x that of the long exposure ones and 1 % reach a factor 3. It is of the uppermost interest to "catch" these images, unspoiled by the contribution of the others: since an $r_0 = 40$ cm means an FWHM $\approx 0."3$, 10 % of the images during such a night would have a resolution (better than or) of $\approx 0."15$ - 0."12 !!

II.b **Some astrophysical problems**

Generally speaking, a step forward by a factor of 2 in resolution is a great improvement since the image blurring has diminished by a factor of 4. Any problem resolvable at a sub-arc second level (= 75 pc at the distance to the Virgo cluster) does not need to wait for the Space Telescope to be tackled. We list here some problems that we have investigated for which high resolution has played a determining factor (in galaxy photometry and imaging).

i. Detailed structures of extragalactic objects

We mention for instance :

- Structure of optical jets (the M87 jet at 0."5 resolution (Nieto and Lelièvre, 1982) and the 3C273 jet at 0."9 (Lelièvre et al., 1983, 1984)).

- Clumpy structures such as the HII regions lying in the central region of NGC 1510 (Eichendorf and Nieto, 1984), or the structure of the central region of hot-spot galaxies (Nieto and Lelièvre, 1981, Nieto and Vidal, 1985).

- Detailed geometry of spheroidal components of galaxies, particularly ellipticals (Nieto and Vidal, 1984a)

ii. Extension of the measurable distance range

The determination of the distance to galaxies relies on different criteria that are usable within a certain distance range whose upper limit is strongly seeing dependent. See an illustration in Nieto and Tiennot (1984).

iii. Central cores of galaxies

Whether galaxies contain nuclear spikes of light or have isothermal cores is a controversial question for normal (i.e. non active) galaxies. The interpretation of the observed profiles may be so seeing dependant that the best possible resolution is necessary. An evidence for a spike of light in NGC 3379 independantly of any model assumed to fit the rest of the galaxy has been put forward by Nieto and Vidal (1984b) who showed that the central brightness of this galaxy increases with increasing resolution.

III - **Improvement on the regular long-exposure resolution**

i. The $D/r_0$ parameter

The degradation of the image is due to wavefront distorsions produced by thermal disturbances in the Earth's atmosphere, (assuming no guiding errors of the observer and tracking defects of the telescope).

In image accumulation, we may distinguish three time scales :

- the very short exposures (a very few tenths of a second), i.e. shorter than the time scale of large scale atmospheric turbulence) yield a speckle structure.

| $D/r_0$ | 76/100 | 10/100 | 1/100 | 1/1000 |
|---------|--------|--------|-------|--------|
| 3       | 2.0    | 2.7    | 2.9   | 3.0    |
| 4       | 2.0    | 3.1    | 3.4   | 3.6    |
| 5       | 1.9    | 3.1    | 3.6   | 4.0    |
| 7       | 1.7    | 2.8    | 3.4   | 4.1    |
| 10      | 1.5    | 2.4    | 3.0   | 3.8    |
| 15      | 1.4    | 2.1    | 2.6   | 3.2    |
| 20      | 1.3    | 1.9    | 2.3   | 3.0    |
| 50      | 1.2    | 1.5    | 1.7   | 2.1    |
| 100     | 1.1    | 1.4    | 1.6   | 1.8    |

Table 1 : Gain in angular resolution for different selection thresholds of short
exposure images.

A third step after i) shortening and ii) selecting the exposures would be to
apply some restoration techniques to the composite image resulting from the
selected short exposures. An illustration of the resulting improvement of the
resolution is shown in Lorre and Nieto (1984) who restored a composite of high-
resolution (FWHM ∾ 0."5-0."7) U and B photographic plates of the M87 jet. The
estimated resolution achieved was estimated to be about 0."2 for the highest
signal-to-noise parts of the image.

ii. How to achieve this ?
Different attempts to reach such a resolution have been undertaken here and  there.
Let us mention first (although it does not concern galaxies), since it is a
central aspect in the topic of another specialized colloquium during this  European
meeting, the observation of the solar granulae by Muller (1983) with few
millisecond long exposures taken continuously, the best images selected leading
to a resolution of .. 0."25.
A more systematic approach has been realized by Thompson and Ryerson (1983) who
have built an "active mirror image stabilizing instrument system" (ISIS) removing
the erratic motion of the centroid of the image by a microprocessor controlled
active plane mirror, capable of repositioning the image every 2ms. An active
shutter eliminates the period of excessive image blur (frequency 6Hz). The first
astrophysical result in galaxy imaging is a V CCD image of the radio-galaxy Cygnus
A whose resolution has improved from 0".9 to about  0."65 (Thompson, 1984).
Other attempts have been made at the CFH with the use of a very high resolution
camera whose shutter opens when the images are better than a given value. Although
some results are quite impressive (FWHM ∾ 0."25 on a 5 sec exposure), there is in
this system no correction for image motion (Christian et al., 1983).
Another technique has been presented by Fort et al. (1984) using the CCD camera  in

a cinematographic mode, that they call cine-CCD. The purpose is to take very short individual exposures (0.1 to 1 sec). In a first step all images are added, resulting in a reference image. Each individual image is afterwards correlated to this reference image, used as a mask image and is recentered by cross-correlation (step ≠ 1). This sum of the recentered individual images is a final image corrected for the erratic image motion and having a better resolution (but a reduced signal-to-noise ratio compared to that of a single image having the same exposure time).

It is easy to improve further the resolution by applying also selection criteria (step ≠ 2) to the individual images, eliminating those whose resolution is worse than an a-priori given threshold.

This means, at variance with the ISIS project, an a posteriori computer study. A project similar to the Cine-CCD, based on the use of a very performing Image Photon Counting System is in preparation : individual images are recorded at the television frame rate (512 x 512 pixels at 30 Hz).There is here certainly a possibility of reaching a resolution better than that of the Cine-CCD made since the individual exposure times are shorter.

The advantage of these two methods relative to systems applying corrections for atmospheric turbulence during the exposures is that they allow to record several images having several resolutions and several signal to noise ratios:

- an integrated one with no correction for atmospheric effects,
- an integrated one where all individual images are corrected for image motion (step ≠ 1),
- several integrated ones where different (more or less drastic) selection criteria for atmospheric blurring have been used (step ≠ 2).

They require however very much computer time. All these images, beside the fact that they will give us relevant information on the different time scales of the atmospheric turbulence during the observations, will allow to undertake step ≠ 3 : the use of image restoration techniques (see for instance Lorre and Nieto, 1984). The best compromise will be then found between resolution and signal-to-noise ratio, yielding the best image that it is possible to derive during the given corresponding observing period.


IV - Conclusion

We have presented briefly results from high-resolution observations either with the CFH telescope or at Pic du Midi Observatory. They mainly correspond to direct acquisition data relying only upon the high resolution of the site.

To improve the resolution, several possibilities have been already or are being exploited. We discuss a few of them, in the light of the expectations derived from atmospheric turbulence theory.

The high resolution range is just starting to be explored. It is very likely that

the number of high resolution studies will increase very soon, since Space
Telescope data will strongly need ground-based high-resolution data in many
respects.

References

Christian,C., Racine,R., Waddell,P., Salmon,D., 1983, in Proc. SPIE 445,
    Instrumentation in Astronomy V, p. 484

Coupinot,G., 1984, private communication

Coupinot,G., Hecquet,J., Heidmann,J., 1982, Month. Not. Roy. Astr. Soc., 199, 451

Eichendorf,W., Nieto,J.-L., 1984, Astron. Astrophys., 132, 342

Fried,D.L., 1966, J. Opt. Soc. Am., 56, 1372

Hecquet,J., Coupinot,G., 1985, Journal of Optics,15, 375

Lelièvre,G., Nieto,J.-L., Wlérick,G., Servan,B., Renard,L., Horville,D., 1983,
    Comptes Rendus Acad. Sci., 296, Série II, 1779

Lelièvre,G., Nieto,J.-L., Horville,D., Renard,L., Servan,B., 1984, Astron.
    Astrophys., 138, 49

Lorre,J.J., Nieto,J.-L., 1984, Astron. Astrophys., 130, 167

Muller,R., 1983, Solar Physics, 87, 243

Nieto,J.-L., Aurière,M., 1982, Astron. Astrophys., 108, 334

Nieto,J.-L., Lelièvre,G., 1981, "Astronomical Photography 1981", J.-L. Heudier and
    M.E. Sim Ed., p. 189

Nieto,J.-L., Lelièvre,G., 1982, Astron. Astrophys., 109, 95

Nieto,J.-L., Tiennot,L., 1984, Astron. Astrophys., 131, 291

Nieto,J.-L., Vidal,J.-L., 1984a, Astron. Astrophys., 135, 190

Nieto,J.-L., Vidal,J.-L., 1984b, Month. Not. Roy. Astr. Soc., 209, 21P

Nieto,J.-L., Vidal,J.-L., 1985, in preparation

Thompson, L.A., 1984, Astrophys. J., 279, L 47

Thompson,L.A., Ryerson,H.R. 1983, in Proc. SPIE, 445, Instrumentation in Astronomy
    V, in press

# SURFACE PHOTOMETRY WITH THE SPACE TELESCOPE

F.D. Macchetto

Space Telescope Science Institute

Homewood Campus, MD 21212

Abstract :


The Space Telescope will be an extraordinary and versatile observatory. Its complement of scientific instruments will allow far reaching observations to be made with a precision which will have no equal in optical astronomy.

The extended wavelength range in the ultraviolet, the high spatial resolution, the large dynamic range and the high signal to noise achievable will make the cameras on board the Space Telescope unique tools to carry out precise surface photometry of a wide range of objects.

This paper describes the key characteristics of the Space Telescope and its scientific instruments with particular emphasis on its possible applications to surface photometry.

*AFTERNOON SESSION II :*


*PROGRESS IN GALAXY PHOTOMETRY*
*AND RELATIONSHIP WITH OTHER WAVELENGTHS*

# ACCURACY AND IMPROVEMENTS IN GALAXY PHOTOMETRY: WHY AND HOW

Massimo Capaccioli

Institute of Astronomy, University of Padova, 35100 Padova, Italy

## I. INTRODUCTION

The first paper on galaxy photometry (Reynolds 1913) appeared before the nature
of the white nebulae was fully understood. Then since, and particularly after
1940, the number of contributions, from plain aperture photometry to detailed 2-D
studies, grew almost exponentially (see the review by de Vaucouleurs 1979).
Typically the photoelectric technique, pioneered by J.Stebbins, was used to
obtain magnitudes and colors through centered diaphragms (see the UBV catalogue
by Longo and de Vaucouleurs 1983) from which standard parameters were calculated
(e.g., de Vaucouleurs et al. 1976); less frequently to build light profiles
(e.g. Miller and Prendergast 1962), but very rarely at high enough resolution
(see Pe3 in de Vaucouleurs and Capaccioli 1979, hereafter dVC). Apart from
sporadic contributions of electronography, for obvious reasons of convenience 1-D
and 2-D photometry was traditionally reserved to the photographic technique until
very recently, when panoramic digital detectors came into regular operation.

From the compilation of Davoust and Pence (1982) we learn that the total number
of galaxies of all types having some kind of detailed photometric information is
about 600. Without duplications this number would have been larger but...
'repetita iuvant', i.e. multiple studies help to establish the quality of the
available sample. In fact the real difficulty with galaxy photometry to-day is
rather the quality than the quantity. In 1961 the Working Group on Galaxy
Photometry and Spectrophotometry of IAU Commission 28 expressed the need for
photometric standards and selected few objects for this purpose. To date only one
standard is available, the elliptical NGC 3379 for which dVC produced an accurate
E-W profile in one color (B-band) and discussed in detail the role of the various
sources of errors (Capaccioli and de Vaucouleurs 1983 = CdV). Another
photometric standard among those listed in 1961 is NGC 3115 (Capaccioli, Held and
Nieto, this volume). In this paper I will discuss some of those areas of
extragalactic astronomy which suffer most from the lack of an adequate accuracy
in the photometric data. I will limit the discussion to elliptical galaxies.

Interest in elliptical galaxies was renewed when astronomers realized that they
were not as well understood as thought before. In particular the discovery that
they may rotate slower than predicted by oblate models flattened by rotation
(Bertola and Capaccioli 1975; see also the reviews by Capaccioli 1979, and
Illingworth 1981) opened the way to theoretical speculations on the coupling of
internal kinematics and intrinsic shape (Binney 1981,1982a). Problems to solve
concern the internal dynamics, the intrinsic shape, the trends of the density

profiles (particularly in the very inner and outer regions), the presence of nuclear singularities, the possible segregation in different families (e.g. giants and dwarfs, and/or oblate and prolate, and/or disk or diskless), the link to other spheroidal subsystems (glubular clusters and bulges of SO's and spirals), the environmental effects (merging, cannibalism, fly-by interactions), etc. The final goal is clearly an understanding of the processes of formation and evolution.

Surface photometry, taking advantage of modern hardware and software techniques which make the work easier and more reliable, is greatly contributing to this enterprise (see the review by Kormendy 1980, and the ponderous 'summa rerum' by the same author, 1982, hereafter JK), both as a complement to the kinematical studies and as an input to genuinely new speculations. Typical examples of this second aspect are the discoveries of nuclear spikes of light in excess of an isothermal (Young et al. 1978, Schweizer 1978, de Vaucouleurs and Nieto 1979a, dVC) and of the twisting of the isophotes (Barbon et al. 1976a,b,1980; Carter 1978,1979).

The organization of the paper is suggested by the consideration that in the photometry of galaxies there are three surface brightness domains facing different scientific problems and requiring different techniques of analysis. A brief discussion on the causes of degradation of resolution (Section II) precedes that on the distribution of the light (Section III) and of the shape of the isophotes (Section IV) in the nuclear regions of ellipticals. Sections V and VI are devoted to considerations about the so called intermediate regions, those going from just the seeing blurred nuclei down to about $\mu(B) = 25$ B-mss. The very faint outer regions will not be considered for reasons of space and because there is little to say about accuracy except that it is needed and rarely achieved so far.

## II. THE POINT SPREAD FUNCTION

The effects of several sources of smearing compound to modify heavily the true shape of the luminosity profiles of galaxies. The atmosphere, the telescope, the detector and the scanning device operate successive convolutions which redistribute the energy of a galaxy image according to a cumulative Point Spread Function (PSF). The latter is not known 'a priori'; in principle it could be a very complicated function of time-varying as well as geometry-luminosity dependent parameters. To avoid complications it is common practice to make the PSF constant over a given image and to identify it with a smoothed analytical interpolation of the observed luminosity profile of a star. Usually circular symmetry is also assumed. Several empirical formulae have been proposed and used (e.g. Brown 1974, Newell 1979, dVC). A single gaussian often represents with sufficient accuracy the so called seeing disk, which is the dominant component in smoothing out the sharp details. In the optimum conditions ground-based observations reach a resolution better than $\sigma = 0"5$. However the PSF does extend much farther than the seeing disk (see Fig. 4 in CdV) and in principle covers the whole of the sky. The outer part (the stellar aureole and the Rayleigh scattering tail) becomes eventually relevant in the faint outskirts of galaxies where the feeble luminosity of the object may hardly exceed the light scattered from the bright nucleus (CdV).

## III. LUMINOSITY PROFILES OF THE CORES

So far very few elliptical galaxies have been accurately mapped in their central regions, and only one or two under exceptionally good seeing conditions (e.g. NGC 3379; Nieto 1983a,b); it is easily foreseeable, however, that the situation will change soon with CCD cameras mass producing top quality photometry of galaxy nuclei in several color bands (see Kent 1983,1984; Lauer 1984). Instead the photographic technique has been largely used (cf. de Vaucouleurs 1979, and Davoust and Pence 1982), but it has some well known limits. First, the linearization of photographic response requires a properly established calibration curve (not so crucial in the low density gradient outer parts of galaxies); unfortunately most of the largest and best telescopes have poorly designed calibration devices (see also de Vaucouleurs 1983) and that almost prevents the use of otherwise splendid plates. Second, the bright nuclei require such short exposures that often the background level does not exceed the chemical fog, causing difficulty in the background subtraction and in the absolute calibration. Moreover, under exceptionally good seeing conditions photographic adjacency effects (such as Eberhard's) may occur in the central spikes (Nieto 1983b); they can be avoided by an adequately large focal plane scale but at the expense of exposure time and therefore resolution. In conclusion, photographic plates may give fair results provided all observational and reduction variables are carefully controlled.

Due to their linear response and high dynamical range digital panoramic detectors are particularly suited to study the bright inner regions of galaxies (Young et al. 1978, Leach 1981, Kent 1983,1984, Lauer 1984). Their traditional limitation is the narrow field, often thought to be a serious problem for the background subtraction. But in fact this need not be the case; the sky can be mapped directly at the telescope almost in real time (due to the very short exposures required) or can be reconstructed later by the standard comparison with aperture photometry.

Once the observed luminosity map of the central region of a galaxy is available at the highest possible level of accuracy (conceivably better that 0.02 mag), it must be deconvolved for seeing before beeing of any astrophysical application. In other words, concepts such as that of 'core radius' $r_c$ introduced by King (1966) as a fitting parameter gain a physical meaning only after deconvolution. Similarly, any discussion about the existence of massive nuclei (Sargent et al. 1978, de Vaucouleurs and Nieto 1979b) is partly dependent upon the knowledge of corrected core profiles.

However it is a fact that on a small enough scale lost information can never be recovered, whatever fancy procedure is devised. Nonetheless two quite different methods can be utilized to retrieve some of that hidden in the data. One is simply the direct deconvolution (for instance the Maximum Entropy algorithm developed by Bryan and Skilling 1980), which however relies upon a very careful (and often risky) treatment of the noise. Another and simpler approach (Schweizer 1978, de Vaucouleurs and Nieto 1979a, dVC) is the straightforward convolution of an analytical model of the galaxy light distribution (luminosity profile and isophotal geometry) whose parameters are adjusted to fit the observations by trial and error. The only obvious disadvantage common to this method too is that the found solution is not unique.

The last consideration brings us to the point of the existence of seeing-convolved spikes of light at the centers of ellipticals as claimed, for instance, by dVC for NGC 3379 and by Young et al. (1978) and de Vaucouleurs and Nieto (1979a) for NGC 4486. This is part of a larger problem concerning the true trend of the light profiles in elliptical cores in relation to empirical as well as theoretical fitting formulae. Let us consider the case of NGC 3379 as an example. According to JK, his observations of the E-W luminosity profile of this galaxy (as reported by dVC, Appendix I) are satisfactorily matched by the King formula with $r_c$ = 2"9. This formula, however, is not much affected by seeing convolution for any reasonable value of $\sigma$ (Schweizer 1978) because it has a quasi-isothermal core profile, and therefore it may also reproduce the deconvolved profile with just a slight change of the core radius to 2"3 (see Fig. 19 in JK). Such result is not surprising since the King formula has a solid dynamical ground and it certainly approximates very reasonably the trend of the luminosity in the mean bodies of ellipticals, i.e. everywhere but the very centers and the faint outer parts (see also Section IV).

On the other hand dVC have proven that an even better fit of the observations can be achieved by convolving a much more peaked model, basically consisting of the $r^{1/4}$ law (de Vaucouleurs 1948,1953,1959,1961) which represents the outer parts of NGC 3379. As dCV noted, this model gives a good match of observed profiles under very different seeing conditions, even when the corrections are large ($\gg$ 1 mag; see also CdV). This is not the case for the King formula, whose central seeing corrections never exceed 1 mag for any practical value of $r_c$ and $\sigma$. This last remark opens the way to the solution of the problem. When we have accurate determinations of the central surface brightness $\mu(0)$ of a galaxian nucleus under very different seeing conditions, we can test the central corrections for the two models and check which provides the best match (cf. de Vaucouleurs and Davoust 1980). Very recently Nieto (1984) has obtained the needed set of data for NGC 3379, including some high resolution plates with the CFH telescope. When his data and the others already existing in the literature are plotted versus the estimated seeing, it becomes clear that, while $\sigma$ decreases from 1"-2" values to about 0"5, the observed central brightness increases by almost one magnitude. This would not be the case if the nucleus of NGC 3379 had an isothermal profile.

The expected central correction applicable to an $r^{1/4}$ law with $r_e = 55\overset{''}{.}5$ (as in dVC model of NGC 3379) at different values of $\sigma$ gives a good fit to the data. In passing, this seems the right place to rectify an historical injustice. Since the light profile of NGC 3379 is peaked similarly to that of M87 (Young et al. 1978, de Vaucouleurs and Nieto 1979a), then if the latter galaxy has a massive nuclear component (Sargent et al. 1978), the same ought to be true for the former, although on a smaller scale.[1] Indeed the trends of the velocity dispersion curves are also quite similar for these two objects (see Fig. 3 in Illingworth 1981).

In conclusion, at least in the case of NGC 3379 we are now able to prove the existence of a nuclear spike of light (i.e. a departure from an isothermal behaviour) consistent with the $r^{1/4}$ law. What about the other ellipticals ? Schweizer (1978) pointed out that the nuclei of E galaxies may be either spiky or not. Such deduction however is still model dependent. More secure information will be available when the central brightness test is applied to a large sample of galaxies (Capaccioli and Nieto 1985). It is also foreseeable that CCD images will soon provide high quality material for this test.

## IV. THE GEOMETRY OF THE INNER ISOPHOTES

It is very clear that convolution affects the _true_ shape of the nuclear isophotes of elliptical galaxies, making them smoother and rounder. Therefore measurements of axis ratios and position angles do not have a unique meaning in the central regions and vary with the image definition. In addition, structures as peculiar isophotal shapes, dust lanes and filaments, jets etc., are weakened or even cancelled. For this particular aspect the restoration based on model convolution has no chance to be successful due to the large number of free parameters required by a 2-D model. More effective and widely used are the digital algorithms which go from the simple minded Convolution Theorem in the Fourier domain to the advanced methods presently used, e.g. the Pixel Compensation Technique (Lorre and Nieto 1984). The latter have been mostly applied to few selected problems only, typically to improve the spatial resolution of the jet of M87 (Arp and Lorre 1976, Perryman 1981), but not yet, to my knowledge, to reconstruct the shapes of the nuclear isophotes of elliptical galaxies, although this is a very interesting subject. For instance, how many ellipticals have nuclear dust lanes ? This question can be answered with multicolor photometry at high resolution and good dynamical range (CCD), provided a proper filtering/restoration technique is applied.

---

[1] This was first argued by de Vaucouleurs, Capaccioli and Young in 1976 in a paper submitted to the Astrophysical Journal, but rejected by a referee who disbelieved this kind of speculations.

From observations we know that ellipticals tend to gain circular shapes near their centers. Is it due to convolution effects or is it rather a property intrinsic to the nature of the objects ? So far we must rely only upon a few high resolution studies of the closest elliptical-like objects, the bulge of M31 (Light et al. 1974) and the compact dwarf M32 (Bendinelli et al. 1981). Another piece of information supporting the idea that the nuclear regions of E galaxies can be elongated was given by Capaccioli and Rampazzo (1981); they showed that the observed ellipticity profile of NGC 3379 is consistent with a model having a constant intrinsic axis ratio. Their prediction has been partly confirmed by measurements of high resolution material (Nieto 1983b).

Convolution effects are also important for many different types of measurements concerning the cores of galaxies, their rotation curves (see Fig. 6 in Capaccioli and D'Odorico 1980) and velocity dispersion profiles (Binney 1982b), etc.

## V. THE INTERMEDIATE REGIONS

For elliptical galaxies with large enough angular sizes the intermediate region is the easiest to be mapped photometrically; it extends from just outside the blurred center to a distance where the surface brightness is about 25 B-mss, i.e. about 1/15 of the typical luminosity of the night sky in a good astronomical site. This light level can be up to 5 mag fainter than that of the effective isophote, that encircling half of the total luminosity (for NGC 3379 the light emitted within the 25 B-mss isophote is already 80% of the total). The practical advantages of working in this region are numerous.

First of all one needs not to worry about either seeing convolution (inner part of PSF) or the light scattered from the nucleus (outer PSF) and the consequent appearance of a spurious corona (crf. CdV). Second, the linearization of the measured quantities (density in photographic work, currents or photon counts in photoelectric work) is not as critical as in the bright core. For instance, if the signal-to-noise ratio is reasonable (which depends mainly on exposure time and sky background luminosity), the flat fielding and/or the background subtraction (including telescope vignetting, which is too often ignored) are not as crucial for the intermediate regions as they are for the very faint outer parts of galaxies (see Fig. 1 in CdV). Moreover it has to be noted that in photographic work the uncertainty in the calibration curve tends to increase with the density, becoming dominant at the galaxy centers. In other words, irrespective of resolution, the largest errors are expected in the brightest regions.

The above considerations are corroborated by a pragmatic argument. As shown by Carter and Dixon (1978) and by Benacchio et al. (1985 = LB), the spread of the results from various studies of the same galaxies, is always minimum in the intermediate regions (apart from zero point errors); but that does not indicate a general consistency or accuracy, even in case of agreement. Let us take the examples of NGC 4374 and NGC 4406, two bright ellipticals at the end of the Markarian Chain in Virgo. The comparison among various studies, made by LB and shown in Fig. 1 in form of residuals from best fitting $r^{1/4}$ laws, is indicative of the state of the art of photographic photometry. Some authors agree reasonably well, within ±0.1 mag over the entire range of 5 mag, but a few depart greatly from the mean. Is this a test for accuracy ? Clearly not, because the mean profiles of these two galaxies from the best studies disagree stronly with the photoelectric aperture photometry (up to a total of 0.4 integrated magnitudes).

The intermediate regions of E galaxies contain a significant fraction of their total luminosity (and therefore, very plausibly, of their total mass), thus they are the right places to investigate the effects of the internal dynamics. Since kinematical studies are greatly hampered by the relatively low surface luminosity of these regions, up to now the easiest approach has been that of surface photometry. In absence of appropriate theoretical models (see however Binney 1982a, van Albada 1982), the strategy is traditionally that of fitting the luminosity profiles with reasonable empirical or semi-empirical formulae. The procedure is somewhat risky expecially when the fitting functions have too many parameters, or when a simple physical meaning is not easily attached to them. A review of the most common formulae was given by dVC. We will concentrate here on the so called $r^{1/4}$ law (de Vaucouleurs 1948), because it is now widely accepted as a good empirical fitting law without free parameters (other than the required scale factors) for the light profiles of elliptical galaxies and spheroidal bulges of lenticulars and spirals, in some cases from the very centers (de Vaucouleurs 1974) to the outermost regions (dVC); exceptions are the tidally truncated profiles of dwarfs (King 1966) and the extended or 'stretched' profiles of some supergiants in rich groups/clusters (Thuan and Romanishin, 1982). The de Vaucouleurs law is a dimensionless function (in units of the scale factors):

$$\log J(\rho) = 3.3307 \ (\rho^{1/4} - 1) \tag{1}$$

where $J(\rho) = I/I_e$ is the relative surface intensity in units of the surface intensity $I_e$ of the effective isophote, that which encompasses half the total light of the galaxy and whose radius $r_e$ (the effective radius) is the unit of radial distance from the center ($\rho = r/r_e$). Clearly equation (1) refers to a specific cross-section of the galaxy image (e.g. the major axis). To eliminate the dependence on the isophotal shape (ellipticity and twisting) it is common practice to transform $\rho$ into the so called equivalent radius $\rho^*$ by circularization of the isophotes ($\rho^* = (ab)^{1/2}$). This procedure is also very useful to improve the signal-to-noise, for instance when the equivalent radius is obtained by a 2-D best fitting of the isophotes with ellipses (Barbon et al. 1976a). In many cases however, particularly in multicomponent galaxies such as lenticulars, it may a source of confusion and should not be used.

Since the good qualities of the $r^{1/4}$ law are well known, we will rather warn against its possible misuses. In the first place, while this law compares

LOG OF EQUIVALENT RADIUS (ARCSEC)

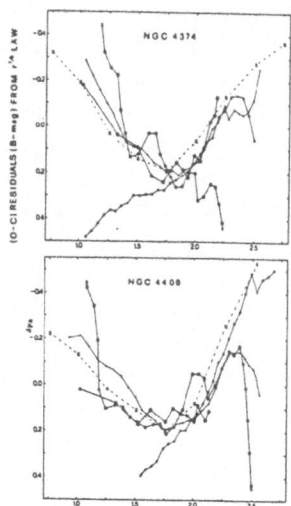

FIGURE 1. Comparison of the equivalent B-band light profiles from various photographic studies (from Benacchio et al. 1985). The main slope has been removed by subtraction of the same $r^{1/4}$ interpolation from all the studies of each object. Data of King (1978), LB and Markarian et al. (1961) are in satisfactory agreement, while those of Liller (1960) and Fraser (1977) depart significantly from the mean.

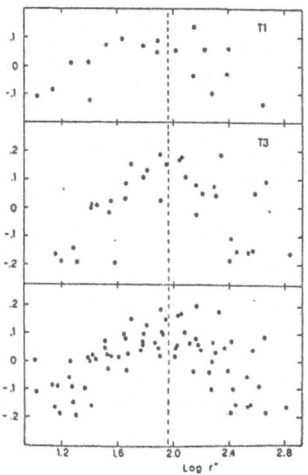

FIGURE 2. Residuals in magnitudes of the light profiles of the E galaxies in King's (1978) sample with respect to the best fitting $r^{1/4}$ laws, plotted versus central distances reduced to the scale of M87. Data for galaxies of the extreme Kormendy's (1977) types T1 and T3 are plotted separately in the two upper panels. The lower panel reports the entire sample (T1 + T2 + T3).

directly with the observations, the value of its scale-length $r_e$ is often uncertain. In fact, what one measures directly is the fourth rooth of $r_e$ and even a very small error is greatly amplified by taking the fourth power. Second, due to the expansion of the abscissa at small distances from the center and the consequent expansion at large radii, the law is not always able to discriminate among the various morphological classes, i.e. it tends to represent too well also the light profiles of lenticulars and even spirals, expecially if the observations are not very accurate and do not extend enough. Third, the $r^{1/4}$ law is difficult to handle mathematically in many instances, for instance when one needs to calculate the spatial deprojection of the luminosity profile (cf. Young 1976). Finally, even the fit to 'bona fide' ellipticals is good to a first order level only; this fact deserves some comments.

As shown by dVC the residuals to the best fitting $r^{1/4}$ law for the E-W profile of NGC 3379 exhibit a weavy pattern with an amplitude of about 0.2 mag (Fig. 3 in dVC). Such residuals, present also in all the other directions (cf. Benacchio 1975, whose results were recently confirmed by the reduction of a long exposure UK Schmidt plate), are _real_ and interesting departures from the $r^{1/4}$ law giving a smooth interpolation of the actual light distribution. Explanations for these waves are lacking. This is a clear example of the need for more numerous and accurate photometric studies of ellipticals in several colors (and not limited to the small frame-size of CCDs).

Another good example of the need for better accuracy, not limited to the central parts, may be given by the study of the light profiles of ellipticals made by Kormendy (1977). Using the sample of King (1978; tested to be externally accurate to ±0.1 mag by LB), he noted an excess over the $r^{1/4}$ law in the outer profiles of E galaxies possessing close companions, and interpreted it as evidence for a tidal stretching of an otherwise almost rigorous $r^{1/4}$ profile. We repeated Kormendy's test with the same sample of objects; all data points with r < 10" were rejected because of convolution by the PSF, and those fainter than 27 B-mss because of their large uncertainty (cf. LB). The remaining data for each galaxy were best fitted by Least Squares with an $r^{1/4}$ law, giving equal weight to all points. The residuals, plotted in Fig. 2 versus the distance from center reduced to the scale of NGC 4486 confirm the different trends of the two extreme galaxy families, i.e. those with and without close companions (or T3 and T1 respectively in Kormendy's notation), and show that all galaxy profiles depart systematically from the $r^{1/4}$ law, with minima located at about the effective radius. This was already noted by de Vaucouleurs (1953) and is confirmed by the galaxy sample of LB. The overall trend of the residuals is that expected if galaxy profiles follow a truncated isothermal model (de Vaucouleurs 1953, Woolley 1954, King 1966). However, because the King formula is not a good representation of the light profiles of giant galaxies in the very outer parts (Thuan and Romanishin 1982), we conclude that the truncated-isothermal scheme, on which this formula is based, possibly holds good for the intermediate regions of ellipticals, but fails in the outermost ones possibly due to the longer relaxation times and/or to the stronger interactions with the environment (or due to the initial conditions).

The representation of the light profiles of S0 and lenticular bulges is a very difficult task which requires extremely accurate observations and proper seeing deconvolution. Rampazzo (1981) noted that, if bulges are expected to extend to large distances, then the $r^{1/4}$ law is not steep enough to vanish in those outer regions where the disk component is still seen to be dominant through the exponential profiles (see also de Vaucouleurs 1958). In this respect a better approximation of the bulge profiles would be an $r^{1/2}$ law. Is this an indication that bulges are not small ellipticals surrounded by a disk ? Possibly, but the final word will come only when two-dimensional decomposition of lenticular and spiral light distribution is available.

## VI. THE INTERMEDIATE REGIONS: GEOMETRICAL PROPERTIES

Variations of the isophotal ellipticity and changes of the orientation of the major axis in E galaxies (first note by Evans 1951 and Liller 1960,1966) have been reported in the past years. Barbon et al. (1966a,b), implementing the Numerical Mapping method of Jones et al. (1967), introduced the technique of analytical ellipse fitting which allows a more objective evaluation of the geometrical parameters (see also Capaccioli 1973, and Barbon et al. 1980,1982). These authors pointed out that the study of the ellipticity and orientation profiles and the loci of the centers of the isophotes, might disclose the presence of substructures in ellipticals. Recently photometric and geometric parameters of 17 systems were reported by Carter (1978), together with some discussion on models of galaxy formation. A large body of data was published by Fraser (1977) and by Strom and Strom (1978a,b). Twisting and ellipticity in early-type galaxies were studied by Williams and Schwarzschild (1979a,b) and Williams (1981). King (1978) gave an extensive photometric analysis of 17 giant E's and S0, detecting an isophotal twisting in four objects. Statistical studies on the flattening profiles of ellipticals were published by di Tullio (1979), Bertola and Galletta (1979), Leach (1979,1981), Barbon et al. (1980), Watanabe (1983) and LB. The above list of references on geometrical studies can be supplemented with the bibliography of Davoust and Pence (1982).

Before we can make any astrophysical use of this large body of data, we must discuss their reliability. The fact that galaxy isophotes appear to twist and ellipticity to change with radius is not questionable; there are some examples which stand out clearly even by visual inspection (e.g., NGC 1600; Barbon et al. 1984). What has to be evaluated is the goodness of the various techniques in coping with sources of errors mimicking the phenomena one wants to measure: distortions (apparent or real) induced by nearby companion galaxies and very bright stars, superposition of foreground and background objects or defects, intrinsic shapes other than elliptical etc. We will not review here the methodologies used to eliminate the disturbances and/or measure the geometrical parameters, but rather comment on the general lack of impersonal methods capable of giving not only the formal parameters for the fit of the analytical isophotal models to the data, but also an estimate of the similarity between the two, i.e. a shape parameter. Usually visual inspection is used as a surrogate, although quite subjective and time consuming. Recently an effort along the direction of a rigorous solution of the problem was made by Cawson (private communication) in the framework of the Fourier approach (Carter 1978).

In the absence of any evaluation of the consistency of the data in the literature, we must rely upon external comparisons among the various studies. The reliability of ellipticity measurements has been discussed by LB. They conclude that in general, with a few remarkable exceptions such as Fraser (1977), flattening measurements are quite consistent. The same conclusion does not apply to the twisting. As an example, let us consider the case of Liller (1960,1966). LB find that her twisting profiles look reliable, but that she failed to detect any twisting in some galaxies where the phenomenon is indeed present (e.g. NGC 4486). The same remarks apply to King (1978), as this author himself noted. However, the external comparison is clearly not a good test. When two orientation profiles from two different studies agree even in revealing similar variations of the position angle, it is still possible the doubt that such variations are a real property of the object because they could be due to some common hidden or unremoved sources of disturbances or shape distortions.

The trend of ellipticity (b/a) with radius a has been used by various authors in an attempt of establishing a correlation with galaxy type (Valentijn 1983) or with environmental properties (di Tullio 1979). So far, however, there is no such correlation for which counter-examples have not been found. The combined analysis of both ellipticity and twisting characteristics seems to be more conclusive. Barbon et al. (1980,1984) have pointed out the existence of correlations between features of the ellipticity and twisting profiles, which are interpreted as an indication in favour of triaxial configurations. Similar conclusions were drawn by Williams (1981) in his study of the complex case of NGC 596 and by Benacchio and Galletta (1981) in their discussion of the statistical correlation between maximum flattening and amount of isophotal twisting (Galletta 1980). However, there is no unanimous agreement toward the interpretation of the ellipticity-twisting correlations and variations in terms of triaxiality alone. For instance, JK stressed the point that at least in some cases these phenomena must be related to tidal interactions. The important question is still open and, as many others, will not be settled until more numerous and accurate photometric observations are produced.

Many of the topics discussed here come from my work in collaboration with Prof. Gerard de Vaucouleurs, to whom I am particularly grateful for many suggestions and for a critical reading of this manuscript. This paper was supported in part by the Consiglio Nazionale delle Ricerche of Italy.

REFERENCES

Arp,H.C., Lorre,J. 1976, Astrophys.J., 210, 58.

Barbon,R., Benacchio,L., Capaccioli,M. 1976a, Mem.Soc.Astron.It., 51, 25.

Barbon,R., Benacchio,L., Capaccioli,L. 1976b, Astron.Astrophys., 51, 25.

Barbon,R., Benacchio,L., Capaccioli,M., De Biase,G., Santin,M., Sedmak,G. 1980, 'Applications of Digital Image Processing to Astronomy', ed. D.A.Elliott, SPIE No. 264, p. 264.

Barbon,R., Capaccioli,M., Rampazzo,R. 1982, Astron.Astrophys., 115, 388.

Barbon,R., Benacchio,L., Capaccioli,M., Rampazzo,R. 1983, Astron.Astrophys., 137, 166.

Benacchio,L. 1975, Dissertation, University of Padova.

Benacchio,L., Galletta,G. 1981, Mont.Not.R.Astron.Soc., 193, 885.

Benacchio,L., Capaccioli,M., De Biase,G., Santin,P., Sedmak,G. 1985, preprint (LB).

Bendinelli,O., Parmeggiani,G., Zavatti,F. 1981, Astrophys.Space Sci., 83, 239.

Bertola,F., Capaccioli,M. 1975, Astrophys.J., 200, 439.

Bertola,F., Galletta,G. 1979, Astron.Astrophys., 77, 363.

Binney,J.J. 1981, 'Structure and Evolution of Normal Galaxies', eds. M.S.Fall and D.Lynden-Bell, Cambridge Univ.Press, p. 55.

Binney,J.J. 1982a, Ann.Rev.Astron.Astrophys., Vol. 20, p. 399.

Binney,J.J. 1982b, Mont.Not.R.Astron.Soc., 200, 951.

Brown,G.S. 1974, Publ.Dept.Astronomy, Univ. of Texas at Austin, No. 11.

Bryan,R.K., Schilling,J. 1980, Mont.Not.R.Astron.Soc., 191, 69.

Capaccioli,M. 1973, Mem.Soc.Astron.It., 44, 417.

Capaccioli,M. 1979, 'Photometry, Kinematics and Dynamics of Galaxies', ed. D.S.Evans, Austin: Univ. of Texas, p. 165.

Capaccioli,M., D'Odorico,S. 1980, 'Astrophysics from Spacelab', eds. P.L.Bernacca and R.Ruffini, Reidel Publ.Co., p. 317.

Capaccioli,M., Rampazzo,R. 1981, Mem.Soc.Astron.It., 51, 491.

Capaccioli,M., de Vaucouleurs,G. 1983, Astrophys.J.Suppl.Ser., 52, 465 (CdV).

Capaccioli,M., Nieto,J-L. 1985, in preparation.

Carter,D. 1978, Mont.Not.R.Astron.Soc., 182, 797.

Carter,D. 1979, Mont.Not.R.Astron.Soc., 186, 897.

Carter,D., Dixon,K.L.. 1978, Astron.J., 83, 574.

Davoust,E., Pence,W. 1982, Astron.Astrophys.Suppl.Ser., 49, 631.

de Vaucouleurs,G. 1948, Annal d'Astrophys., 11, 247.

de Vaucouleurs,G. 1953, Mont.Not.R.Astron.Soc., 113, 134.

de Vaucouleurs,G. 1958, Astrophys.J., 128, 465.

de Vaucouleurs,G. 1959, Hand.der Phys., Vol. 53, p. 331.

de Vaucouleurs,G. 1961, 'Problems of Extragalactic Research', ed. G.C..McVittie, Nev York: MacMillan, p. 3.

de Vaucouleurs,G. 1974, 'Formation and Dynamics of Galaxies', ed. R.Shakeshaft, Reidel Publ.Co., p. 335.

de Vaucouleurs,G. 1979, 'Photometry, Kinematics and Dynamics of Galaxies', ed. D.S.Evans, Austin: Univ. of Texas, p. 1.

de Vaucouleurs,G. 1983, 'Astronomy with Schmidt type telescopes', ed. M.Capaccioli, Reidel Publ.Co., p. 367.

de Vaucouleurs,G. de Vaucouleurs,A., Corwin,H.C. 1976, 'Second Reference Catalogue of Bright Galaxies', Texas University Press.

de Vaucouleurs,G., Nieto,J-L. 1979a, Astrophys.J., 230, 697.

de Vaucouleurs,G., Nieto,J-L. 1979b, Astrophys.J., 231, 364.

de Vaucouleurs,G., Capaccioli,M. 1979, Astrophys.J.Suppl.Ser., 40, 631 (dVC).

de Vaucouleurs,G., Davoust,E. 1980, Astrophys.J., 239, 783.

Di Tullio,G. 1979, Astron.Astrophys.Suppl.Ser., 37, 591.

Evans,D.S. 1951, Mont.Not.R.Astron.Soc., 111, 526.

Fraser,C. 1977, Astron.Astrophys.Suppl.Ser., 29, 161.

Galletta,G. 1980, Astron.Astrophys., 81, 179.

Illingworth,G. 1981, 'Structure and Evolution of Normal Galaxies', eds. M.S.Fall and D.Lynden-Bell, Cambridge Univ.Press, p. 27.

Jones,W.B., Obitts,D.L., Gallet,R.M., de Vaucouleurs,G. 1967, Publ. Dept. Astronomy, Univ. of Texas at Austin, Ser. II, Vol. I, No. 8.

Kent,S. 1983, Astrophys.J., 266, 562.

Kent,S. 1984, preprint.

King,I. 1966, Astron.J., 71, 64.

King,I. 1978, Astrophys.J., 222, 1.

Kormendy,J. 1977, Astrophys.J., 214, 359.

Kormendy,J. 1980, 'ESO Workshop on Two-dimensional Photometry', eds. P.Crane and K.Kjar, ESO: Geneva, p. 191.

Kormendy,J. 1982, 'Morphology and Dynamics of Galaxies', XII Advanced Course of the Swiss Society of Astronomy and Astrophysics, eds. L.Martinet and M.Mayor, Sauverny: Geneva Obs., p. 75 (JK).

Lauer,T.R.. 1984, Dissertation, University of California at Santa Cruz.

Leach,R. 1979, 'Photometry, Kinematics and Dynamics of Galaxies', ed. D.S.Evans, Austin: Univ. of Texas, p. 75.

Leach,R. 1981, Astrophys.J., 248, 485.

Light,E.S., Danielson,R.E., Schwarzschild,M. 1974, Astrophys.J., 194, 257.

Liller,M. 1960, Astrophys.J., 132, 306.

Liller,M. 1966, Astrophys.J., 146, 28.

Longo,G., de Vaucouleurs,A. 1983, The University of Texas, Monographs in Astronomy, No. 3.

Lorre,J., Nieto,J-L. 1984, in preparation.

Miller,R.H., Prendergast,K.H. 1962, Astrophys.J., 136, 713.

Newell,B. 1979, 'International Workshop on Image Processing in Astronomy', eds. G.Sedmak, M.Capaccioli and R.J.Allen, Miramare: Trieste Obs., p. 100.

Nieto,J-L. 1983a, Astron.Astrophys.Suppl.Ser., 53, 287.

Nieto,J-L. 1983b, Astron.Astrophys.Suppl.Ser., 53, 343.

Nieto,J-L. 1984, preprint.

Perryman,M.A.C. 1981, 'Optical Jets in Galaxies', ESA Sp-162, p. 25.

Rampazzo,R. 1981, Dissertation, Univ. of Padova.

Reynold,G.H. 1913, Mont.Not.R.Astron.Soc., 74, 132.

Sargent,W.L.W., Young,P.J., Boksenberg,A., Shortridge,K., Lynds,C.R., Hartwick,F.D.A. 1978, Astrophys.J., 221, 731.

Schweizer,F. 1978, Astrophys.J., 220, 98.

Strom,S.E., Strom,K.M. 1978a, Astron.J., 83, 732.

Strom,S.E., Strom,K.M. 1978b, Astron.J., 83, 1293.

Thuan,T.X., Romanishin,W. 1982, Astrophys.J., 248, 439.

Valentijn,E.A. 1983, Astron.Astrophys, 118, 123.

van Albada,D. 1982, Mont.Not.R.Astron.Soc., 201, 939.

Watanabe,M. 1983, Ann. Tokyo Astron. Obs., Ser. II, 19, No. 2.

Williams,T.B. 1981, Astrophys.J., 244, 458.

Williams,T.B., Schwarzschild,M. 1979a, Astrophys.J., 227, 56.

Williams,T.B., Schwarzschild,M. 1979b, Astrophys.J.Suppl.Ser., 41, 209.

Woolley,R. 1954, Mont.Not.R.Astron.Soc., 114, 191.

Young,P.J. 1976, Astron.J., 81, 807.

Young,P.J., Sargent,W.L.W., Boksenberg,A., Lynds,C.R., Hartwick,F.D.A.   1978,
Astrophys.J., 222, 450.

DISCUSSION


G.Galletta: I have just a comment about the maximum flattening-twisting
correlation found for ellipticals. It is important to have new good photometry of
E galaxies since, even if the correlation could be due in part to photometric
errors, nonetheless some real effects might exist. It is difficult to ascribe to
photometric errors a twisting of 40° in a E4 galaxy.

# COMPARISON OF PHOTOGRAPHIC AND CCD SURFACE PHOTOMETRY OF GALAXIES

Laurent VIGROUX[1], Jean-Luc NIETO[2]


(1) Service d'Astrophysique, CEN Saclay, 91191 Gif-sur-Yvette,
Cedex, France


(2) Observatoires du Pic du Midi et de Toulouse, 14 Av. Edouard
Belin, 31400 Toulouse, France

All classical works on galaxy surface photometry have been performed with photographic plates. During the seventies, several other panoramic detectors such as electronographic cameras or intensified televisions have been used. The gain in speed and accuracy of these new detectors was counterbalanced by numerous defects such as geometrical distorsion and calibration difficulties. They never challenged, in this particular field of surface photometry, the photographic plates. The appearance of CCDs in late seventies has reversed the situation and despite their small size, they are ideally suited for surface photometry. In this paper we will first review the good and bad sides of plates and CCDs for surface photometry, and then compare the results obtained with the two methods on selected galaxies.

## I - PHOTOGRAPHIC PLATES

The photographic plates suffer from several well known defects : first of all they are not linear. The density-intensity calibration is still the major problem in photographic surface photometry. It can be calibrated on an empty place on the plate by spot sensitometer but one must assume that the calibration curve is identical everywhere else. Moreover plate and microdensitometer have a small dynamic range which prevents to determine on a single plate the surface brightness of a galaxy from its center to its edges. One must take several plates of the same galaxy with different exposure times. Plate uniformity is also a problem for sky subtraction -and plate non-uniformity, even at a level of a few per cent in intensity can mimic faint extensions of galaxies. Finally photographic plates have quite a poor quantum efficiency, of the order of a few per cent, and they do not offer simple quick-look analysis.
Several studies have been done in the past to solve these problems and, with care,

it turns out possible to obtain a good accuracy in photographic surface photometry of galaxies. They remain unique in two domains : their overall size and their resolution. On a good plate it is possible to obtain meaningful intensity determinations on a 10-15 μ scale. Since they can have very large dimensions, up to 20 centimeters for a prime focus plate of a 4 meter telescope, they allow the observation of very large fields, up to 1 degree on a prime focus.

A last advantage of plates is their low cost. They do not require very expensive high technology systems around the telescope itself. Their analysis can be done after in another laboratory which offers all the hardware and software needed for image analysis. To be analysed, plates must be first digitalised with a microdensitometer. Unfortunately microdensitometers did not follow the same improvements that photographic plates had and they are certainly the main cause of the limited dynamic range of photographic surface photometry.

## II - CCDs

The quality and defects of CCDs are exactly opposite to those of photographic plates. They are absolutely linear, they have a very good quantum efficiency, up to 80 % for thin backside illuminated CCDs ; they have a wide dynamic range, of the order of $10^4$ and since they give numerical intensity maps, it is easy to obtain quick look information. But they have small sizes, typically of the order of 1-2 cm and large pixel sizes, 20-30 μ. The next generation of large size CCDs, 6 cm x 6 cm announced to be available in 1985 must solve this problem of size but even with present CCDs it is possible to increase the observed field by a scanning method (Boroson et al 1983). In this mode the telescope drifts along the column direction of the CCD and the CCD is read out with a speed with corresponds to the crossing time of an object across the CCD. This allows to observe very long strips of the sky.

The outstanding photometric properties of CCDs make surface photometry much easier than with plates. In practice only 2 effects limit the accuracy of photometry : dark current and flat fielding. In general CCDs are operated at sufficiently low temperatures, below 170K, for the thermal dark current to be negligible even for exposures as long as 1 hour, relative to the number of photoelectrons coming from the nightsky alone. But other effects such as photoemission by the on chip amplifier or edge effects can produce large scale patterns. These patterns are not always exactly proportional to exposure time and may be difficult to remove with an accuracy better than 1 %. However these patterns can be suppressed in most CCDs by an appropriate setting of the clock voltages. Unfortunately this is not always done for the CCDs available on telescopes. Accurate flat fielding is also somewhat troublesome. A flat field correction much better than 1 % is needed if one wants to measure the faintest parts of galaxies. Then one must obtain a flat field with the highest possible signal to noise ratio. A good recipe is to obtain from 5 to 10 uniform frames at an intensity of half the saturation of the CCD. This can be done in several ways. A very telescope time consuming method, but the best one is to make several long exposures of empty fields ; then remove the stars and galaxies in each image and take their mean. Another possibility consists of observing twilights, but twilight color temperature is not identical to that of nightsky. This may lead to some residuals in the flat field correction with broadband filters. Or one can use a continuum lamp which illuminates a diffusing zone inside the dome with a filter which increases the color temperature of the lamp to mimic that of nightsky. Accuracy of 1 % may be easily obtained and with some care it is possible to reach an accuracy better than $5 \times 10^{-3}$.


III - <u>COMPARISON OF RESULTS</u>


The comparison of surface photometry data of a galaxy obtained with different detectors and even by different authors is never an easy task. Each author has his own way to determine intensity profiles and to present the results. In the worst case one finds only a simple plot in arbitrary units. Moreover when all published data are homogeneized, results are in general very disappointing. Differences as large as 1 magnitude can be found. Capaccioli (this conference) has given good examples of these types of problems. Meaningful comparisons can be made only on 3 standard galaxies : NGC 3379, 4486, and 3115. However since they have very large sizes which do not easily fit CCD fields, comparisons can be made on their central parts only.

For NGC 3379 we have compared the photographic data discussed in Nieto (1983a,b) and the INAG-CEA CCD data obtained with the CFH telescope in the blue light. The CCD is a 320 x 512 pixels RCA thin and backside illuminated. It was set behind the Cassegrain focal reducer. The pixel size is 0.83 arc second, the field is 4.4 arc min EW and 7.1 arc min NS. The CCD image was obtained with a 10 min exposure through isophotes with a 0.2 magnitude step. In figures 1 and 2 are superimposed the CCD data (dots) on the original figures of Nieto (1983a, b). All magnitudes are given by differences between actual measurements and the $r^{1/4}$ law which gives the best fit to the data. Figure 1 represents all published results compiled by Nieto (1983a). The CCD data were not reduced to the same zero point, and there is a ∿ .07 mag systematic difference. But relative variations are very small. The rms deviation between our data and the mean of all the others is less than 0.02 mag. Figure 2 represents in the same way the CCD data (dots) superimposed to profiles obtained from several individual plates (Nieto 1983b). The general agreement is excellent. However the noise of each individual photographic profile is much larger than that obtained with a single CCD exposure. To obtain an accuracy with plates such as that with CCDs one must smooth the photographic profile and take the mean of several plates as it was done in Figure 1.

For NGC 3379, we also compared the R CCD profile obtained in the same conditions and another R CCD profile obtained by Kent (1984). Figure 3 represents the magnitude differences between both sets of data. With the exception of the 3 inner arc second, which are always affected by seeing and pixel sampling effect, the agreement is very good and never exceeds 0.05 magnitude.

We were not able to obtain CCD data for NGC 4486 but detailed comparisons were made between photographic, SIT, CCD and scanning CCD results by Boroson et al (1983). While they found a reasonably good agreement between SIT, CCDs and the scanning CCD, the agreement with the photographic results is much worse. With the exception of the de Vaucouleurs and Nieto (1979) measurements which agree at a 0.05 magnitude level with the CCDs results, the others exhibit systematic trends where differences reach ± 0.2 magnitude.

This discussion shows that CCD surface photometry is at least of the same quality than the photographic one. For objects which fit the small field of a CCD frame (typically less than 5 X 5 (arcmin)$^2$), it really saves telescope time to use CCDs since observing the same object with plates will require the acquisition of several plates with much longer exposures. Moreover CCDs allow the use of more specific filters centered on selected spectral features which give better insights of the galaxy content (see Vigroux, this meeting). On the other hand the large plate size allows the study of objects in much larger fields, and they may be preferred for studies of very large objects or for the study of large amounts of objects in a large field.

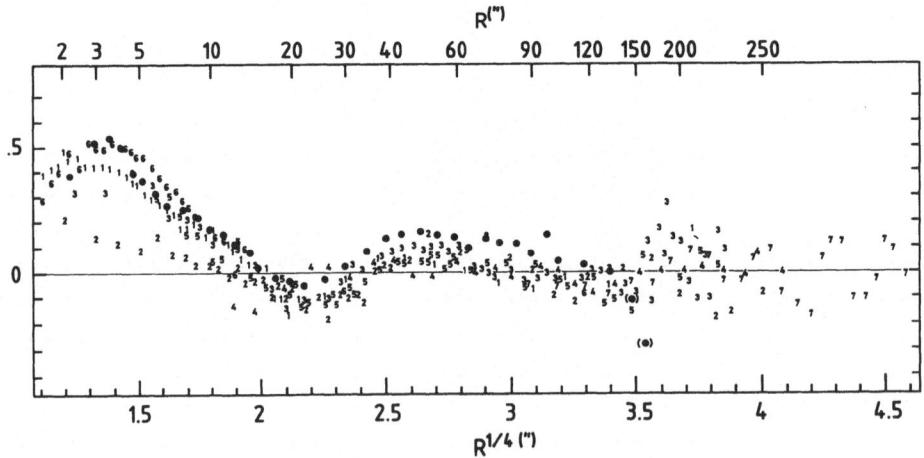

Figure 1

Differences (obs - r$^{1/4}$) between all published data on the EW profile of NGC 3379 and the r$^{1/4}$ law adopted by de Vaucouleurs and Capaccioli (1979) (photographic results : numbers, CCD data : dots).

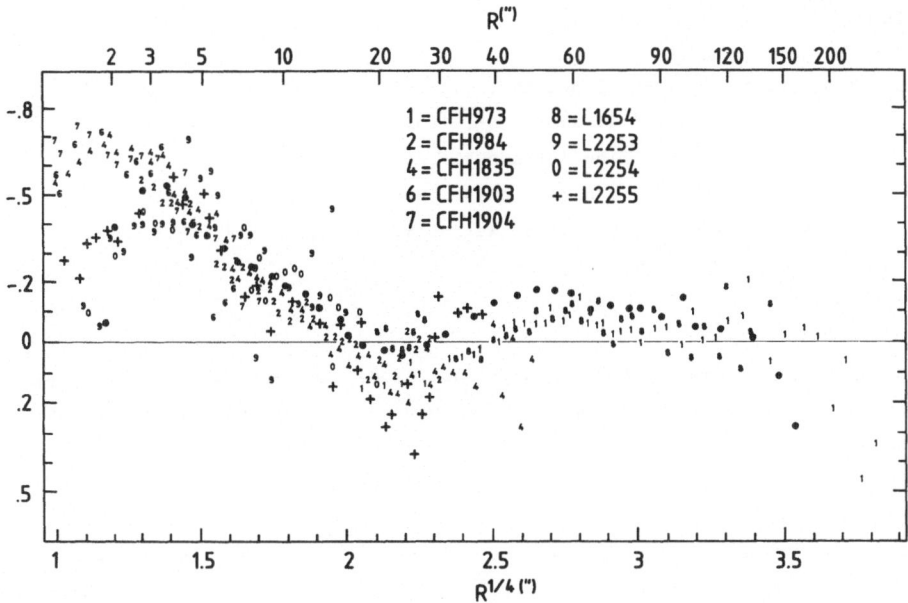

**Figure 2**
Same as Fig. 1 but with unfiltered data from single photographic plates
(Nieto, 1983b).

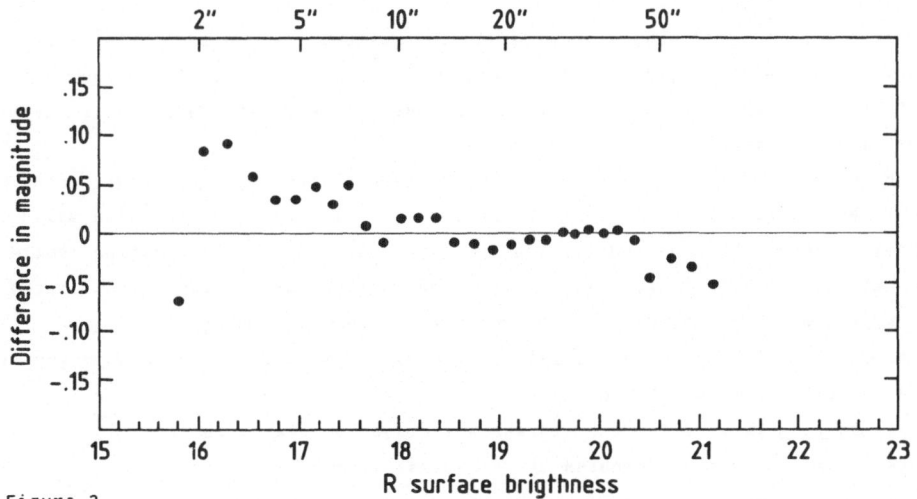

**Figure 3**
Differences (in magnitudes) between the INAG-CEA CCD E-W profile of NGC 3379
and Kent (1984) CCD profile (R light).

References

Boroson, T.A, Thompson, I.B., Schectman, S.A., 1983, Astrophys. J. 88, 1707.
Kent, S.M., 1984, Astrophys. J. Suppl. 56, 105.
Nieto, J.-L., 1983a, Astron. and Astrophys. Supp. 55, 247.
Nieto, J.-L., 1983b, Astron. and Astrophys. Supp. 55, 383.
de Vaucouleurs, G., Nieto J.-L., 1979, Astrophys. J., 230, 697.

Discussion

D. Malin : A comment rather than a question, to enlarge on some of the figures which have been presented at this meeting regarding properties of the photographic plates.

1. With careful work, flat fielding is not necessary to levels of about 0.5 % of the night sky brightness.

2. The dynamic range of IIaO plates is 10,000:1, not the 1000:1 value often quoted, which refers to IIIaJ emulsions, especially designed to have high contrast.

3. In this context the apparent quantum efficiency of a well-hypersensitized IIIaJ, exposed in a nitrogen atmosphere can reach 5 %, not the 1 % often quoted. It should also be noted that hypersensitizing and $N_2$ exposure also eliminates reciprocity failure.

4. There is no loss of observing time due to readout, a considerable factor in reducing the overall observing efficiency with CCDs, especially large ones.

5. The storage capacity and density of plates is enormous and their archival properties, after some early alarm are excellent.

M. Capaccioli : I disagree with your statement that CCD pictures cost much less telescope time than conventional photographs. It seems to me that a more fair comparison should be done considering the total amount of information collected per unit times : this is certainly very large for Schmidt or prime focus plates.

L. Vigroux : For single objects which fit into CCD fields I believe it is better to use CCDs. However as I said previously in my talk, I agree with you than there is still room for photographic plates in large field studies.

S. Djorgovski : As for the storage advantages of plates vs. CCDs : in order to use a plate, for most purposes it is necessary to digitize it ; then you have many magnetic tapes, just as with CCDs. But, there is another advantage of CCDs, which nobody mentioned so far : they are Poissonian, white noise detectors, and thus very friendly for any image processing and image restoration. This is not the case with any photography - based detectors.

# PHOTOGRAPHIC PHOTOMETRY OF 16000 GALAXIES ON ESO BLUE AND RED SURVEY PLATES

Andris LAUBERTS
European Southern Observatory

Edwin A. VALENTIJN
Kapteyn Laboratorium, Groningen
European Southern Observatory

Since the completion of the ESO/Uppsala survey for southern galaxies 1982 a follow-up study of that survey has been prepared. The main purpose of the original survey was to find and classify galaxies on the ESO Quick Blue Survey (QBS). The copy plates were visually inspected and all parameters, such as position, size and morphology, were determined with the help of the human eye. The number of galaxies was restricted by including only objects with angular diameters larger than 1'. This limit roughly corresponds to 14.5 magnitude and had the advantage that the detected systems showed enough structure to classify them morphologically. 14000 galaxies passed the diameter limit and together with about 2000 peculiar galaxies, 1000 star clusters and 1000 planetary nebulae they were assembled in a single volume (ESO/ Uppsala Catalogue, A. Lauberts, 1982).

As soon as ESO initiated the new red survey on IIIa-F emulsion, plans developed to extract for all 16000 galaxies all possible photometric and morphological parameters from the complete set of B and R survey plates. At that time less than 10% of the ESO catalogue galaxies had published magnitudes and almost no red photometry existed for these objects. Detailed photometry was known for perhaps 100 of the brightest objects. Here we describe an extensive project that aims at calibrating both the B and R survey plates and extracting automatically both photometric and morphological parameters for all the galaxies present in the catalogue. A flow chart linking the different steps of the project is given in Fig. 1 (Lauberts and Valentijn, 1983). Essentially the project contains the following parts:

1. Scan the 16000 galaxies with the PDS on 606 Blue and 606 Red original ESO survey plates.

2. Bring together existing photometry in a catalogue and add complementary photometry using own measurements at the ESO 1 meter telescope.

3. Calibrate the plates and determine automatically the properties of the galaxies using the ESO VAX computers.

4. Produce a catalogue on paper, magnetic tape and possibly on video disk.

5. Investigate the data base scientifically.

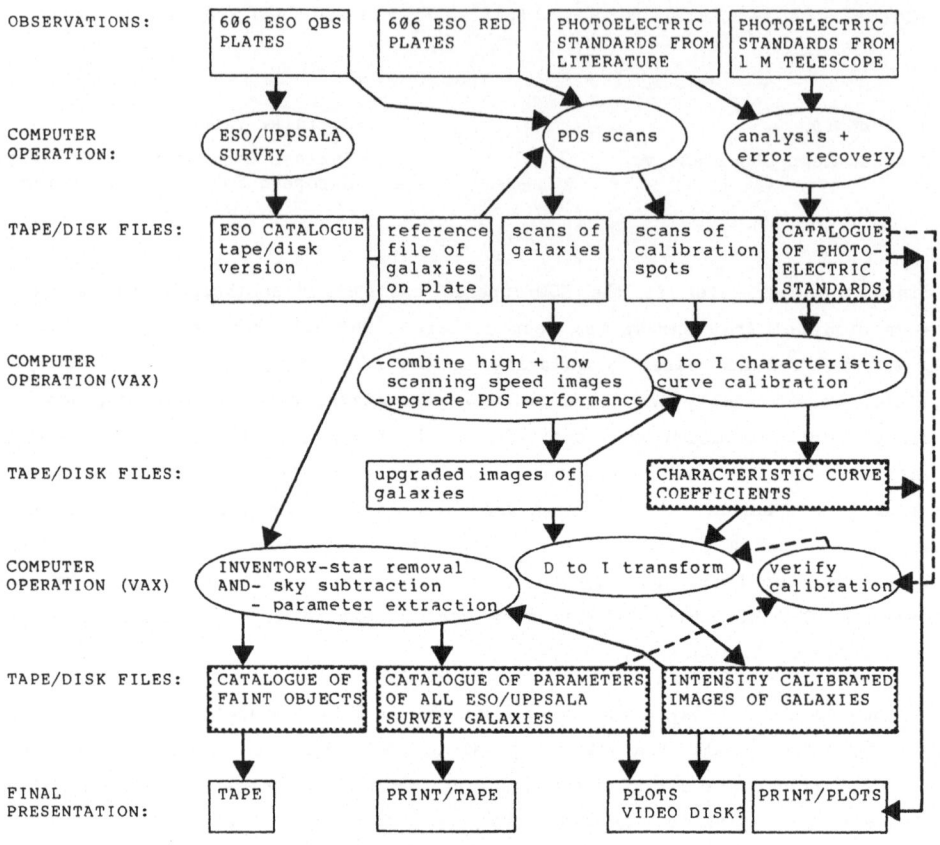

OBSERVATIONS:

- 606 ESO QBS PLATES
- 606 ESO RED PLATES
- PHOTOELECTRIC STANDARDS FROM LITERATURE
- PHOTOELECTRIC STANDARDS FROM 1 M TELESCOPE

COMPUTER OPERATION:

- ESO/UPPSALA SURVEY
- PDS scans
- analysis + error recovery

TAPE/DISK FILES:

- ESO CATALOGUE tape/disk version
- reference file of galaxies on plate
- scans of galaxies
- scans of calibration spots
- CATALOGUE OF PHOTO-ELECTRIC STANDARDS

COMPUTER OPERATION(VAX):

- -combine high + low scanning speed images -upgrade PDS performance
- D to I characteristic curve calibration

TAPE/DISK FILES:

- upgraded images of galaxies
- CHARACTERISTIC CURVE COEFFICIENTS

COMPUTER OPERATION (VAX):

- INVENTORY-star removal AND- sky subtraction - parameter extraction
- D to I transform
- verify calibration

TAPE/DISK FILES:

- CATALOGUE OF FAINT OBJECTS
- CATALOGUE OF PARAMETERS OF ALL ESO/UPPSALA SURVEY GALAXIES
- INTENSITY CALIBRATED IMAGES OF GALAXIES

FINAL PRESENTATION:

- TAPE
- PRINT/TAPE
- PLOTS VIDEO DISK?
- PRINT/PLOTS

Fig. 1: Flow chart of the automatic parameter extraction listing the computer operations and the resulting disk files. Marked disk files indicate that they provide a useful data base to the astronomical community.

By October 1984 all 606 B plates and over 400 R plates have been scanned, a catalogue of photometric UBVRI standards has been compiled and after thorough tests on ESO field 358 satisfactory versions of the software are in operation.

In the following we give more detailed information on the different steps and present some results of the test field, covering the Fornax cluster.

Digitizing the Plates

The plates are positioned emulsion up in the PDS machine, set to zero density at the plate fog level. From a reference file the HP computer selects the objects to be scanned. An area as large as three times the visible size (about 25 mag B surface brightness) of the object is scanned at high speed. A second scan of the central area of presumably much higher density is performed at a lower speed. Later the two images are combined into one. The artifacts of the logarithmic amplifier of the PDS (prior to 1984) which produced asymmetric profiles at high densities are mainly recovered by a 2-dim interpolation. January 1984 a much faster amplifier with a 12 bits A-D converter was installed resolving most of the previous problems. The scanning procedure is straightforward although much care must be taken for plate alignment and focussing.

Photometric Calibration

During three runs (1982-1983) at the ESO 1 meter telescope, multi-aperture photoelectric U,B,V,R,I measurements of 191 standard elliptical and S0 galaxies have been acquired, selecting one object per survey field (Lauberts, 1984). Together with existing measurements 7000 entries have been created in a disk file for the photometric calibration of the survey plates (Lauberts and Sadler, 1984). Polynomial fits to the observed magnitude versus the log of the aperture used have been made for every standard galaxy. To recover any mishaps in the observations, a plotted version of the photometric catalogue was made by drawing the radial distribution of the V, U-B, B-V and V-R measurements together with the fits.

The characteristic curve $I = I(D)$ is found by fitting $\log I = A*\log(D-Dfog) + B*\log(Dsat-D) + C$ to both the standard galaxy aperture photometry and the measured calibration spots, the latter being used for the lower intensities. This formula has been adopted from Llebaria and Figon in "Proceedings on Astronomical Photography", Nice 1981. The resulting transformation coefficients are stored in disk files. At present we have data to calibrate about 400 Blue and 250 Red original plates.

Only one standard galaxy is used for the determination of the characteristic curve, leaving the other standards for a verification of the residuals of the photoelectric versus photographic measurements. In Fig. 2, representing the data for about 20 galaxies in field 358, we see a clear division in three parts of surface brightness. The calibration was given the highest weight in the middle portion with the B surface brightness ranging from 21 to 23. On the low intensity side the residuals fluctuate since the photoelectric observations themselves become inaccurate at levels just above the sky brightness. On the high intensity side the residuals grow systematically reflecting the lower accuracy of photographic methods, mostly due to emulsion saturation at high densities (D>3 on Blue IIa-O, D>4 on Red IIIa-F). To some extent these errors may be recovered in the final analysis by combining the results from different plates.

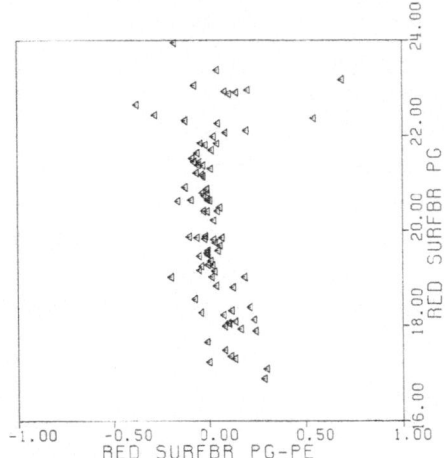 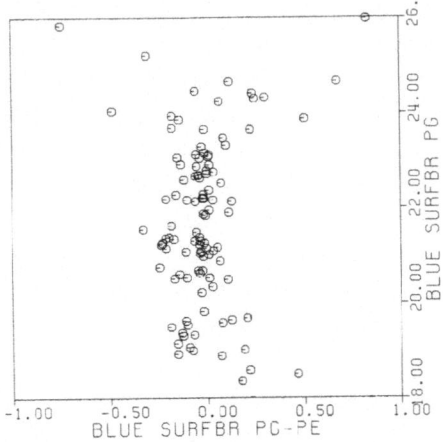

Fig. 2: The residuals of photographic and photoelectric measurements versus the red and blue surface brightness are given in the right and left diagram respectively. The diagrams include data for about 20 galaxies in ESO field 358.

Providing a catalogue with the photometry of standard galaxies and the list of coefficients of the characteristic curve may be considered as the first two products of our project.

The Automatic Parameter Extraction

The images of the galaxies are converted from density into intensity, normalized to sky units, and are then ready for further analysis.

First a version of the programme INVENTORY is run which detects and classifies all objects present in a single frame of a target galaxy and whose central surface brightness exceeds that of the sky by a factor 1.5. On average 20 such objects are found per frame and their positions, magnitude and classification are calculated using a reference point spread function. The data are stored in a separate disk file for later investigation. In the future this data bank will be used to search for

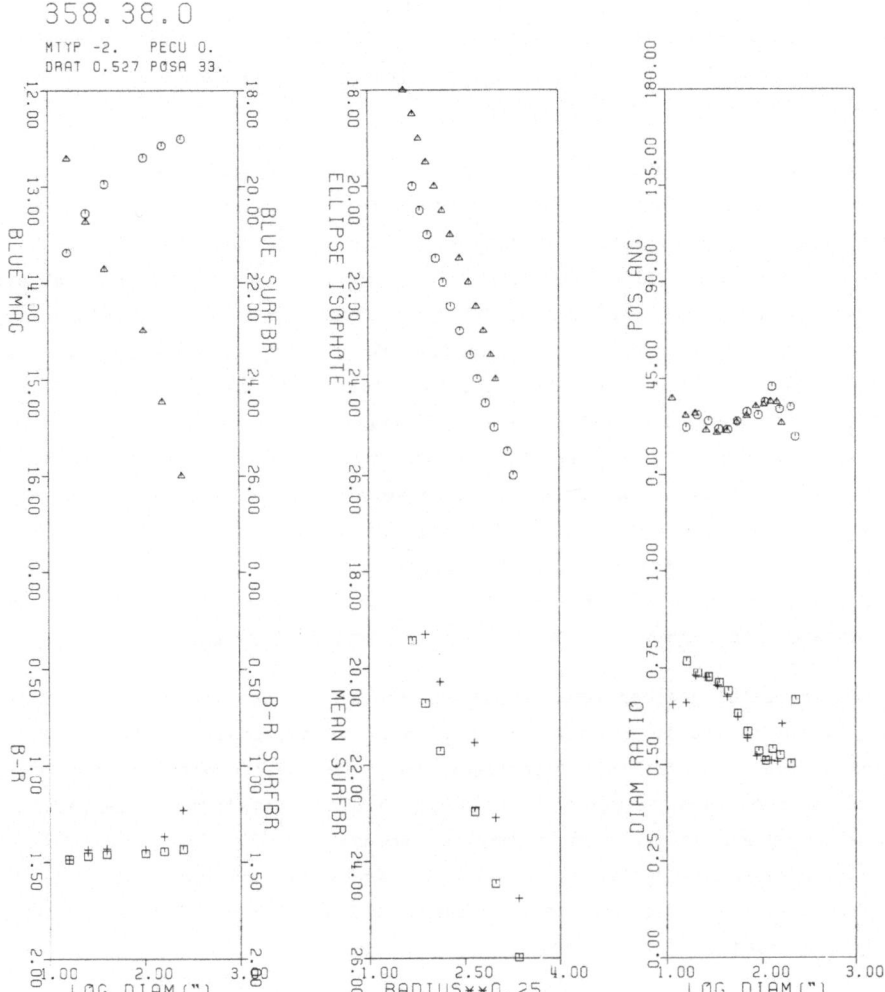

**Fig. 3:** Photographic properties derived for the SO galaxy ESO 358-G38 in the Fornax cluster. BLUE MAG vs LOG DIAM (circle): Integrated blue magnitude versus log major diameter (arcsec), sampled in concentric ellipses with same position angle and flattening. For blue surface brightness < 22 the photographic values are replaced by, if available, more accurate photoelectric values in circular apertures. BLUE SURFBR vs LOG DIAM (triangle): Mean blue surface brightness versus log major diameter of the outer ellipse of two bounds. B-R vs LOG DIAM (square): Integrated B-R colour versus log major diameter of the outer ellipse. B-R SURFBR vs LOG DIAM (cross): Local B-R colour versus major log diameter of the outer ellipse. MEAN SURFBR vs RADIUS**0.25 (square: B, cross: R): Mean surface brightness versus major radius**0.25 (arcsec) of the outer ellipse. ELLIPSE ISOPHOTE vs RADIUS**0.25 (circle: B, triangle: R): Ellipse approximation to isophotes on smoothed image versus major radius**0.25. POS ANG vs LOG DIAM (circle: B, triangle: R): Position angle (degrees) versus log major diameter of concentric ellipses fitted to isophotes. DIAM RATIO vs LOG DIAM (square: B, cross: R); Ratio minor to major diameter versus log major diameter of ellipses fitted to isophotes.

peculiar objects, such as quasars and novae, in the neighbourhood of the target galaxies. For instance, an automatic survey for objects with a certain colour excess will be feasible. The INVENTORY programme finally creates an image with all neighbouring objects subtracted from the input frame.

Next, as a first step in a string of our routines, which we have called AND, the sky brightness distribution is approximated by a plane using 8 surrounding subregions. After the subtraction of the sky we are finally set to extract the photometric and structural information from the B and R images, aligned by superposing common stars. Radial B, R and B-R profiles are stored in the disk catalogue together with overall elongations and position angles. For elliptical and SO galaxies concentric ellipses are fitted to the isophotes determining the radial change of their eccentricity and position angle (isophotal twisting). We aim to obtain all these parameters for the 16000 survey galaxies and to publish them on paper. However, a plotted version of the catalogue seems also very useful as is illustrated by Fig. 3 which presents one page of such a catalogue. The SO galaxy NGC 1389/ESO 358-G38 has a total B = 12.5, a central B-R = 1.5 and a bluer halo colour B-R = 1.3 (Cousins systems), follows an r 1/4 law profile and shows, consistently on B and R plates, clear evidence for isophotal twisting and has varying axial ratios.

Once the catalogue has been completed, many items can of course be studied. We are highly interested in studying the influence of the environmental conditions of galaxies on their fundamental properties. Since the sample contains galaxies in all sorts of environments, ranging from purely isolated systems to members of rich clusters or superlusters, this objective becomes fesible. Systematic correlative studies between galaxy radii, colours, colour gradients and isophotal twisting will be possible. At the bottom of the flow chart in Fig. 1 we have indicated one way of presenting the acquired data base.

## References

Lauberts, A., Valentijn, E.A.: 1983, The ESO Messenger 34, 10.
Lauberts, A.: 1984, Astronomy & Astrophysics Suppl. 58, 249.
Lauberts, A., Sadler, E.: 1984, ESO Scientific Report, in press.

## DISCUSSION

F. Schweizer:  How do you combine the fast and slow scans? Do you simply insert the smaller slow scan into the larger fast scan?

A. Lauberts:  Yes, I do.

PHOTOMETRY AND RADIO CONTINUUM OBSERVATIONS OF GALAXIES

R.-J. Dettmar[1], R. Beck[2], R. Wielebinski[1]

1) Max-Planck-Institut für Radioastronomie
Auf dem Hügel 69, D-5300 Bonn 1, FRG

2) Max-Planck-Institut für Kernphysik
Saupfercheckweg 1, D-6900 Heidelberg, FRG

## 1. Introduction

Astronomical observations can now be made in a wide range of the electromagnetic spectrum. Comparable angular resolution is achievable in radio, infrared, optical, ultraviolet and X-ray ranges.

The radio continuum emission of galaxies in the cm to m wavelength range is due to two different physical processes: the thermal (free-free) emission from HII regions ionized by the UV-radiation of high mass stars and the nonthermal (synchrotron) radiation from cosmic ray electrons bound by magnetic fields. We do not discuss the central activity and the weak radio continuum of early type galaxies and therefore we will confine the following discussion to spiral galaxies only. For these types of galaxies the radio continuum traces the high mass end of the stellar mass range. The young OB stars are indirectly responsible for the thermal emission while the nonthermal radiation is most probably due to electrons accelerated in supernova remnant shock fronts. The comparison of optical and radio data opens new insight into the physics of active star formation regions and the interstellar medium.

## 2. The Distribution of Radio Continuum

The radio continuum emission from galaxies can be separated into three distinct components: (1) the nucleus, (2) the "thin disk" connected with the young population and (3) the "thick disk" or flattened halo.

The nucleus is a prominent region in radio continuum in almost every galaxy. Its predominance varies with galaxy type. The most intense central sources are found in elliptical and starburst galaxies like M82 and NGC 253. Statistical arguments by Hummel (1981) indicate however that the radio intensity of the nucleus has no correlation with that of the disk.

In most galaxies more than 90% of the total radio power originates in the total disk (Hummel, 1981). An important contribution to the radio intensity comes from the spiral arms. In high resolution maps radio continuum peaks are associated with the dust lanes. This is generally believed to be a confirmation of the existence of density waves. In these dust lanes we expect compressed magnetic fields which contribute to the enhancement of the synchrotron radiation. However, since in some galaxies the radio continuum peaks do not coincide with dust lanes, this interpretation is not generally applicable.

Observations of edge-on galaxies indicate that cosmic rays and magnetic fields have a much larger scale height in z than the young population. The disk emission could be separated into a "thin" and a "thick" component for NGC 891 and the Galaxy (Allen and Hu, 1984).

Studies of larger samples of galaxies (e.g. Gioia et al., 1982; Harnett, 1982; Israel et al., 1984) showed that the total flux density spectrum is straight (power law) over a large frequency range. Furthermore the spectral index is $\alpha = 0.74$ ($S_\nu \propto \nu^{-\alpha}$) with a remarkably small deviation. Bearing in mind that thermal emission is known to be present in galaxies from optical studies, this value is the lower limit for the nonthermal spectral index. Taking a nonthermal spectral index $\alpha_{nt} = 0.8$ Gioia et al. showed that their sample of galaxies had a thermal component of less than 40% at 10.7 GHz.

Multifrequency maps allow studies of the distribution of the spectral index and are very valuable in enabling us to separate the thermal and nonthermal components. The spectral index variations are not large, and therefore maps of the same angular resolution at widely spaced frequencies are required. Attempts to determine the thermal fraction have been made by numerous authors (e.g. Beck and Gräve, 1982; Klein et al., 1982). The nonthermal spectral index was found to be around 0.8 with a steepening of about 0.2 towards the edges of the galaxies. Assuming this spectral index to be constant across the inner disk of the galaxy the variation of the thermal fraction could be determined.

More detailed discussions about the radio continuum emission of galaxies are given by Ekers (1981), Sancisi and van der Kruit (1981), Wielebinski (1983), and Beck and Reich (1984).


3. Origin of Nonthermal Radio Continuum

Cameron (1971) showed first that the total optical luminosity and the radio power correlate rather well. This implies that the radio power increases with the number of stars. More detailed studies of radio emissivity, integrated colours and Hα fluxes (Klein, 1982; Kennicutt, 1983) indicated that nonthermal emission is related to the young stellar population and hence the birth rate of massive stars. Further evidence that massive stars are responsible for the nonthermal emission is given by the strong correlation between radio continuum at 6 cm and the IRAS 60 μm and 100 μm fluxes (Klein et al., 1984a).

This result seems incompatible with the scale lengths of synchrotron radiation which are several kpc longer than those of HII regions (Berkhuijsen and Klein, 1984). As synchrotron emission reflects the combined distribution of cosmic ray electrons and interstellar magnetic fields, both of these distributions must be broad while the energy may be provided by sources in the spiral arms, most probably supernovae.

Single SNR's are strong radio sources, but their contribution to the total radio emission of a galaxy is small. After acceleration in the shock fronts of SNR's, cosmic ray electrons diffuse into the hot interstellar medium (see model by Bogdan and Völk, 1983). This view is supported by the determination of the mean synchrotron spectral index which is significantly steeper than that of supernova remnants indicating diffusive losses of cosmic rays in the interstellar medium. The understanding of cosmic ray propagation may also benefit from the observation that the spectral index in the "thick disk" of edge-on galaxies is steeper than near the plane (Allen and Hu, 1984).

More detailed models of the particle acceleration in SNR's require more information about the evolution of SNR's with different explosion energies and with different properties of the surrounding interstellar medium. Progress on the question of the propagation of cosmic rays can be expected from the observation of the variation of synchrotron intensities and radio spectral indices and from the distribution of massive stars, Hα emission and SNR's in spiral galaxies.


4. Origin of Thermal (free-free) Emission

In external galaxies the thermal fraction is small even at cm wavelengths. Radio measurements reveal the total thermal emission unaffected by extinction. Therefore this thermal flux density is usually higher than that expected from Hα observations. The thermal flux density yields the number of Lyc-photons $N_{Lyc}$ produced by the ionizing stars. This gives information on the upper end of the PDMF and SFR if $N_{Lyc}$ is compared to the number of OB stars (Berkhuijsen, 1983).

The comparison of the thermal radio emission with the calibrated Hα emission yields an estimate of extinction and its variation within a galaxy (e.g. Berkhuijsen, 1983; Klein et al., 1984b). The galaxies studied so far show some indication for varying extinction with radius which may correlate with the variation of other

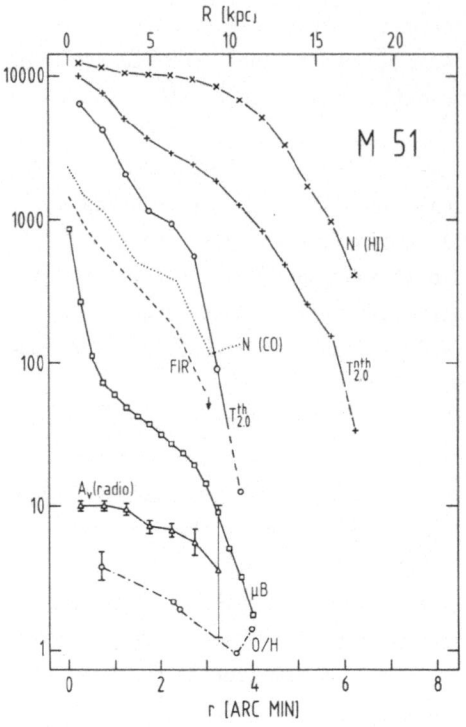

Fig. 1: Radial distributions of several constituents in M51: the blue light, the far infrared luminosity, the CO column density, the thermal and nonthermal radio brightness temperature at 14.7 GHz and the optical extinction. Relative units are used. (From Klein et al., 1984b)

quantities like metallicity (see Figure 1).

If the extinction is known from measurements of the Balmer decrement, an independent test for the separation of thermal and nonthermal emission is possible. This is necessary to investigate whether the nonthermal spectral index is in fact almost constant across the inner parts of a galaxy.

## 5. Magnetic Fields and Dust

Magnetic fields are known to be present in galaxies from optical polarisation measurements (for a review see Elvius, 1978). The optical results indicate fields aligned along the spiral arms. The detection of radio polarization in M51 and M81 by Segalovitz et al. (1976) confirmed the existence of large-scale magnetic fields in nearby spiral galaxies. The 'E' vectors were perpendicular to the spiral arms with a rather uniform distribution. Some more galaxies are studied until now showing again 'E' vectors at 90° to the arms and dust lanes.

A more general pattern of magnetic fields in galaxies has now emerged. The magnetic fields are aligned along the spiral arms with a high degree of uniformity in those galaxies where density waves are thought to be strong. The mean ratio of the uniform to random component of the field shows some correlation with the absolute luminosity of a galaxy (Beck and Reich, 1984). The magnetic field strengths (using the energy equipartition argument) are in the range 3 - 10 µG. These values imply that magnetic fields should be considered as an important parameter in the evolution of spiral structure.

Gas motion and differential rotation can drive a dynamo to maintain the interstellar magnetic field in a torus-like configuration, as observed in M31 (Beck, 1982). Dynamo theory gives predictions for the radial variation of the magnetic field strength (Ruzmaikin and Shukurov, 1981) which can be tested if the distribution of cosmic rays is known. The interpretation of magnetic fields in terms of a bisymmetric open configuration was given by Tosa and Fujimoto (1978). Observations of the rotation measures can distinguish between both theories. The questions of the small-scale structure of magnetic fields, the occurrence of reversals etc. remain open.

The polarized synchrotron emission depends on the strength, orientation and alignment of interstellar magnetic fields while polarized starlight, as a result of the orientation of dust particles (Davis-Greenstein-mechanism), mainly depends on the size, shape and composition of the particles. Coordinated polarization observations in the radio and optical range could help investigating the magnetic field structure and properties of the interstellar dust. More detailed optical polarization measurements are therefore highly desirable and within the possibilities of modern detectors (McLean et al., 1983).

## 6. Future Prospects

Beside the desired observations of Pop I objects and optical polarization, observations of blue compact galaxies are of special interest since they lack any ordered pattern but nevertheless exhibit high star formation rates occurring in bursts (Klein et al., 1983). Their spectra are significantly flatter than those of spiral galaxies. The nonthermal emission expected from SNe seems to be very small. The observed nonthermal emission is probably a superposition of single SNRs and the relativistic electrons are not stored in these galaxies. Blue compact galaxies offer the chance to study a small portion of interstellar medium without the effect of fast rotation, density waves and dynamos.

We wish to thank Drs. E. Berkhuijsen, E. Hummel and U. Klein for very helpful discussions and G. Breuer for typing the manuscript.

References

Allen, R.J., Hu, F.X.: 1984, this conference
Beck, R.: 1982, Astron. Astrophys. 106, 121
Beck, R., Gräve, R.: 1982, Astron. Astrophys. 105, 192
Beck, R., Reich, W.: 1984, in: IAU Symp. No. 106 "The Milky Way Galaxy", (ed. H. van
    Woerden), D. Reidel Publ. Co., Dordrecht, in press
Berkhuijsen, E.M.: 1983, Astron. Astrophys. 127, 395
Berkhuijsen, E.M., Klein, U.: 1984, in: IAU Symp. No. 106 "The Milky Way Galaxy",
    (ed. H. van Woerden), D. Reidel Publ. Co., Dordrecht, in press
Bogdan, T.J., Völk, H.J.: 1983, Astron. Astrophys. 122, 129
Cameron, M.J.: 1971, Monthly Notices Roy. Astron. Soc. 152, 403
Ekers, R.D.: 1981, in: "The Structure and Evolution of Normal Galaxies", (eds. S.M.
    Fall and D. Lynden-Bell), Cambridge
Elvius, A.: 1978, Astrophys. Space Sci. 55, 49
Gioia, I.M., Gregorini, L., Vettolani, G.: 1982, Astron. Astrophys. 116, 164
Harnett, J.I.: 1982, Australian J. Phys. 35, 321
Hummel, E.: 1981, Astron. Astrophys. 93, 93
Israel, F.P., Hulst, J.M. van der: 1983, Astron. J. 88, 1736
Kennicutt, R.: 1983, Astron. Astrophys. 120, 219
Klein, U.: 1982, Astron. Astrophys. 116, 175
Klein, U., Beck, R., Buczilowski, U.R., Wielebinski, R.: 1982, Astron. Astrophys.
    108, 176
Klein, U., Gräve, R., Wielebinski, R.: 1983, Astron. Astrophys. 117, 332
Klein, U., Jong, T. de, Wielebinski, R., Wunderlich, E.: 1984a, Astron. Astrophys.,
    submitted
Klein, U., Wielebinski, R., Beck, R.: 1984b, Astron. Astrophys. 135, 213
McLean, I.S., Aspin, C., Reitsema, H.: 1983, Nature 304, 243
Ruzmaikin, A.A., Shukurov, A.M.: 1981, Sov. Astron. 25, 553
Sancisi, R., Kruit, P.C. van der: 1981, in: IAU Symp. No. 94 "Origin of Cosmic Rays",
    (eds. G. Setti, G. Spada and A.W. Wolfendale), D. Reidel Publ. Co., Dordrecht,
    p. 209
Segalovitz, A., Shane, W.W., Bruyn, A.G. de: 1976, Nature 264, 222
Tosa, M., Fujimoto, M.: 1978, Publ. Astron. Soc. Japan 30, 315
Wielebinski, R.: 1983, 18th Intern. Cosmic Ray Conf., Bangalore 061-1, p. 161

# GALAXY PHOTOMETRY IN THE INFRARED

A.F.M. Moorwood
European Southern Observatory
D-8046 Garching bei München

## 1. Introduction

Both the scope and the opportunities for obtaining infrared photometric data on galaxies have increased substantially during the last
few years following the provision by most of the major observatories
of common user infrared photometers on 3-4m class telescopes. Included
amongst these are the 3m NASA IRTF and 3.8m UKIRT dedicated infrared
facilities on Mauna Kea. Not only the quantity but also the quality of
infrared photometry has also improved as a result of both instrumental
developments and the greater effort devoted to achieving accurate,
well calibrated data. In the first section of this review I will
briefly elaborate on some of these aspects by way of an introduction
to the infrared photometric system and its capabilities which I hope
will prove useful to non-infrared specialists.

Within the space allocated to this contribution it is impossible to
attempt anything approaching a comprehensive review of the, now
extensive, applications and results of infrared photometry in galaxy
research. By drawing primarily on recent work in the areas of star
formation and cosmology however I will attempt to emphasize both some
unique aspects of infrared photometry and, in line with the theme of
this session, its complementarity with photometry at shorter
wavelengths.

## 2. The Groundbased Infrared Photometric System

Infrared photometry from the ground is restricted to a number of
atmospheric transmission windows which are shown on Fig. 1 together
with the designations and centre wavelengths of the most commonly used
broad-band filters. The best windows are J, H, K and N where the
extinction is typically $\lesssim$ 0.1 mag./air mass. As can be seen on the
figure, transmission at $\lambda$ > 18$\mu$m is critically dependent on atmospheric $H_2O$ content and observations in this range are consequently
limited to excellent mountain altitude sites.

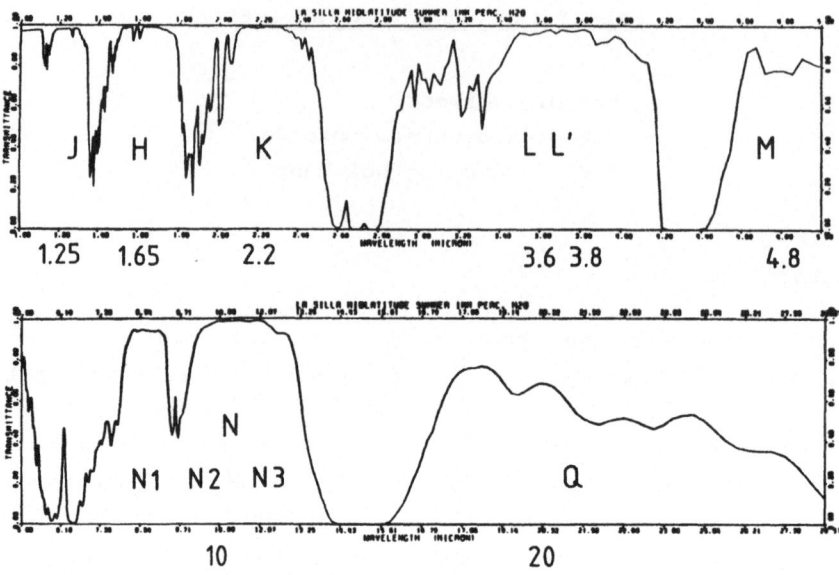

1.25  1.65    2.2                                          3.6 3.8              4.8

10                              20

Fig. 1. Atmospheric transmission for 1mm of precipitable $H_2O$ above a site at 2400m altitude (computed by G. Finger at ESO using the LOWTRAN program). Letter designations and centre wavelengths are shown for the principal IR photometric bands.

Airglow and thermal emission from the atmosphere and telescope determines the limiting performance of broad band photometry ($\Delta\lambda/\lambda > 0.1$) over most of this wavelength range. At $\lambda > 2\mu m$, usually called the thermal infrared, this background radiation exceeds that from faint astronomical sources by many orders of magnitude. In order to isolate the source signal therefore it is usually necessary to employ a combination of sky chopping and beam switching i.e. a suitable optical element (usually the telescope secondary) is driven such that the detector alternately views two fields projected on the sky (at ~ 10 Hz) while the telescope is moved periodically ($\lesssim$ .1 Hz) so as to centre the astronomical object alternately in the two fields. Using this technique, both the sky emission and any signal resulting from differences in telescope background between the two chopped fields can thus be cancelled. Shot noise on the background radiation cannot be cancelled of course and this determines the achievable magnitude limits. Using a 4m class telescope and diaphragms of a few arc sec., practical magnitude limits are around 20 at J and H, 18 at K, 12 at L, 8 at N and 4 at Q. In relating these to optical photometry it should be noted that the stellar light from normal galaxies has

V-K ≃ 3 at zero redshift and ≃ 6 at z ≃ 1. Large samples of galaxies are thus now accessible for near infrared photometry and the detection limit for distant galaxies at K is competitive with optical CCD's. At longer infrared wavelengths, direct starlight can only be observed in bright, nearby, galaxies and most programmes in the thermal infrared are directed towards active and star burst galaxies which exhibit infrared excess emission above the stellar continuum.

Standardization of near infrared photometry has improved considerably in recent years and several lists of standard stars have now been published. Curiously, these originate mostly from Southern Hemisphere observatories: AAO (Allen and Cragg, 1983), CIT/CTIO (Elias et al., 1982), ESO (Engels et al., 1981; Koornneef, 1983), SAAO (Glass, 1974; Carter, 1984, in preparation). These systems exhibit internal consistencies in the 0.01-0.03 mag. range and colour equations for transforming between them are discussed in the above references. They all yield close to zero magnitude in all bands for α Lyr and thus provide a natural extension of the optical photometric system. For the mid-infrared, a limited list of standards useable around 10μm is given in Thomas et al. (1973) and the 20μm system has recently been discussed by Tokunaga (1984).

3. Some Astronomical Applications of IR Galaxy Photometry

i) Overview. The sensitivity of currently available instrumentation is now such that all known classes of extragalactic objects can be observed at least in the near infrared. This includes not only visually identified objects but also certain radio galaxies with no detectable or extremely faint optical counterparts (Rieke et al., 1979; Lilly and Longair, 1984). Early results from the IRAS satellite all sky infrared survey indicate that several thousand galaxies have been detected at λ ~ 100μm and that a substantial fraction of these are radiating the bulk of their luminosity in the infrared (Soifer et al., 1984). The subject of infrared galaxy photometry is thus an extensive one and one that is growing rapidly to include most areas of extragalactic research.

In terms of radiation mechanisms, the infrared region is unique in providing information on thermal emission from dust grains heated by recently formed massive stars or by the non-thermal central sources in active galaxy nuclei. The discovery of infrared sources of this type and the possibility of also detecting pure non-thermal emission in AGN's in fact established and then dominated infrared galaxy photometry in the early 1970's. This is still an important field of research which continues to generate considerable debate on the nature and origin of the infrared emission from QSO's, BL Lacs, Seyfert galaxies and LINERS. IRAS data at longer wavelengths may have a considerable impact. Given the restricted length of this contribution however I will avoid further discussion of this issue by referring to the excellent review by Rieke and Lebofsky (1979) and to some more recent papers (Glass and Moorwood, 1984; Glass et al., 1982; Allen et al., 1982; Hyland and Allen, 1982; Ward et al., 1982). In order to illustrate some of the specific features of infrared photometry I will discuss in somewhat more detail a number of recent developments in the areas of star formation and cosmology which are probably less widely known.

ii) <u>Star Formation</u>. The presence of hot, massive stars in galaxies has traditionally been detected by searching for blue visible-ultraviolet colours and/or optical emission lines from ionized gas. Infrared surveys of spiral galaxy nuclei however indicate that it may be more common for the luminosity of recently formed OB stars to be absorbed and re-radiated at mid-far infrared wavelengths by dust associated with their parent molecular cloud/HII region complexes. Adopting emission at 10μm as an indicator for example, Rieke and Lebofsky (1978) concluded from a survey of nearby galaxies that recent star formation has probably occurred in ~ 40% of all spiral galaxy nuclei. Becklin et al. (1984) suggest that this figure may be closer to 100% for spirals in the Virgo cluster and a similarly high efficiency for star formation induced by galaxy interactions has been claimed by Joseph et al. (1984) on the basis of a 3.8μm survey.

A number of galaxies exhibiting extremely prominent thermal infrared spectra and luminosities of typically $\gtrsim 10^{10}$ $L_\odot$ have been studied in more detail with the conclusion that intense bursts of massive star formation have indeed occurred recently in their nuclei (e.g. M82 and NGC 253 (Rieke et al., 1980), IC 342 (Becklin et al., 1980), Circinus and NGC 4945 (Moorwood and Glass, 1984), NGC 6240 and Arp 220 (Rieke et al., 1984)). The latter two represent the most luminous examples known with $L_{IR} \sim 10^{12}$ $L_\odot$ as revealed by IRAS observations. Fig. 2 is a plot of the ultraviolet-radio energy distribution of the Circinus galaxy showing the characteristically steep rise through the mid and

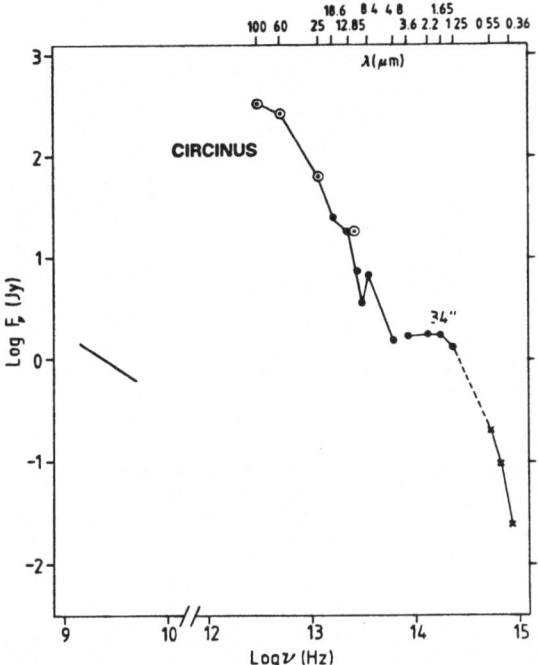

Fig. 2. Energy distribution from ultraviolet to radio wavelengths for the Circinus galaxy (Moorwood and Glass, 1984). Solid dots represent groundbased IR observations while the dots in circles are from IRAS.

far infrared. In several of these cases however, caution must be exercised before attributing all of the infrared luminosity to star formation. Some for example exhibit Seyfert 2 or LINER rather than the intense HII region type optical spectrum which characterizes the prototype starburst galaxy NGC 7714 as defined by Weedman et al. (1981). The Seyfert 1 galaxy NGC 7469 provides an example of star formation in a region surrounding an obviously non-thermal active nucleus (Cutri et al., 1984). Whether Seyfert and starburst activity sometimes co-exist by chance, whether they are related in a physical or evolutionary way and whether Seyfert and/or LINER type optical spectra can result from starbursts alone in the absence of non-thermal sources of ionization are all important questions for future work.

The contribution of near infrared photometry to star formation studies stems curiously enough largely from the fact that near infrared colours are dominated by the older stellar population and are relatively insensitive to the presence of hot stars. As pointed out by Struck-Marcell and Tinsley (1978) however, this means that the

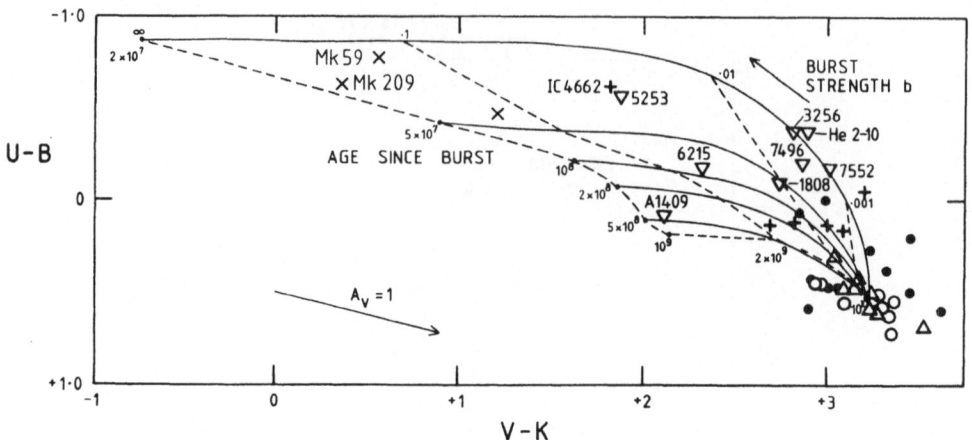

Fig. 3. Observed galaxy colours compared with model predictions (see text). Symbols are: O - ellipticals, Δ - lenticulars, • - spirals, + - emission line spirals, ∇ - emission line spirals with IR excesses (Glass and Moorwood, 84) and × - blue compact dwarf galaxies (Thuan, 83).

position of a galaxy in the U-B, V-K two colour diagram should be extremely sensitive to its star formation history. Their own model predictions for a starburst lasting for $2 \cdot 10^7$ years in an old stellar population are shown in Fig. 3 together with recent observational data for various galaxy types as identified in the caption. What is observed is a steady progression in the direction of mainly burst strength (fractional galaxy mass involved in the burst) from normal galaxies to ones with optical emission lines to ones with optical emission lines plus infrared excesses to the most extreme blue compact dwarf galaxies. Significant colour evolution can be seen to be present already when only ~ 0.1% of the galaxy mass is involved in the burst. Highly obscured regions of star formation such as those discussed above are not of course revealed in this diagram. A nice example of the complementarity with mid-infrared photometry however is provided by the E/S0 galaxy NGC 5253. Its position in the U-B, V-K diagram indicates recent star formation and is consistent with the presence of a very obvious and extended visible HII region complex. Its mid-infrared properties however suggest that massive stars are forming now in a compact, visually obscured, region at the centre of the galaxy (Moorwood and Glass, 1982). Further work is now required in this area to include colour effects due to the presence of supergiants and ionized gas not included in the original models (Huchra, private communication, Terlevich, comment appended to this contribution).

The possibility of combining UBV and near infrared photometry to study star formation occurring under entirely different conditions from those discussed so far has been demonstrated recently by Valentijn and Moorwood (1984). They obtained infrared photometry as a function of radial distance out to ~ 100 kpc from the nucleus of the central cD galaxy in the A496 cluster which appears to be accreting intracluster gas in an X-ray cooling flow. Visible-infrared colour indices were then determined by integrating digitized U, B, V electronographic images under simulated apertures corresponding to those used in the infrared. Population synthesis models constructed to fit the observed colour gradients lead to the conclusion that the main sequence stars are characterized by an upper mass cutoff which increases from ~ 0.5 $M_\odot$ at the centre to ~ 1.6 $M_\odot$ at 100 kpc and the contribution from giants increases with radial distance from the centre. Support for the idea that the currently observed stars may have all formed in the cooling flow comes independently from the X-ray data. These yield a radial dependence of the Jeans mass which mimics the derived upper mass cutoff and a total accreted gas mass equal to the total mass of stars required in the population synthesis.

iii) <u>Cosmology</u>. The fact that the near infrared emission from most galaxies appears to be dominated by late type giants makes this an attractive spectral region for several cosmological applications. Aaronson and Mould (1983) for example have shown that the Tully-Fisher relationship is much tighter and probably less type dependent if infrared H (1.6µm) rather than B magnitudes are used. This relationship between magnitude and hydrogen 21cm line width reflects the underlying luminosity-mass relationship and the smaller dispersion probably reflects both the fact that H magnitudes are a better measure of the old galaxy population and that the extinction at this wavelength is far lower than in the blue. Observations of ~ 300 galaxies have now been used to determine distances to several clusters and lead to a value of $H_o \approx 80$ km s$^{-1}$ Mpc$^{-1}$. So far, calibration of this relationship is based on existing distance calibrators. Independently however, McGonegal et al. (1983) have now undertaken a redetermination of the Local Group distance scale based on infrared photometry of Cepheids. This is expected to yield a more reliable result than previous optical work because the period-luminosity relationship in the infrared appears to again be tighter and suffer less from extinction, temperature, metallicity etc.

Another consequence of the fact that the near infrared emission from galaxies is dominated by giants is the lack of colour evolution. Models due to Bruzual (1983) predict no significant colour evolution in the near infrared out to $z \sim 2$. Observations now seem to confirm this out to $z = 0.6$ for an optically selected sample (Ellis and Allen, 1983) and to $z = 1.2$ for a radio selected sample (Lilly and Longair, 1984). An interesting potential application of infrared photometry therefore is the determination of redshifts for distant galaxies. While confirming the absence of infrared colour evolution, both of these studies have revealed strong visible-infrared colour evolution which appears to reflect increased star formation activity in the past.

Lilly and Longair have also used their data to construct a K ($2.2\mu m$), z Hubble diagram with the aim of determining $q_o$. Although promising, further work now seems to be required. Assuming no luminosity evolution, the value obtained is $\sim 3$. Alternatively it can be argued that a value of $< 0.5$ is consistent with luminosity evolution at K amounting to $\simeq 1$ mag. at $z = 1$. If this is the case however then this reflects the evolution of the underlying old stellar population and is thus likely to be much better behaved than that at shorter wavelengths where an additional and, apparently, random contribution due to bursts of star formation is also present.

4.  Future Prospects

There is no shortage of astrophysically interesting problems which infrared photometry can tackle in the coming years. Until now however infrared photometers have, of necessity, employed only single element detectors and have thus been capable of providing only limited spatial coverage (by mapping) and resolution. This situation is now changing rapidly with the development of various panoramic CCD, CID and other detectors which are sensitive in the groundbased infrared region. Although not yet in routine astronomical use, devices with up to $32 \times 32$ pixels have already been used for galaxy imaging (McCreight, 1984) and the further development of these devices should clearly increase not only the data rate but also the scope for infrared galaxy photometry in the future.

## 5. References

Aaronson, M., Mould, J.: 1983, Ap.J., 265, 1.

Allen, D.A., Cragg, T.A.: 1983, Mon. Not. R. astr. Soc., 203, 777.

Allen, D.A., Ward, M.J., Hyland, A.R.: 1982, Mon. Not. R. astr. Soc., 199, 969.

Becklin, E.E., Gatley, I., Matthews, K., Neugebauer, G., Sellgren, K., Werner, M.W., Wynn-Williams, C.G.: 1980, Ap.J., 236, 441.

Becklin, E.E., Devereux, N.A., Capps, R.W., Scoville, N.Z. Young, J.: 1984, Bull. Am. Astr. Soc. - Proceedings of 162nd AAS meeting, Minneapolis.

Bruzual, G.: 1983, Ap.J., 273, 105.

Cutri, R.M., Rudy, R.J., Rieke, G.H., Tokunaga, A.T., Willner, S.P.: 1984, Ap.J., 280, 521.

Elias, J.H., Frogel, J.A., Matthews, K., Neugebauer, G.: 1982, Astron. J., 87, 1029.

Ellis, R.S., Allen, D.A.: 1983, Mon. Not. R. astr. Soc., 209, 685.

Engels, D., Sherwood, W.A., Wamsteker, W., Schultz, G.V.: 1981, Astron. Astrophys. Suppl., 45, 5.

Glass, I.S.: 1974, Mon. Not. astr. Soc. S. Africa, 33, 53.

Glass, I.S., Moorwood, A.F.M., Eichendorf, W.: 1982, Astron. Astrophys., 107, 276.

Glass, I.S., Moorwood, A.F.M.: 1984, Mon. Not. R. astr. Soc. (in press).

Hyland, A.R., Allen, D.A.: 1982, Mon. Not. R. astr. Soc., 199, 943.

Joseph, R.D., Meikle, W.P.S., Robertson, N.A., Wright, G.S.: 1984, Mon. Not. R. astr. Soc., 209, 111.

Koornneef, J.: 1983, Astron. Astrophys. Suppl., 51, 489.

Lilly, S.J., Longair, M.S.: 1984, Mon. Not. R. astr. Soc. (submitted).

McCreight, C.R.: 1984, Proceedings of IAU Coll. 79 (edited by M.-H. Ulrich and K. Kjär, ESO).

McGonegal, R., McAlary, C.W., McLaren, R.A., Madore, B.F.: 1983, Ap.J., 269, 641.

Moorwood, A.F.M., Glass, I.S.: 1982, Astron. Astrophys., 115, 84.

Moorwood, A.F.M., Glass, I.S.: 1984, Astron. Astrophys., 135, 281.

Rieke, G.H., Lebofsky, M.J.: 1978, Ap.J., 220, L37.

Rieke, G.H., Lebofsky, M.J., Kinman, T.D.: 1979, Ap.J., 232, L151.

Rieke, G.H., Lebofsky, M.J.: 1979, Ann. Rev. Astron. Astrophys., 17, 477.

Rieke, G.H., Lebofsky, M.J., Thompson, R.I., Low, F.J., Tokunaga, A.T.: 1980, Ap.J., 238, 24.

Rieke, G.H., Cutri, R.M., Black, J.H., Kailey, W.F., McAlary, C.W., Lebofsky, M.J., Elston, R.: 1984, Steward Observatory Preprint 541.

Soifer, B.T., et al.: 1984, Ap.J., 278, L71.

Struck-Marcell, C., Tinsley, B.M.: 1978, Ap.J., 221, 562.

Thuan, T.X.: 1983, Ap.J., 268, 667.

Thomas, J.A., Hyland, A.R., Robinson, G.: 1973, Mon. Not. R. astr. Soc., 165, 201.

Tokunaga, A.T.: 1984, Astron. J., 89, 172.

Valentijn, E.A., Moorwood, A.F.M.: 1984, Astron. Astrophys. (in press).

Ward, M., Allen, D.A., Wilson, A.S., Smith, M.G., Wright, A.F.: 1982, Mon. Not. R. astr. Soc., 199, 953.

Weedman, D.W., Feldman, F.R., Balzano, V.A., Ramsey, L.W., Svanek, R.A.: 1981, Ap.J., 248, 105.

## Discussion

R. Terlevich: I would like to mention some recent results of work that will be appearing in MNRAS in the near future and that is a result of the collaboration with Alison Campbell from Cambridge. We performed infrared photometry on a sample of 14 violent star forming regions. Their IR colours show that most of the 2µm flux in these objects is produced by ionized gas and cool evolved stars. The detection of a deep CO index ($\lambda$ = 2.3µm) in two objects reveals that the cool stars are in fact young, luminous, red supergiants and not old red giants.

L. Gouguenheim: You mentioned the infrared Tully-Fisher relation and argued that IR luminosities give a better estimation of total luminosities than the blue ones. However, it must be stressed that total magnitudes are needed (corrected for aperture).

# GALAXY PHOTOMETRY AND X-RAY ASTRONOMY

A.C. Fabian

Institute of Astronomy

Madingley Road

Cambridge CB3 OHA

England

## Introduction

Many nearby normal galaxies have been detected as X-ray sources by the Einstein Observatory. The X-ray emission from the normal stellar component is presumably negligible but populations of X-ray binaries and supernova remnants as well as any hot interstellar medium were observable. The X-ray luminosity of a galaxy is typically about 0.01 per cent of its visual luminosity. The emission from late-type galaxies, which is mostly due to X-ray binaries, is discussed here, then a more detailed account is given of the discovery and implications of substantial amounts of $\sim$ 10 million degree interstellar gas in early-type galaxies. The density of much of this hot gas is high enough that a cooling flow may occur with consequent high-pressure star formation.

## Spiral and Irregular Galaxies

The X-ray appearance of our own Galaxy may be typical of many spiral galaxies. Two major populations of luminous X-ray binaries are evident. Those in the disc of the galaxy are mainly associated with OB stars and are due to accreting neutron star or black hole binary companions (e.g. Cen X-3 and Cyg X-1). The Population II sources concentrated in the 'bulge' and in the globular clusters are in orbit about lower mass stars. The most luminous sources are not much more powerful than $10^{38}$ erg s$^{-1}$, which is the Eddington limit for 1 $M_\odot$ objects. Galactic supernova remnants appear to be at least a factor of 10 - 100 times less luminous.

The Magellanic Clouds are much richer, per unit mass, in luminous Population I X-ray binaries than our Galaxy (Clark et al. 1978). M31 is much richer in X-ray luminous globular cluster and central bulge sources (van Speybroeck et al. 1979). There are no clear answers to these differences, although abundance variation may be an important factor. A number of other nearby spiral galaxies have been resolved in X-rays (Long & Van Speybroeck 1983 and references therein).

Some irregular galaxies in which star formation is particularly active have ratios of X-ray to visual luminosity of between $10^{-1}$ and $10^{-2}$ (Fabbiano et al. 1982; Stewart et al. 1982). It is presumed that many massive X-ray binaries are formed in these regions, although there is no firm evidence.

## Elliptical and SO Galaxies

Extended X-ray emission from a number of early-type galaxies in the Virgo cluster was discovered soon after the Einstein Observatory was launched (Forman et al. 1979). This was in addition to the well-known diffuse X-ray source centred on M87. The X-ray 'plume' to M86 (Forman et al. 1979) suggested that much of the emission was due to hot gas, which in that case was being pushed out of M86 by the ram pressure of the intergalactic (or circum-M87) gas. Diffuse X-ray emission was also found in Centaurus-A (NGC 5128) by Feigelson et al. (1981), who argue convincingly that the emission is most unlikely to be due to stars or to X-ray binaries.

Further observations by Bechtold et al. 1983, Biermann & Kronberg 1983, Nulsen et al. 1984 and, in particular, Forman & Jones 1984, have produced a list of 30 or more X-ray detected elliptical and SO galaxies. Although many of these are in the Virgo cluster, there appears to be no strong environmental influence provided they are not in a rich cluster. Reasonably isolated elliptical galaxies have similar X-ray to optical luminosity ratios ($L_X/L_V$) to those in groups and loose clusters. The dependence of $L_X$ on $L_V$ is steeper ($L_X \propto L_V^{\alpha}$, where $3/2 \lesssim \alpha \lesssim 2$) than for spiral galaxies, where $\alpha \approx 1$. This argues further against a binary source origin (Forman & Jones 1984).

The X-ray spectra of early-type galaxies is best fit by radiation from hot gas with a temperature of about $10^7$K. The emission extends for at least 10 kpc in most cases and may extend much further in some well-studied cases (Forman & Jones 1984). If we adopt the simple view-point that the X-rays are due to hot gas, then there are a number of interesting consequences.

First, we note that the mass of interstellar gas in elliptical galaxies is at least $10^8 M_{\odot}$ within about 10 kpc, and may be considerably more at larger radii. Early-type galaxies have a substantial interstellar medium.

NGC 4472

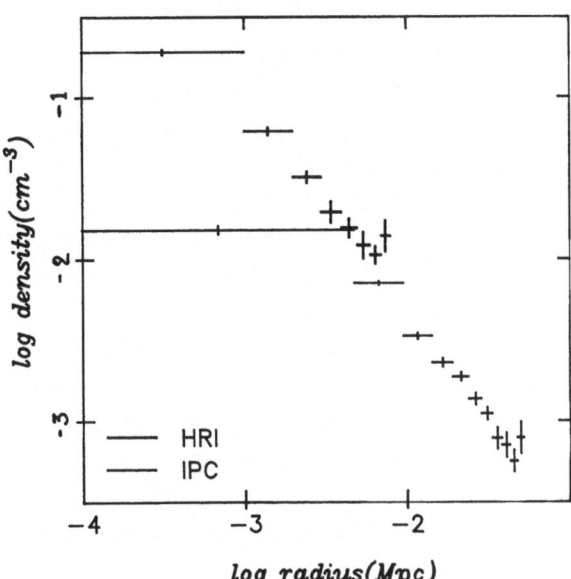

Figure 1. Electron density profile in the Virgo elliptical galaxy, NGC 4472, produced from a deprojection of the X-ray surface brightness by K.A. Arnaud. The X-ray data (Forman & Jones 1984) are from the High Resolution Imager and the Imaging Proportional Counter of the Einstein Observatory.

Second, it is clear that much of this gas must be close to hydrostatic equilibrium. If its motion approaches the sound speed, then the mass flow rate must exceed ~ 10 $M_\odot$ yr$^{-1}$ - far too high for either an outflow or an inflow. Galactic winds (see Johnson & Axford 1971, Mathews & Baker 1971) do not occur in most galaxies. The energy input from supernovae cannot be large.

Third, the equation of hydrostatic support can be used to obtain accurate mass profiles for elliptical galaxies from X-ray observations. Basically

$$\frac{dP_{gas}}{dr} = - \rho_{gas} \, g(r)$$

where $g(r)$ is the acceleration due to gravity $(GM(r)/r)$. The X-ray surface brightness profile and gas temperature distribution give $\rho_{gas}(r)$ and $P_{gas}(r)$ and thus $M(R)$. So far the weak link in this reasoning is the measurement of gas temperature profiles; the Einstein data suggest a roughly isothermal gas with T ≃ 10$^7$K. The gravitating mass of a galaxy M is then proportional to radius;

$$M \approx 10^{12} \left(\frac{T}{10^7 K}\right) \left(\frac{R}{20 \ kpc}\right) M_\odot \ .$$

Finally, the relatively high density of the interstellar gas means that its radiative cooling time $t_{cool}$ is less than the Hubble time within R $\sim$ 20 - 30 kpc. $t_{cool} \lesssim 10^8$ yr within $\cong$ 10 kpc (Nulsen et al. 1984). It is likely that a cooling inflow of gas occurs (for a review see Fabian et al. 1984) involving about 0.1 - 1 $M_\odot$ yr$^{-1}$. A rough steady-state may be reached within the galaxy where stellar mass-loss into the interstellar medium is compensated for by cooling (see also White & Chevalier 1984).

The cooled gas presumably forms stars. The conditions are however different from spiral galaxies and star formation may there proceed with a different initial-mass-function (IMF) than is inferred for our Galaxy (Jura 1977; Fabian et al. 1982; Sarazin & O'Connell 1983). In particular, the pressure is 100 - 1000 times higher than in the local interstellar medium so that the Jeans mass (at a given low temperature of say 10 K or less) is 10 times smaller. The bulk of the star formation may therefore involve low-mass stars which contribute little optical light. This is particularly necessary for the large-scale cooling flows observed in clusters of galaxies and involving hundreds of solar masses per year (see Fabian et al. 1984). The implication of this is that the star formation rate in elliptical galaxies is only slightly less than it is in spiral galaxies. The major difference lies in the stellar IMF. As inflow is involved - individual gas elements probably do not move in more than a factor of two or so in radius - the structure of early-type galaxies must be continuing to evolve. The implied steep IMF for cooling flows provides a source of baryonic matter with a high mass-to-light ratio. The cooling flow may be the 'tail-end' of galaxy formation.

There are, of course, further consequences of such inflows for nuclear activity in elliptical galaxies, for the propagation of radio jets, and for the signs of star formation observed at ultraviolet wavelengths (e.g. Bertola et al. 1979). Unfortunately, further X-ray observations must await the launch of ROSAT in two or three years time.

## References

Bechtold, J., Forman, W., Giacconi, R., Jones, C., Schwarz, J., Tucker,W. & van Speybroeck, L., 1983. Astrophys. J., 265, 26.
Bertola, F., Capaccioli, M., Holm, A.V. & Oke, J.B., 1980. Astrophys. J. 237, L65.

Biermann, P. & Kronberg, P., 1983. Astrophys. J. Lett., 268, L69.
Clark, G., Doxsey, R., Li, F., Jernigan, F.G. & Van Paradijs, J., 1978. Astrophys. J. Lett., 221, L37.
Fabbiano, G., Feigelson, E. & Zamorani, G., 1982. Astrophys. J., 256, 397.
Fabian, A.C., Nulsen, P.E.J. & Canizares, C.R., 1982. Mon. Not. R. astr. Soc., 201, 933.
Fabian, A.C., Nulsen, P.E.J. & Canizares, C.R., 1984. Nature, 310, 733.
Feigelson, E.D., Schreier, E.J., Delvaille, J.P., Giacconi, R., Grindlay, J.E. & Lightman, A.P., 1981. Astrophys. J., 251, 31.
Forman, W., Schwarz, J., Jones, C., Liller, W. & Fabian, A., 1979. Astrophys. J. Lett., 234, L29.
Forman, W. & Jones, C., 1984. Preprint of talk given at Bologna meeting on X-ray Astronomy.
Johnson, H.E. & Axford, W.I., 1971. Astrophys. J., 165, 381.
Long, K. & Van Speybroeck, L., 1983. in Accretion-Driven Stellar X-ray Sources, (ed. Lewin & van den Heuvel), CUP.
Mathews, W. & Baker, J., 1971. Astrophys. J., 170, 241.
Nulsen, P.E.J., Stewart, G.C. & Fabian, A.C., 1984. Mon. Not. R. astr. Soc., 208, 185.
Sarazin, C.L. & O'Connell, R.W., 1983. Astrophys. J., 268, 552.
Stewart, G.C., Fabian, A.C., Terlevich, R.J. & Hazard, C., 1982. Mon. Not. R. astr. Soc., 200, 61P.
Van Speybroeck, L., Epstein, A., Forman, W., Giacconi, R., Jones, C., Liller, W. & Smarr, L., 1979. Astrophys. J., 234, L45.
White, R.E. & Chevalier, R.A., 1983. Astrophys. J., 280, 561.

## DISCUSSION

M. Capaccioli : Both the asymmetry and the distortion presented by the X-ray map of M86 have been detected in the optical image of the galaxy. From a collection of (weak) evidence D. Malin, J.L. Nieto and myself are inclined to believe that there is a physical but transient inter-action between M86 and M84, which took place in tne past $10^8$ years and caused the optical and radio phenomena observed in the two objects, including the supernova explosion which occurred in between them.

J. Nieto : Concerning the mass of M87 determined from X-rays and the underlying hypotheses, I would like to mention that other mass determinations relying on completely different hypotheses highly confirm a mass larger than a few times $10^{12}$ $M_\odot$ beyond 5' - 7' (Nieto and Monnet, 1984, A. & A., to be published; Huchra. 1984, ESO Workshop on the Virgo cluster of galaxies).

O. Gerhard : In order to work out the gas mass involved in the cooling flow, could you give the median inferred gas mass in the ellipticals sample, and the radius at which the cooling time exceeds a Hubble time?

A. Fabian : The unpublished work of Forman & Jones suggests a median gas mass of $\sim 10^{10}$ $M_\odot$ for large ellipticals ranging down to < $10^8$ $M_\odot$ for smaller objects. The cooling time $\approx H_0^{-1}$ at $\sim 30$ kpc in N4472. For N1395 (wnich may not be very different), Nulsen et al. (MNRAS, 208, 185) obtain 5 x $10^8$ $M_\odot$ within 10 kpc.

E. Khachikian : Have you any observational evidences of the gas flow to the centre of any galaxy?

A. Fabian : No direct evidence. The flows are highly subsonic, so we do not expect to easily observe the inflow velocity (but see HI result by Crane et al. in IAU 97 Proceedings).

A. Bosma : What would happen to the cooling flow if a small galaxy falls into a big elliptical?

A. Fabian : Not very much (I think). It will depend upon whether the galaxy motion is supersonic or not; the gas will settle back in a sound crossing time.

A. Moorwood : Valentijn and myself have observed radial visible-IR colour gradients out to 100 kpc from the centre of the A496cD galaxy which are consistent with star formation in a cooling flow. Population synthesis indicates a main sequence upper mass cutoff which increases from $\sim$ .6 $M_\odot$ at the centre to $\sim$ 1.6 $M_\odot$ at 100 kpc plus an increasing contribution from giants with increasing radius. The mass cutoff of the MS is also consistent with the radial variation of the Jeans mass as deduced from X-ray data.

# PROBLEMS OF GALACTIC AND EXTRAGALACTIC CO PHOTOMETRY

F.P. Israel

Sterrewacht Leiden

The Netherlands

## 1. Basics

In the past decades, tremendous insight into the global properties of galaxies has been gathered from (atomic) HI observations. The importance of HI for studies of galaxies lies in the fact that the observed 21-cm line emission allows a determination of both HI column densities and velocities as a function of position. Thus, mass distribution, kinematics and dynamics of galaxies can be determined out to the visibility limits of HI in each galaxy, usually well beyond visual boundaries.

The study of CO line emission, particularly at 115 GHz (2.6 mm), has developed to the point where it significantly adds to the knowledge gained from HI studies. This has become possible because of the simultaneous development of very sensitive detectors (overall system temperatures of ~ 100 K typically) and large, dedicated mm telescopes (10-m class single dishes, mm interferometers). The scientific importance of CO lies in the fact that it preferentially traces, unlike HI, high density gas (because CO needs densities of order $10^3$ $cm^{-3}$ or higher for its excitation) in the form of clouds of molecular hydrogen. CO is the most abundant molecule after $H_2$; its abundance with respect to $H_2$ is $10^{-4}$, and it is at least $10^2$ times more abundant than other molecules. Moreover, it is easily observable contrary to $H_2$ itself which is, with some effort, observable in far-UV (near 1100 Å) absorption against bright early type stars and in near-infrared emission in localized hot (T $\sim$ 2000 K regions) (Shull and Beckwith 1982).

Thus, extragalactic CO studies yield information on the distribution and kinematics of dense and usually large molecular cloud complexes as well as information on the overall molecular content of galaxies. At least in our Galaxy, $H_2$ is an important constituent of the interstellar medium: within the Solar Circle the inferred mass of $H_2$ roughly equals the mass of HI. Since the emphasis of this volume is on photometry, we will in the following concentrate on observations of CO brightness, and largely ignore the kinematical information also present.

In the interpretation of observed CO intensities complications occur. These are in part due to assumptions that are usually made, explicitly or implicitly. If $\tau(^{12}CO) \gg 1$, the observed antenna temperature yields the excitation temperature $T_{ex}(CO)$. If the space density of CO is sufficiently high (assumed to be the case) then $T_{ex} = T_{kin}$. At the same time, the assumption that $\tau(^{13}CO) \ll 1$ and $T_{ex}(^{13}CO) = T_{ex}(^{12}CO)$ yields the column density $N(^{13}CO)$. One way of determining whether $^{12}CO$ is optically thick or not lies in observing different rotational transitions of CO. If, for instance, $^{12}CO$ is optically thin, the $J=2-1$ transition at 1.3 mm will be brighter by up to a factor of four than the $J=1-0$ transition at 2.6 mm. A practical problem that arises here, and elsewhere, is that of correctly applying a variety of instrumental correction factors e.g. whether the result is expressed in $T_A^*$ or $T_R^*$; this problem is fully described in Kutner and Ulich (1981). Frequently, the correction procedure described is at odds with actually published values.

From Copernicus far-UV absorption measurements, Bohlin et al. (1978) found $N(HI+2H_2+p) = 6 \times 10^{21}$ E(B-V) for gas clouds in the Solar Neighbourhood. Dickman (1978) found $A_V = 4 \times 10^{-16} N(^{13}CO)$, so that $N(H_2)/N(^{13}CO) = 4 \times 10^5$ (assuming $A_V/E(B-V)=3.1$), with an uncertainty of 50%. In the meantime, Frerking et al. (1982) have presented evidence for a more complicated relation, although the order of magnitude is similar. Several authors have attempted to determine (empirically) a more direct relation between $H_2$ mass and observed velocity-integrated $^{12}CO$ strength:

$$M(H_2)/(M_\odot) = 4 \times 10^5 \int_A T_A^* \, dV \, D^2 \quad (K \; km \; s^{-1} \; Mpc^2)$$

again with an uncertainty of 50% and valid only for an ensemble of clouds in the Galaxy (see e.g. Israel and Rowan-Robinson, 1984, and references therein). Finally, comparison of galactic CO data with results from the COS-B $\gamma$-ray satellite strongly suggests that throughout the Galaxy $N(H_2) = (2.5-3.0) \times 10^{20} \int T_A^* dV$ (CO) (molecules $cm^{-2}$) although this should be regarded as an upper limit due to the (probably small) contribution of unresolved $\gamma$ ray point sources (Lebrun et al., 1983; Bloemen et al., 1984, 1985). One should also bear in mind that, except for the last determination, all these estimates really only apply in our Galaxy within a few kpc from the Sun. An important question is therefore: to what extent is the Solar Neighbourhood CO representative of all galactic CO and of CO in other galaxies. In general, problems of the interpretation of CO observations are related to problems of CO formation, destruction (among those the questions of fractionization and species dependent selfshielding) and excitation which are at present insufficiently resolved.


2. CO in the Galaxy.


Since 1975, a large number of CO surveys have been carried out, both surveys of particular types of objects (HII regions, dark clouds, etc.) and general surveys of the Galaxy usually limited to the Galactic plane. The most important of these are the following.

| Telescope | Beamsize | Species | Region | Reference |
|-----------|----------|---------|--------|-----------|
| NRAO | 1.1' | $^{12}CO$, $^{13}CO$ | Centre; 1st quadrant | Burton & Gordon, 1978<br>Liszt et al., 1981<br>Liszt et al., 1984 |
| UMass<br>NRAO<br>UMass<br>BTL | 0.8'<br>1.1'<br>0.8'<br>1.7' | $^{12}CO$, $^{13}CO$ | 1st, 4th quadrant | Sanders et al., 1984<br>Solomon et al., 1983 |
| GISS | 8.0' | $^{12}CO$ | 1st, 2nd quadrant | Cohen & Thaddeus, 1977<br>Cohen et al., 1980<br>Dame, 1983<br>Myers et al., 1985 |
| CSIRO | 2.8' | $^{12}CO$ | 4th quadrant | Robinson et al., 1983 |
| BTL | 1.7' | $^{12}CO$ | 1st quadrant | Knapp et al., 1985 |

Several major conclusions have resulted from these surveys (not considering kinematical and dynamical aspects).

a. Within the Solar Circle, a strong CO signal is observed throughout the Galaxy. Use of the Solar Neighbourhood (SN) relations mentioned above indicates an $H_2$ mass within the Solar Circle of 1-3 x $10^9 M_o$ (Sanders 1981; Dame 1983). b.The molecular gas, as traced by CO, not only has a distribution markedly less diffuse than HI, but, it seems to exist almost exclusively concentrated in dense (n $>$ $10^3$ cm$^{-3}$) clouds, known as giant molecular clouds (GMC's). GMC's are usually defined as clouds having a size d $>$ 20 pc and a mass $M(H_2)$ $>$ $10^5$ $M_o$. They tend to be elongated with typical dimensions of 20x100 pc (e.g. Stark and Blitz, 1978). The existence of these GMC's was completely unexpected; their discovery is one of the major triumphs of observational molecular astronomy.

c. The extent to which GMC's delineate a spiral pattern is still a controversional issue. The question, at present, centers on the existence of 'interarm' CO clouds. For arguments regarding CO spiral arms, see Blitz, 1983; Liszt, 1985. Because of their discrete nature, and their frequent association with nebulae and young stars, CO clouds provide very useful

means to determine the Galactic rotation curve, especially <u>outside</u> the Solar
Circle (Blitz, 1979; Brand, 1985).

d. <u>GMC's play an important role in star formation</u>; virtually all known regions
of star formation are associated with GMC's. Probably all OB stars in the
Galaxy are formed through collapse processes in GMC's (c.f. Habing and
Israel, 1979). As is the case with other young populations, the
distribution of GMC's throughout the Galaxy shows strong preferences. They
are concentrated towards galactic plane with scale heights of about 60-80
pc; most are found between distances of 4 and 8 kpc from the Galactic
Centre. Some GMC's are found at the Galactic Centre, but few are present
outside the Solar Circle (Solomon et al., 1983; see, however, Kutner and
Mead, 1981)

## 3. Properties of Giant Molecular Clouds

That almost all molecular material in the Galaxy resides in GMC's and
particularly in massive GMC's is clear from the following.

| Mass Interval $M(H_2)/M_\odot$ | Percentage of Total Mass $M(H_2)$ in Interval | |
|---|---|---|
| | Dame (1983) | Sanders (1981) |
| $10^4-10^5$ | 23 | 16 |
| $10^5-10^6$ | 73 | 39 |
| $10^6-10^7$ | – | 45 |

<u>These cloud complexes are very strongly clumped.</u> Most observing has been
done in the $^{12}CO$ line, but because this line is almost always optically thick,
it is mostly sensitive to temperature variations and shows little contrast
between high-density and moderate-density regions. Thus, GMC's observed in $^{12}CO$
tend to look like large, rather featureless emission regions with a few peaks.
Observations of $^{12}CO$ with <u>high linear</u> resolution, and especially $^{13}CO$
observations sensitive to column density clearly show the extremely clumped
nature of most molecular cloud complexes (Bally and Israel, 1985). Other
indications for strong clumping are a) observations of dense clump nuclei in
line emission from high density tracers such as $NH_3$ (e.g. Ungerechts et al.,
1982; Myers and Benson, 1983) and b) the large volume filling factors (of order
10 or more) often derived for GMC's.

Typical clump parameters are a linear size of order one parsec and a mass
of order a few hundred solar masses. It follows that a GMC may contain several
thousand clumps. <u>The concept of strongly clumped molecular cloud complexes</u>

implies physical properties rather different from the often held concept of a more or less homogeneous cloud complex. The simplest version of the first concept is a two-component model: dense clumps $(n > 10^3 \text{cm}^{-3})$ embedded in a much more tenuous, preferably non-molecular medium $(n \sim 1\text{-}10 \text{ cm}^{-3})$, with abrupt density changes. Obviously, a clumped model has a much higher (molecular) surface-to-volume ratio than a homogeneous model.

Several authors have attempted to establish the mass/size distribution of clouds (c.q cloud complexes in the Galaxy). The usual assumption is a power law distribution of the type $N(M) \propto M^{-n}$ or $N(d) \propto d^{-m}$, where $n=(m+2)/3$ if the clouds all have the same density. Results are summarized below.

| Reference | Mass distribution $n$ | Size distribution $m$ |
|-----------|-----------------------|-----------------------|
| Stark (1979) | 1.2 | 1.8 |
| Rowan-Robinson (1979) | 1.0 | 1.0 |
| Liszt et al. (1981) | 1.8 | 3.3 |
| Sanders (1981) | 1.5 | 2.5 |
| Dame (1983) | 0.5 | -0.5 |
| Casoli et al. (1983) | 1.25 | 1.75 |

It should be noted that each of these distributions was determined from a different data set, covering a different mass interval. Combining these results, Drapatz and Zinnecker (1984) concluded that in the Galaxy the size distribution of a molecular cloud complex can be represented by a half log normal distribution in the size interval 0.1-100 pc. Very little work has so far been done on the distribution of clump sizes and masses within a GMC. A preliminary observation of the mass distribution of clumps in the modest S255 molecular cloud complex $(M_{tot} \simeq 1.5 \times 10^5 M_\odot)$ yields $n=1.0 \pm 0.2$ (corresponding to $m=1.0$) for 31 clumps with masses ranging from $10^2$ to $10^4$ $M_\odot$ (Bally and Israel 1985).

## 4. CO in other Galaxies

Statistically useful numbers of CO detections in galaxies have only become available in the last few years, because extragalactic CO signals tend to be very weak $(T_A^* \text{ (CO)} \lesssim 1.0 \text{ K}$ for all but the strongest, while often $T_A^* \text{ (CO)} < 0.1$ K). This is due to generally severe beam dilution and large filling factors (the observing beam often being of a size similar to that of the observed galaxy), and frequently also to intrinsically weak CO signatures. The situation up to about 1981 has been summarized by Morris and Rickard (1982). Since that time, the available sample of CO galaxies has been enlarged, in particular by observers using the Bell Telephone Labs and the FCRAO telescopes. Most of this material is summarized and referenced by Verter (1983; 1985). The following conclusions can be reached (see Morris and Rickard, 1982, and Verter, 1983).

a) A few galaxies (IC342, M82, NGC 253) are bright in CO ($T_A^* > 1.0$ K), but such systems are exceedingly rare. It is interesting to note that the CO emission of the last two galaxies appears to include a significant optically thin $^{12}CO$ component (as determined from a comparison of the $^{12}CO(1-0)$ and $^{12}CO(2-1)$ transitions), which is also quite unusual, and which is most likely related to the intense nuclear activity in these galaxies.

b) Integrated CO luminosities tend to peak for galaxies of intermediate type (Sb-Sbc). For these galaxies, the ratio CO/HI is also highest. The observed peak values correspond to typical molecular gas contents of order $M(H_2)$ = $2 \times 10^9$ $M_\odot$ and molecular gas fractions $M(H_2)/M(H_2+HI)$ = 0.35, if the Solar Neighbourhood CO to $H_2$ conversion is assumed to be applicable.

c) CO surface brightness of spiral galaxies appears to be correlated with radio continuum brightness, This suggests that both are tied in with the young-star population in these spiral galaxies (Israel and Rowan-Robinson, 1984). Similarly, CO emission appears to be correlated with far-infrared emission. Correlations with near-infrared and HI emission are generally less clear or absent.

d) Magellanic-type galaxies and compact blue dwarf galaxies generally have extremely weak CO signals (Elmegreen et al., 1980; Israel et al., 1985) not commensurate with their radio continuum, far-infrared, or HI emission.

The above conclusions appear to be relatively well-established. Most attempts to arrive at more specific conclusions are, however, hampered by selection effects. These arise from a). the usually poor linear resolution (of order 1 kpc or more) obtainable. b). incomplete coverage of a galaxy: often only the center of a galaxy is observed, sometimes a major axis map is obtained, and only rarely a galaxy is completely mapped. c). the uncertainty in the CO to $H_2$ conversion mentioned earlier; this is particularly disturbing, because on galactic scales, and between different galactic types (possibly large) systematic effects might be expected.

Several of the conlusions reached by the FCRAO observers (see. e.g. Young, 1983; Scoville, 1984, and references therein) are subject to these selection effects. The first problem arises when a sample of single beam CO measurements, centered on galactic nuclei, and refering to galaxies at different distances and with different sizes is used to determine general properties of these galaxies. The problem is severe, because different fractions of the galactic disks are covered and the CO contributions of the disk and the nucleus cannot be separated. Yet the disk and the nucleus behave in different and unrelated ways: there are e.g. indications that galaxies with large optical bulges have centers deficient in or devoid of CO (Young, 1983; Scoville, 1984). Moreover, the nature of CO emission in the center of galaxies can be radically different from that of CO emission in galactic disks (see e.g. Rickard and Blitz, 1985), especially in cases like M82 and NGC 253 where the strong $^{12}CO$ signal appears to be due largely to optically thin $^{12}CO$, in contrast to the usual optically thick $^{12}CO$ emission from the disk. This problem can only be partly solved by observing galaxies at the same distance and thus with the same linear resolution, such as galaxies in the Virgo Cluster (Young et al., 1985) because the actual influence of the nucleus is still unknown. Another possibility lies in the use of observations with the very high resolution (of order arcseconds) obtainable with mm interferometers (e.g. Lo et al., 1985).

Thus, single-point measurements of galaxies should only be used with great caution to draw generalized conclusions on the total CO content of galaxies; at the same time it is especially hazardous to derive $H_2$ surface densities from CO observations that contains emission from both galactic nuclei and disks. This last point can be illustrated by refering to our own Galaxy. The strong CO signal observed toward the galactic nucleus would imply an impressive $H_2$ concentration. However, gamma-ray observations fail to substantiate this (Blitz et al., 1985).

Even when the CO distribution of galaxies is mapped, or properly modeled (by using actual CO observations as input, rather than taking the optical-light distribution as representative), and when care is taken to exclude the nucleus, it is dangerous to use standard CO to $H_2$ conversions. Direct evidence for this statement is given by Rickard and Blitz (1985) who find great variations in the $^{12}CO/^{13}CO$ ratio within and between disks of galaxies. The problem of the CO-to-$H_2$ conversion is reviewed in detail by Lequeux (1981), who concludes: 'our knowledge of the amount of $H_2$ in external galaxies belongs almost entirely to the realm of speculations'.

## 5. Properties of Extragalactic CO clouds.

The total amount and distribution of $H_2$ in a galaxy are, as we have seen, of great interest because of the important role of this molecule in our Galaxy. Since, unfortunately, direct measurements are practically out of the question, a certain amount of speculation appears inavoidable.

An important clue is the weak to extremely weak CO emission of the irregular dwarf galaxies (Elmegreen et al., 1980; Israel et al., 1985). Two effects may conspire to produce such low signal strengths in these galaxies. First, these galaxies are characterized by low metallicities. For the abundance of CO the C abundance, and the C/O ratio are of importance. In the LMC and especially the SMC, these are quite low compared to the Solar Neighbourhood (see

Dufour, 1984). It is reasonable to assume that the CO abundance roughly scales with the C abundance; this would, however, only explain weak $^{12}$CO signals if the CO abundance is so low that even $^{12}$CO would be optically thin. This is not likely to be the case, except perhaps in extreme cases such as the SMC where the C abundance is down by a factor of 20 compared to the Solar Neighbourhood, and the nuclear regions of some galaxies (e.g. M82 and NGC 253).

A weak $^{12}$CO signal can be explained by a second effect (which is not quite independent of the first). Low metallicity galaxies also have a low dust content, so that molecules are less shielded against destructive UV radiation (shortwards of 1120 Å). Blue dwarf galaxies have local UV radiation fields comparable to, or stronger than that in the Solar Neighbourhood. Thus, CO will be photo-dissociated, and even more effectively when the CO clouds are strongly clumped (e.g. section 3). The combined effect of low metallicity and low shielding can be strong enough to decrease the signal strength of optically thin lines such as $^{13}$CO by large factors. Signal strengths of optically thick lines such as $^{12}$CO would also decrease (but to a lesser extent) because photodissociation of clumps would decrease the total CO geometrical cross-section, hence increase the beam filling factor. A complicating factor is that CO might be self-shielding, but to what degree is not known. However, even in that case, low CO abundances would significantly diminish the importance of self-shielding so that optically thick $^{12}$CO emission would still be weakened. Simple modeling shows that at least the weak $^{12}$CO emission of the Magellanic Clouds can be explained in this way (Israel et al., 1985).

Several other observations, both in our Galaxy and in other galaxies can, at least qualitatively, be explained by the combined effects of metallicity and photodissociation in clumped molecular cloud complexes. Examples are:

a. The Galactic Center has a strong CO signature, but gamma-rays indicate a relatively low $H_2$ content. This could be due to a relatively higher CO abundance, to a high $T_{ex}$ or to both (Blitz et al., 1985).

b. Gamma-ray observations indicate that the ratio between $H_2$ column density and the integral $^{12}$CO line intensity shows no strong variations throughout the Galaxy (Bloemen et al., 1985); however, the $^{13}$CO/$^{12}$CO ratio shows a strong gradient (Burton et al., 1984) which resembles the N and O gradients (Shaver et al., 1983). Assuming that in the 'molecular ring' the effects of a higher dust-to-gas ratio (shielding) and a higher UV radiation field cancel each other, the (optically thick) $^{12}$CO emissivity would be more or less constant, while the (optically thin) $^{13}$CO emissivity would still show the effects of a metallicity gradient.

c. Observed CO gradients in normal spiral galaxies do not necessarily indicate $H_2$ gradients. For instance, Blitz (private communication) has shown that the CO gradient in M101 matches, perhaps fortuitously closely, the O/H gradient; a similar result is found for M31 (Blitz, 1985).

d. Strong CO signals observed in IRAS 'starburst' galaxies (IC4524, NGC 6240) may not be due to an $H_2$ excess (Young et al., 1985) but to a combination of low $^{12}$CO opacity and high $^{12}$CO luminosity as appears to be the case in M82 (Bally and Stark, 1985).

## 6. Summary.

CO observations provide a powerful tool to study the distribution and kinematics of dense molecular gas present in galaxies in the form of highly clumped, giant molecular clouds. Derivation of quantitative information on total molecular ($= H_2$) column densities and masses is severely hampered by the great uncertainty in the relevant $CO/H_2$ ratio. This ratio is influenced by (often related) factors such as opacity, metallicity, dust-to-gas ratio and ambient UV field intensity. The observations can generally be modeled in a qualitative way, but quantitative modeling needs more systematic CO (and where possible $H_2$) observations, more extensive observations of metallicity (in particular the C and O abundances) and dust content, as well as a better understanding of the CO formation and destruction processes.

## References

Bally, J., Stark, A.A., 1985, in preperation

Bally, J., Israel, F.P., 1985, in preparation

Blitz, L., 1979, Astrophys. J. Lett., _231_, L115 (also _234_, L172)

Blitz, L., 1983 in "Surveys of the Southern Galaxy", Eds. W.B. Burton and F.P. Israel, Reidel Publ. Co., p. 117

Blitz, L., 1985, Astrophys. J., in press

Blitz, L., Bloemen, J.B.G.M., Hermsen, W.J., Bania, T.M., 1985, Astron. Astrophys., in press

Bloemen, J.B.G.M., and 8 other authors, 1985, Astron. Astrophys. submitted

Bloemen, J.B.G.M., and 6 other authors, 1985, Astron. Astrophys. _139_, 37

Brand, J., 1985, in preparation

Burton, W.B., Gordon, M.A., 1978, Astron. Astrophys. _63_, 7

Casoli, F., Combes, F., Gérin, M., 1983 in "Surveys of the Southern Galaxy", Eds. W.B. Burton and F.P. Israel, Reidel Publ. Co., p. 181

Cohen, R.S., Thaddeus, P., 1977, Astrophys. J. Lett. _217_, L155

Cohen, R.S., Cong, H., Dame, T.M., Thaddeus, P., 1980, Astrophys. J. Lett. _239_, L53

Dame, T.M., 1983, Ph. D. Thesis Columbia University (U.S.A.)

Dickman, R.L., 1978, Astrophys. J. Suppl. _37_, 407

Drapatz, S., Zinnecker, H., 1984, Mon. Not. Roy. Astron. Soc. 210, 11P

Dufour, R.J., 1984, in "Future of Ultraviolet Astronomy", 3$^{rd}$ IUE Symposium,

Elmegreen, B.G., Elmegreen, D.M., Morris, M., 1980, Astrophys. J. 240, 455

Frerking, M.A., Langer, W.D., Wilson, R.W., 1982, Astrophys. J. 262, 590

Habing, H.J., Israel, F.P., 1979, Ann. Rev. Astron. Astrophys. 17, 345

Israel, F.P., Rowan-Robinson, M., 1984, Astrophys. J. 283, 81

Israel, F.P., De Graauw, Th., Van der Stadt, H., De Vries, C., 1985, Astrophys. J. submitted

Jenkins, E.B., Savage, B.D., 1974, Astrophys. J. 187, 243

Knapp, G.R., Stark, A.A., Wilson, R.W., 1985, Astrophys. J. Suppl. in press

Kutner, M., Ulich, B.L., 1981, Astrophys. J. 250, 341

Kutner, M., Mead, K., 1981, Astrophys. J. Lett. 249, L15

Lebrun, F., and 14 other authors, 1983, Astrophys. J. 274, 231

Lequeux, J., 1981, Comments on Astrophysics 9, 117

Liszt, H.S., Xiang, D., Burton, W.B., 1981, Astrophys. J. 249, 532

Liszt, H.S., Burton, W.B., Xiang, D.L., 1984, Astron. Astrophys. 140, 303

Liszt, H.S., 1985, in IAU Symposium 106, in press

Lo, K.Y., and 11 other authors, 1985, Astron. Astrophys. submitted

Morris, M., Rickard, L.J., 1982, Ann. Rev. Astron. Astrophys. 20, 517

Myers, P.C., Benson, P.J., 1983, Astrophys. J. 266, 309

Myers, P.C., and 6 other authors, 1985, Astrophys. J. submitted

Rowan-Robinson, M., 1979, Astrophys. J. 234, 111

Rickard, L.J., Blitz, L., 1985, Astrophys. J. Lett. in press

Robinson, B.J., McCutcheon, W.H., Manchester, R.N., Whiteoak, J.B., 1983, in "Surveys of the Southern Galaxy, Eds. W.B. Burton and F.p. Israel, Reidel Publ. Co., p. 1

Sanders, O.H., 1981, Ph. D. Thesis SUNY, Stony Brook (U.S.A.)

Sanders, D.B., Solomon, P.M., Scoville, N.Z., 1984, Astrophys. J. 276, 182

Scoville, N.Z., 1984, in "Star Formation Workshop", ROE, p. 199

Shaver, P.A., McGee, R.X., Newton, L.M., Danks, A.C., Pottasch, R.S., 1983, M.N.R.A.S. 204, 53

Shull, J.M., Beckwith, S., 1982, Ann. Rev. Astron. Astrophys. 20, 163

Stark, A.A., 1973, Ph. D. Thesis, Princeton University (U.S.A.)

Stark, A.A., Blitz, L., 1978, Astrophys. J. Lett. 224, L15

Solomon, P.M., Stark, A.A., Sanders, D.B., 1983, Astrophys. J. Lett. 267, L29

Ungerechts, H., Walmsley, C.M., Winnewisser, G., 1982, Astron. Astrophys. 111, 345

Verter, F., 1983, Ph. D. Thesis, Princeton University (U.S.A.)

Verter, F., 1985, Astrophys. J. Suppl. in press

Young, J.S., 1983, in "Survey of the Southern Galaxy", Eds. W.B. Burton and F.P. Israel, Reidel Publ. Co., p. 253

Young, J.S., Scoville, N.Z., Brady, E., 1985, Astrophys. J., in press

# A STUDY OF STAR FORMATION IN SHAPLEY-AMES SPIRAL GALAXIES BASED ON IRAS FAR-INFRARED OBSERVATIONS

K. Brink and T. de Jong
Astronomical Institute, University of Amsterdam,
Roetersstraat 15, 1018 WB Amsterdam
Netherlands

SUMMARY

IRAS far-infrared observations of an optically complete sample of galaxies are used to compare star formation rates in spiral galaxies in the field, in small groups and in the Virgo cluster.

INTRODUCTION

Optical and radio observations have shown that star formation is often enhanced in interacting galaxies (e.g. Sulentic 1976; Larson and Tinsley 1978). This effect is most evident for galaxies in close pairs or groups, but may also be present in wider systems through the generation of grand-design spiral structure (Elmegreen and Elmegreen 1982, and references therein). On the other hand, galaxies may have consumed a significant fraction of their gas after one or more encounters, so that their present star formation activity is low. In clusters of galaxies, interaction with an intracluster medium may lead to additional depletion of gas. Thus in regions of high galaxy density, where encounters are frequent, a relatively large spread in star formation rates might be expected.

IRAS has detected a large number of galaxies at far-infrared wavelengths, providing a database of unprecedented quality and size for statistical studies of star formation in other galaxies than our own. Using IRAS data, Soifer et al. (1984) have reported an overrepresentation of interacting galaxies in the infrared selected 'minisurvey' sample, indicating enhanced star formation in these galaxies.

To investigate more systematically the effect of environment on star formation we compare in this paper the infrared properties of 'field' galaxies with those of galaxies in small groups and in the Virgo cluster. We restrict ourselves to optically selected samples, chosen from the

Revised Shapley Ames Catalogue of Bright Galaxies (RSA, Sandage and
Tammann 1981).

RESULTS

The group sample consists of the RSA galaxies occurring in the catalogue
of group galaxies of Huchra and Geller (1982), and belonging to a group
with ten members or less. The field sample contains all RSA galaxies not
belonging to one of the Huchra-Geller groups. The Virgo cluster sample
consists of the RSA galaxies in Huchra-Geller group 41.

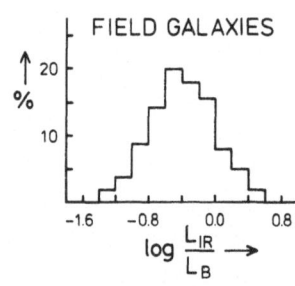

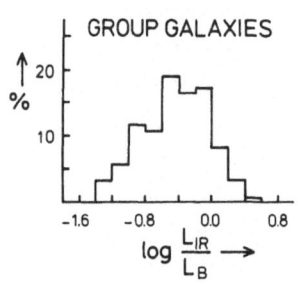

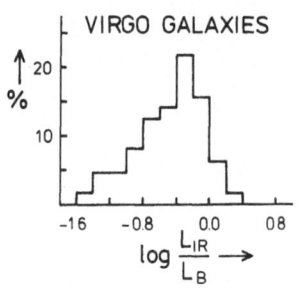

To ensure optical completeness for all samples
we have selected galaxies having an integrated
blue magnitude, corrected for galactic and
internal absorption, $B_T^0 < 12$. Nearby galaxies,
which are often resolved by the IRAS long
wavelength detector beams, are removed by
imposing a lower limit on the radial velocity
of a galaxy with respect to the Local Group,
$v_0 > 1000$ km s$^{-1}$. To retain a sufficiently
large sample, for the Virgo cluster a limit of
500 km s$^{-1}$ is used.

Figure 1 shows the normalised distributions of
the far-infrared/blue luminosity ratio of the
spiral galaxies in the three samples. The
infrared luminosity was calculated from the
IRAS 60 and 100 μm fluxes as described by de
Jong et al. (1984). The optical luminosity was
derived from the corrected blue magnitude $B_T^0$.

Non-detections are not included in figure 1.
Of 161 field spirals, 4 were not detected by
IRAS; the same holds for 2 out of 121 group
spirals and 6 out of 64 Virgo spirals. Including these not-detected galaxies we find

Fig.1. Normalised frequency distribution of the ratio of far-infrared
to blue luminosity for spiral galaxies in the field , in groups
and in the Virgo cluster.

a fraction of 'infrared-weak' galaxies ($L_{IR}/L_B < -0.80$) of 17% for the
field sample, 24% for the group sample and 28% for the Virgo cluster.

No large global differences between field, group and Virgo cluster galaxies are apparent in the distributions shown in figure 1. There may exist
a weak trend for an increasing number of infrared-weak spirals going from
field to Virgo cluster galaxies.

The absence of obvious differences between the three samples studied
could be understood if spirals are converted into galaxies of other
morphological type when their gas is consumed by star formation or de-
pleted by other means. In this respect it is noteworthy that the total
field sample contains 7 lenticulars versus 161 spirals, whereas the group
sample has 16 lenticulars versus 121 spirals and the Virgo cluster 14
versus 64. A more detailed study, using different methods of analysis and
including additional galaxy samples, is presently underway.

KB is supported by the Netherlands Foundation for Astronomical Research
(ASTRON) with financial aid from the Netherlands Organisation for the
Advancement of Pure Research (Z.W.O.).

REFERENCES

Elmegreen, B.G., and Elmegreen, D.M. 1982, Mon. Not. R. Astr. Soc 201,
    1021.
Huchra, J.F., and Geller, M.J. 1982, Astrophys. J., 257, 423.
de Jong, T., Clegg, P.E., Soifer, B.T., Rowan-Robinson, M., Habing,
    H.J., Houck, J.R., Aumann, H.H., and Raimond, E. 1984, Astrophys. J.
    (Letters), 278, L67.
Larson, R.B., and Tinsley, B.M. 1978, Astrophys. J., 219, 46.
Sandage, A., and Tammann, G.A. 1981, A Revised Shapley-Ames Catalog of
    Bright Galaxies (Washington,D.C.: Carnegie Institution of Washington),
    Publication 635 (RSA).
Soifer, B.T., Rowan-Robinson, J.R., Houck, J.R., de Jong, T.,
    Neugebauer, G., Aumann, H.H., Beichman, C.A., Boggess, N., Clegg,
    P.E., Emerson, J.P., Gillett, F.C., Habing, H.J., Hauser, M.G., Low,
    F.J., Miley, G., and Young, E. 1984, Astrophys. J. (Letters), 278,
    L67.
Sulentic, J.W. 1976, Astrophys. J. Suppl.Ser., 32, 171.

DISCUSSION

A. Moorwood : Are your results in conflict with those of Soifer et al.
who claim that interacting galaxies are overrepresented in the infrared
selected sample in the IRAS minisurvey ?

K. Brink : No. The idea is that galaxies may exhibit strongly enhanced
star formation rates during close encounters, and thus close pairs will
be overrepresented in infrared selected samples.
We have studied (relatively wide) groups as a whole, to get information
on the star formation history of a typical group member. Close pairs are
present in all three samples, though more frequent in groups and in the
Virgo cluster, as could be expected from the group selection criteria.
They indeed show enhanced star formation, but their number is too small
to produce large effects.

IRAS OBSERVATIONS OF THE MAGELLANIC CLOUDS

P.B.W. Schwering

Sterrewacht Leiden

Huygens Laboratorium

Postbus 9513

2300 RA Leiden

Apart from its main task, making an Infra Red all sky survey, the Infra Red Astronomical Satellite observed several objects to enhance the spatial resolution or increase its sensitivity. One of its major targets: the Magellanic Clouds.

Because of their relative proximity  and high ecliptic latitude the Clouds were
ideal objects to study with IRAS. Maps resulting from a special observing technique
(perpendicular scanning) have angular resolutions of 1' (20 pc at 60 kpc). This is
high compared to the available radio data, and also good compared to radio data of
other galaxies (WSRT 21 cm M31: 100 pc). A review about dust and IR in the MC's
was given by Israël (1983).

Infra Red maps at 60 micron are shown in figure 1 for the LMC and figure 2 for the
SMC. The low extended background, although present in both maps, was suppressed in the
first one to show the separate IR spots clearly. These areas correspond very well
with the H$_\alpha$ emission regions (Davies, Elliott and Meaburn, 1976). Also there is a very
good resemblence with the radio continuum (McGee, Brooks and Batchelor,1972; Broten,
1972). The correlation with the HI column density (McGee and Milton, 1966; Rohlfs et
al, 1984) is clearly present but less tight. In the SMC map the background can be seen.
The same correlations were done for the SMC farIR with H$_\alpha$ (DEM,1976), Radio continuum
(Broten, 1972), HI (Hindman, 1966; McGee and Newton, 1981).
Because the maps in figures 1 and 2 have a somewhat uncertain calibration no intensity
levels are given, and no direct flux comparisons were made. That will be done in the
near future.

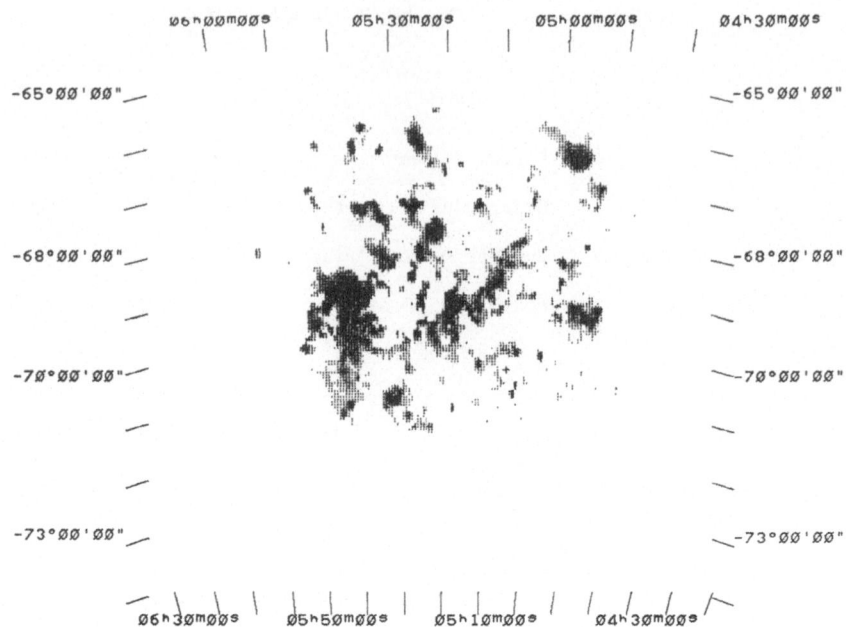

Fig 1: IRAS 60 micron map of the LMC.

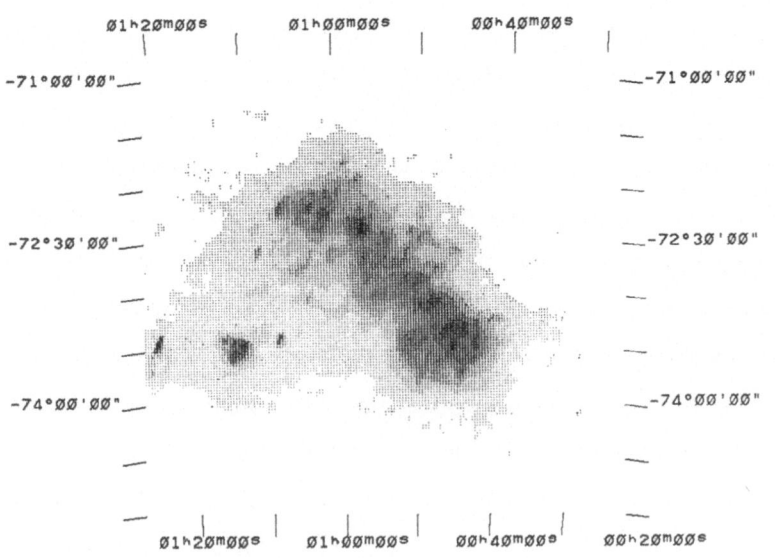

Fig 2: IRAS 60 micron map of the SMC.

References:

Broten,N.W.:Aust. J. Phys. (1972) 25   599-612

Davies,R.D.,Elliott,K.H.,Meaburn,J.:Mem. R. astr. Soc. (1976) 81   89-128

Hindman,J.V.: Aust. J. Phys. (1967) 20   147-171

Isräel,F.P.:1983, IAU symp. 108 eds S. van den Bergh & K.S. de Boer

          (dordrecht: Reidel) 319-332

McGee,R.X.,Milton,J.A.:Aust. J. Phys. (1966) 19   343-374

McGee,R.X.,Brooks,J.W.,Batchelor,R.A.:Aust. J. Phys. (1972) 25   581-597

McGee,R.X.,Newton,L.M.:Proc. ASA (1981) 4(2)   189-195

Rohlfs,K.,Kreitschmann,J.,Siegman,B.C.,Feitzinger,J.V.; Astron. Astrophys. (1984)

          137   343-357

AFTERNOON SESSION III :


GALAXY PHOTOMETRY, MASS
DISTRIBUTION AND DYNAMICS

# CONTEMPORARY DYNAMICAL PROBLEMS - A POSSIBLE
# CONTRIBUTION TO THEIR SOLUTION BY GALAXY PHOTOMETRY

Louis Martinet

Geneva Observatory

1290 Sauverny

Switzerland

Extensive reviews on the main current problems in galactic dynamics have been published very recently (Binney, 1982, on hot systems; White, 1983, on mergers; Toomre, 1983, on warps; Athanassoula, 1984, on spiral structure). In this article some of the most recent developments concerning specific questions in the field will be discussed, in particular those for which the contribution of galaxy photometry may be useful in the future. The study of isophotes or of dust rings in ellipticals, for instance, could provide information on the structural shape. Also, disc stability may be investigated by simultaneous kinematic and photometric observations. Detailed surface photometry of SB galaxies will allow to specify the length, the size, the mass and the angular velocity of bars. Edge-on and face-on galaxy photometry may yield indications on the distribution of masses in various components of spiral galaxies. Population gradients and bursts of star formation as well as the existence of shells could turn out to be constraints for dynamical friction and merger formation processes.

The paper is divided into three parts:  I) Triaxiality in elliptical galaxies,  II) Stability of disks and structure of bars,  III) Interactions between components of galaxies.

## I) Triaxiality

Twisting isophotes and non-zero motions on the apparent minor axis are the only observational indications in favour of triaxiality in elliptical galaxies at the present time. Statistical attempts to determine the exact shape of E's have yet to produce a definite result for want of sufficient photometric and spectroscopic data. One hope lies in the gas in and around galaxies but no one-to-one correspondence exists between the appearance of a dust lane and the shape of the system.

Concerning theoretical aspects, several pioneer papers were published in recent years on the dynamics of systems with 3 degrees of freedom. For such systems, the distribution function F must generally depend on three isolating integrals E, $I_2$, $I_3$ in order to maintain the necessary anisotropic velocity distribution. The nature of $I_2$ and $I_3$ is not well known. Are they necessarily isolating? (Some 3-D systems devoid of certain symmetries seem to have lost one or two isolating integrals, as shown by Martinet and Magnenat (1981)). In his thesis de Zeeuw (1984) considered the perfect ellipsoid model to which corresponds the "perfect elliptical galaxy" having the following properties:  a) the density is stratified in concentric, nearly ellipsoidal surfaces $\rho = \rho_0 (1+m^2)^{-2}$ with $m^2 = x^2/a^2 + y^2/b^2 + z^2/c^2$; most individual orbits have three explicitly known isolating integrals and can be analytically described as a consequence of the fact that the potential is of Eddington type. These properties represent a serious advantage, even if, as de Zeeuw emphasizes, real ellipticals are only nearly perfect: their radial brightness profiles do not rigorously fit that of the perfect ellipsoid, which furthermore has neither twisted isophotes nor rotation. But it is not excluded that separable models exist which come closer to being realistic than the perfect ellipsoid. In fact, the regular character of tube and box orbits found in integrable triaxial systems is preserved when parti-cular forms of potential perturbations occur which seem consistent with observations of ellipticals (Gerhard, 1984):  1) $\cos m\phi$ pertur-bations (m even);  2) modest ellipticity gradients;  3) small figure of rotation. Observations of isophotes yielded that some bulges of spirals (particularly SB) seem to be triaxial. They rotate rapidly and their dynamics is more complicated than that of slowly rotating ellipticals. It appears that some bars could also be triaxial (for the bar of NGC 936, Kormendy, 1983, finds an axis ratio 1:4:10). SA and SB galaxies are systems in which the conditions of regularity mentioned above are not always met. Irregular (semi-ergodic) behaviour of orbits can be non negligible and consequently a secular evolution may be going on in such galaxies. In particular, a new phenomenon character-istic of galactic systems with 3 degrees of freedom, the complex instability of periodic orbits (Magnenat, 1982), can play a role in the evolution of central regions which should not be neglected.

## II) Structure and evolution of bars

Among early Hubble types (SO to Sb) the percentage of SB is larger
than that of SAB; however, for later types (Sbc to Scd) it is the
contrary. The percentage of SAB and SB is slightly decreasing from
70% to 60%, from SO to Sc. Small bars or ovals clearly appear in IR
studies of spiral galaxies. Thus, bars are not rare. In order to
succeed in constructing a coherent scenario of bar formation and
evolution in the future, we need some additional insight on characteristic parameters of bars such as length, angular speed and size.

Results of the blue and the IR survey of galaxies obtained by
Elmegreen (1984) show that early type galaxies seem to have larger
bars relative to the galactic size than late type galaxies. Further,
the correlation between the sizes of bulges and the length of bars in
SO-Sb galaxies found by Athanassoula and Martinet (1980) seems to be
confirmed by Baumgarte (1984). Also, Sellwood (1981) has suggested, by
means of N-body simulations, that the lengths of bars are determined
by the distribution of matter in the spheroidal (bulge) component.

If the bar ends near the corotation radius $r_c$, the measurement of the
circular velocity $V_{rot}$ $(r_c)$ leads in principle to the angular speed of
the bar $\Omega_p$. Only little observational evidence exists for corotation
near the end of the bar: the direction of HI gas flowing in NGC 1365
(van der Hulst et al., 1983) or the dust lane position in NGC 1300
(Roberts, Huntley, van Albada, 1979). Recently, Tremaine and Weinberg
(1984) put forward an original method for determining $\Omega_p$ without any
particular dynamical method, but using galaxy photometry. In a flat
disc where tracers (stars or gas) are neither created nor destroyed,
$\Omega_p$, if well defined, can be obtained as a function of observable
quantities such as the inclination i, the surface brightness $\Sigma$ and the
radial velocity in different points of the projected disc. In spite of
observational difficulties, attempts of applications to SBO's could be
successful.

Data on shapes of bars are also essential for understanding the formation and the evolution of these structures. No extensive statistics
on the shape of bars in different Hubble types exist. Axial ratios
larger than 5 seem hard to find. This fact is explained by means of
numerical studies on orbital behaviour in barred systems, trying to

fix a limit for the existence of strong bars (Athanassoula et al., 1983; Teuben and Sanders, 1984). For example, for a ratio $M_B/M_D \sim 0.1$ of bar to disc mass, if the axis ratio is too large ($\geqslant 5$), the main direct periodic orbit elongated along the bar becomes unstable on large ranges of energy and cannot trap matter around it. The onset of semi-ergodicity can limit the strength of the bar. Semi-ergodic orbits fill equipotential contours which are rounder than the bar. An extension of these calculations to the 3-dimensional case (triaxial bar) confirms this effect and shows that the bar growth is limited by increasing vertical and plane instabilities when the axis ratio is increased (Pfenniger, 1984). One of the instability strips is probably responsible for the box shape of some bulges.

Turning to the question of global disc stability, we observe that a third of the galaxies do not show bar or oval-like perturbations. However, in N-body simulations, a large robust bar forms in the early stages of evolution and generally persists. Several factors likely to prevent the bar formation have been considered: the existence of a non-responsive component such as a bulge or a halo, a hot disc, a fraction of retrograde orbits, or the presence of an Inner Lindblad Resonance (for details, see a review by Athanassoula, 1984). But so far an infaillible guide to the global stability of realistic models has not been found. One need not to recall the misunderstanding resulting from the introduction of the parameter t by Ostriker and Peebles (1974) (see the elaborate criticism by Toomre, 1981): only the mass of the halo inside corotation, where a bar could develop, is important for the disc stabilisation with respect to bar-like perturbations. At the present time, no estimation of this mass in typical galaxies of different Hubble types is available. Until now, one essentially concentrated on the determination of the external mass of the halo, responsible for the flat rotation curve (see for ex. van der Kruit, 1983).

In fact, all relevant effects are not yet incorporated in the existing global stability criteria. In particular, the Efstathiou et al. (1982) criterion, which concerns only very specific models, does not take account of the center-edge variation of the velocity dispersion and is based on epicyclic approximation. Nevertheless, a hot disc does not seem to be the least agent of stabilisation. Sellwood and Carlberg (1984) propose that disc galaxies overcome the bar forming instability

through a combination of both large velocity dispersion and some
modest fraction of bulge and halo mass.

This leads us to mention two processes which may play a role in this
context: disc heating and star formation. Lacey (1984) shows that there
are difficulties in interpreting the heating of discs in terms of
scattering of stars by massive objects, either giant molecular clouds
in the disc or massive black holes in the halo. These processes lead
to contradictions with kinematical observations in the solar neighborhood. The results favour the role played by transient spiral waves in
stellar heating. On the contrary, star formation may contribute to the
cooling of the disc. Carlberg and Sellwood (1984) suggest that the
morphological family of disc galaxies (SA, SAB or SB) is a reflection
of the rate at which the disc is built up after the formation of the
bulge and the halo. Distinctions between Hubble types will probably
have to be introduced in quantitative scenarios.

Conclusions drawn from observational data on the velocity field in
barred galaxies might cause a certain amount of confusion with respect
to the question of disc stability. For instance, from the flat
rotation curve of NGC 936 (SB0), Kormendy (1983) deduces that "barred
galaxies do not appear to lack massive halos". On the other hand, the
disc of NGC 936 is very hot ($Q = 7^{+5}_{-3}$ at $r = r_B/2$, where $r_B$ is the bar
radius). Ironically, this strongly barred galaxy would seem to contain
two ingredients which traditionally work against the bar growth! In
fact, such a high value of Q precisely corresponds to the heating of
the disc by the bar and the flat portion of the rotation curve does
not give any information on the force exerted by a hypothetical halo
inside corotation. It is recommended to be very careful in connecting
observational facts with theoretical views on stability! Also, coordinated programmes of kinematic and photometric measurements of barred
and unbarred galaxies are urged in order to clarify this problem.

N-body simulations do not yet permit an understanding of the bar
evolution in time. First of all, four "external" effects can influence
the results: 1) softened gravity which can simulate velocity
dispersion, 2) graininess due to the small number of particles,
3) a rather severe truncation of discs preventing a secondary growth
of the bar which occurs whenever outer disc material is available to
accept angular momentum from the bar, 4) the time during which the

evolution is followed up. Furthermore, the evolution of the bar may be complicated by the effect of dynamical friction studied by Tremaine and Weinberg (1984). A torque on the bar arises entirely from near resonant stars. If the angular speed is slowly changing, a reversible feedback could stabilize or destabilize the rotation speed and a permanent capture of near resonant stars into librating orbits may occur.

III)   Interactions between galactic subsystems

In this section the characteristics of the response of some galactic components to the gravitational action of others are examined more particularly.

a)   Disc-bulge:   Barnes and White (1984) used a N-body code based on a low-order multipole method to study the response of a proto-bulge to the field of an exponential disc. One makes use of the fact that if the time scale of disc formation is much larger than the orbital period in the bulge, adiabatic invariants make sure that the response of the bulge depends only on the final state of the disc. The accretion of a disc induces modest changes in the ellipticity of a bulge and minor departures from a purely elliptical shape. The bulge is not pulled into a box-shape, which would rather be explained by 3-D resonance effects (Pfenniger, 1984). The addition of a disc does not explain why bulges appear to be rotationally supported. There is very poor observational evidence of mean age differences between discs and bulges. The broad band colour and line strength index study by Caldwell (1983) and SO's is not decisive.

b)   Halo-disc:   Various authors have discussed how a massive galactic halo affects the persistence of warps in galactic discs (see for ex. Sparke, 1984 and references therein). If the disc is to warp at 3-5 scale lengths, then the halo mass must lie in the range $2 M_D < M_H < 12 M_D$. Forcing mechanisms such as triaxiality in the massive halo or dynamical friction against halo stars can pump energy into the warping mode over many rotation periods so that the bending grows slowly in amplitude (Binney, 1981). Tests by way of warp observations in stellar components would be useful.

N-body calculations in which both disc and halo stars are treated
fully self-consistently (Sellwood, 1980) indicate that there is very
little interaction between these populations while the disc remains
axisymmetric; but a strong bar is able to transfer angular momentum
from the disc to the halo.

c) Bar-disc: The response of a gaseous or stellar disc to a bar-like
perturbation has been studied in many papers. Spiral arms or rings are
well known results of the process. From orbit calculations in a
galactic potential with a rather realistic bar (Athanassoula et al.,
1983) we found that the region between corotation and the Outer
Lindblad Resonance is more and more populated by semi-ergodic orbits
if the strength of the bar increases (axis ratio > 4 for $M_B/M_D$ ~ .1).
It is possible that at long term this region could become depopulated
because semi-ergodic orbits spend most of their time in the outer
parts of the galaxy. This depopulated region can spread beyond the
Outer Lindblad Resonance (Contopoulos, 1983). But beyond the simple
1/1 outer resonance, there are again near circular orbits that trap a
ring of matter around them. The existence of these rings has to be
proved through deep photometric observations. Further, Schwarz (1984),
searching for how the response of a gaseous or a stellar disc depends
on the strength of the bar-like or oval distortion and on its rotation
rate, found that the most apparent change with increasing bar strength
is the density contrast across the arms. Spiral arms are almost non
existent for bar to axisymmetric force ratio q = 0.01. For q = 0.25,
the arms are broadest as well as strongest. In conclusion, recent and
older N-body simulations and orbit calculations show that, in principle,
photometric observations of rings and arm-interarm density contrasts
could put some constraints on the strength and rotation rate of bars
in SB galaxies.

d) Bar-bulge: If the bulge is formed before the disc is accreted, a
bar is younger than the bulge. If, in addition, it becomes strong
enough, it could influence the bulge dynamics. N-body simulations are
planned in order to clarify that question.

e) Resonant coupling in 3-D systems: The motion of stars perpendicular
to a galactic plane is given by Mathieu's equation: $\ddot{z} + [a + 2q \cos \tau]z = 0$.
According to Binney (1981), orbits in principal planes are unstable to

the development of vertical oscillations as a consequence of resonant coupling in z-direction, either with epicyclic motion ($a = 4\omega_z/\kappa$) in axisymmetric and triaxial potentials, or with orbital motion in non-axisymmetric (bar) potentials. Such processes explain the instability of periodic orbits around the middle axis of a triaxial galaxy, the vertical structure of discs, the stability of planar orbits in a nearly spherical potential in connection with galactic warps, the high velocity dispersion of old stars perpendicular to the galactic plane, difficult to explain by transient spiral feature action only. However, a study of cooperative effects is needed (by way of N-body simulations) in order to confirm the efficiency of the processes.

## Concluding remark

This review treats some of the most recent problems relative to the internal dynamics of galaxies for which we need the galaxy photometry contribution. For lack of space we did not deal with environmental effects which certainly play an essential role in the galactic evolution. Clusters of galaxies are thus the privileged laboratory for studies of galactic evolution. Statistics on differences between populations of low and high density regions will be very rich in information on the sensitivity of evolution to the environment. In particular, it is of high importance to know if the key role is played by very early environment or by subsequent evolution.

## References

Athanassoula, E., 1984, Physics Reports, in press

Athanassoula, E., Martinet, L., 1980, Astron. Astrophys. 87, L 1a

Athanassoula, E., Bienaymé, O., Martinet, L., Pfenniger, D., 1983, Astron. Astrophys. 127, 349

Barnes, J., White, S.D.M., 1984, preprint

Baumgarte, J., 1984, private communication

Binney, J., 1981, Monthly Notices R.A.S. 196, 455

Binney, J., 1982, in "Morphology and Dynamics of Galaxies", 12th Saas-Fee Course, p.3, Eds. L. Martinet and M. Mayor, Geneva Observatory

Caldwell, N., 1983, Astrophys. J. 268, 90

Contopoulos, G., 1983, Astron. Astrophys. 117, 89

Efstathiou, G., Lake, G., Negroponte, J., 1982, Monthly Notices
    R.A.S. 199, 1069

Elmegreen, B.G., Elmegreen, D.M., 1984, preprint

Gerhard, O.E., 1984, preprint

Kormendy, J., 1983, Astrophys. J. 275, 529

Lacey, C.G., 1984, Monthly Notices R.A.S. 208, 687

Magnenat, P., 1982, Thesis University of Geneva

Martinet, L., Magnenat, P., 1981, Astron. Astrophys. 96, 68

Ostriker, J.P., Peebles, P.J.E., 1974, Astrophys. J. 186, 467

Pfenniger, D., 1984, Astron. Astrophys. 134, 373

Roberts, W.W., Huntley, J.M., van Albada, S.D., 1979,
    Astrophys. J. 233, 67

Schwarz, M.P., 1984, Monthly Notices R.A.S. 209, 93

Sellwood, J., 1980, Astron. Astrophys. 89, 296

Sellwood, J., 1981, Astron. Astrophys. 99, 362

Sellwood, J., Carlberg, R.G., 1984, Astrophys. J. 282, 61

Sparke, L.S., 1984, Astrophys. J. 280, 117

Teuben, P.J., Sanders, R.H., 1984, preprint

Toomre, A., 1981, in "The structure and evolution of normal
    galaxies", Eds. S.M. Fall and D. Lynden-Bell, p. 111,
    Cambridge Univ. press

Toomre, A., 1983, in "Internal Kinematics and Dynamics of Galaxies",
    IAU Symposium no. 100, p. 177, Reidel, Dordrecht, Publ.

Tremaine, S., Weinberg, M.D., 1984a, Monthly Notices R.A.S. 209, 729

Tremaine, S., Weinberg, M.D., 1984b, Astrophys. J. (Letters) 282, L5

van der Hulst, J.M., Ondrechen, M.P., van Gorkom, J.H., Hummel, E.,
    1983, in "Internal Kinematics and Dynamics of galaxies",
    IAU Symposium no. 100, p. 233, Reidel, Dordrecht, Publ.

van der Kruit, P.C., 1983, Proceedings Astron. Soc. of Australia 5, 136

White, S.D.M., 1983, in "Internal Kinematics and Dynamics of galaxies",
    IAU Symposium no. 100, p. 337, Reidel, Dordrecht, Publ.

de Zeeuw, T., 1984, Thesis, Leiden University, The Netherlands

## DISCUSSION

S. Djorgovski : There is some kinematical evidence for similarity of bulges and similar (small) luminosity ellipticals. However, the bulges often show box- or peanut-shaped isophotes, and the low luminosity ellipticals do not. Could that be due to their different dynamical evolution, e.g. due to the presence of a bar?

L. Martinet : Yes. 3-D calculations (N-body simulations and orbits) suggest that such isophote shapes can be a consequence of the presence of a triaxial bar.

E. Athanassoula : J. Sellwood and I have considered the stability of hot disks. Using N-body simulations we have shown that velocity dispersion can reduce very substantially the growth rate of instabilities. In this way we constructed models with arbitrarily small growth rates of the bar instability in the complete absence of any halo. Several cases, with Q profiles flat, increasing or decreasing with radius, were tried. High Q near the center of the disk proved much more effective at reducing the growth rates than increased random motions further out.

# THE THREE-DIMENSIONAL SHAPE OF GALAXIES

F. Bertola
Institute of Astronomy
University of Padova (Italy)

I - Introduction

Since the notion that the universe is populated by agglomerates of
stars was developed during the eighteenth century, the presence on the
sky of the Milky Way suggested the idea that the stellar system we li-
ve in, and, by analogy, all the stellar systems are flattened.
John Herschel seems to be the first who interpreted the shape of our own
galaxy as due to the effect of its fast rotation. Measurements of the
rotational velocity of the Milky Way and of spiral galaxies confirmed
this interpretation.
Contrary to spirals, whose rotation curves are rather  easily derived
from their emission lines, the velocity field of elliptical galaxies,
derived from absorption lines, was difficult to obtain, due to the rapid decrease outwards of the surface brightness. Although rotational
data were lacking for elliptical galaxies, they were supposed to be
flattened by rotation, like the spirals. Several models were constructed
to account for the observed properties of elliptical galaxies on the assumption that they are rotationally supported. The last ones were those
of Gott (1975), Larson (1975) and Wilson (1975).
Based on spectra taken with the image tube spectrograph attached to
the 200-inch Palomar telescope, the first rotation curve of an elliptical galaxy became available in 1972 (Bertola, 1972). It revealed in
the E5 galaxy NGC 4697 a much lower rotational velocity than expected
from the above models. These data were enough to cast severe doubts on
them (Gott, 1977b).
Successive observational works (Bertola and Capaccioli, 1975; Illingworth, 1977; Peterson, 1978; Schechter and Gunn, 1978) confirmed that
elliptical galaxies were slow rotators.
These results prompted Binney (1978) to make new suggestions on the
three-dimensional shape of elliptical galaxies. His argument was the
following: if it is not the rotation to cause the intrinsic shape of
elliptical galaxies, why should we accept "a priori" that their shape

is oblate? The fact that the isophotes of elliptical galaxies are ellipses indicates that the surfaces of constant light density, and consequently of mass density if the mass-to-light ratio is constant, are
spheroids, either oblate or prolate, or, in the most general case, triaxial ellipsoids, as demonstrated by Contopoulos (1956) and Stark (1977).
Therefore the possibility that elliptical galaxies are either prolate,
oblate or triaxial bodies was left open.

Binney's suggestion, stimulated both the construction of theoretical
models on elliptical galaxies and, at same time, the search for observational tests of their intrinsic shape. In the following we shall review these latters, as well as the indications of departure from oblat-
ness in disk galaxies.

## II - Elliptical Galaxies

### a) studies based on photometric data.

Marchant and Olson (1979), Richstone (1979), de Vaucouleurs and Olson
(1981) and Merrit (1982) performed a statistical test based on the mean
surface brightness of a sample of elliptical galaxies. If the galaxies
are oblate ones, those seen face-on should appear with a lower  surface
brightness than those seen edge-on. The contrary is true for prolate
galaxies. Uncertainties in the photometric data make this text inconclusive.

Noerdlinger (1979), Binggeli (1980) and Binney and de Vaucouleurs
(1981) studied the distribution of apparent flattening in elliptical
galaxies. Again the results are inconclusive since the data are at same time consistent with different intrinsic shapes.

The fact that in several cases the orientation of the isophotal ellipses of elliptical galaxies does not remain constant with the distance
from the center (Carter, 1978; King, 1978; Bertola and Galletta, 1979;
Williams and Schwarzschild, 1979a,b) is interpreted as an indication of
triaxiality. A light distribution constituted by dissimilar and coaxial
triaxial ellipsoid, when viewed not along the principal planes, gives
rise to the observed twisting of the isophotes. Detailed modelling of
the twisting in NGC 596 has been performed by T.B. Williams (1981).
Statistical properties of the twisting and flattening have been interpreted in terms of triaxiality (Galletta, 1980); Benachhio and Galletta,
1981).

b) studies_based_on_kinematical_data.

Using the tensor virial theorem, Binney (1978) has derived the relation between $v_m/\sigma$ ($v_m$ is the maximum velocity of rotation and $\sigma$ is the central velocity dispersion) and the intrinsic ellipticity $\varepsilon$ for oblate, prolate and triaxial galaxies with different degree of velocity anisotropy. While in the diagram $v_m/\sigma$ versus $\varepsilon$, representative points (affected by projection effects) of high luminosity ellipticals ($M_B<-20.5$) lie below the line representing oblate galaxies with velocity isotropy, ellipticals of low luminosity ($M_B>-20.5$) are fast rotators and lie on that li ne. Since projection effects move representative points of oblate isotropic ellipticals along this line, it follows that the shape of low lu minosity ellipticals is consistent with that of oblate spheroids (Davies et al., 1983). Whether high luminosity ellipticals are also oblate, but with velocity anisotropy, is matter of opinion.

Lake (1979), Merrit (1982) and Capaccioli, Fasano and Lake (1984) studied the correlation between ellipticity and central velocity dispersion at a given luminosity for non rotating galaxies. Pole-on prolate galaxies would show longer velocity dispersion than equatorially viewed ones, while the contrary is true for oblate galaxies. These tests seem to favour marginally the oblate shape.

The detection of remarkable velocity gradients along the apparent minor axis of elliptical galaxies (Jenkins and Scheuer, 1980; Bertola et al., 1983) is not consistent with rotating flattened spheroids.

Bertola and Galletta (1978) suggested that the dust lanes crossing elliptical galaxies along the minor axis can be used to deduce the three dimensional shape of the stellar body. Hawarden et al. (1981) pointed out the existence of ellipticals with dust lanes also along the major and intermediate axes. Those ellipticals with warped dust lanes along the minor axis allow interesting considerations on their shapes (Caldwell, 1984; Bertola et al., 1984). The fact that the dust lane is not interrupted in the center excludes that the dust and gas are captured in the polar orbits of an oblate galaxy. The merging phenomenon is suggested by the fact that in three out of four cases so far studied the motions in the warps are prograde with respect to the stellar motions, while a stationary situation would imply retrograde motions. On the other hand the relative high values of $v_m/\sigma$ found along their major axis, which is due mainly to streaming than tumbling motion, seem to exclude a purely prolate structure. Fast tumbling would prevent the dust and gas to settle into the observed plane. Therefore one is left with a triaxial structure for these galaxies. These galaxies are viewed approximately along their intermediate axis. Therefore the intrinsic major and minor axis are directly measured on the image, while lower and upper limits for the intermediate axis can be estimated from the $v_m/\sigma$ versus $\varepsilon$ relation and from the warping angle respectively. Elliptical galaxies with the dust lane along the major axis seems to be much less frequent than those with the dust lane along the minor axis. Does this mean that almost prolate figures are predominant over almost oblate ones among elliptical galaxies. In Fig. 1 we show what is probably the best case of an elliptical with a dust lane along the major axis, although this classification can be somewhat controversial. The presence of quite pronounced warps has to be remarked.

Kinematical and morphological decoupling of stars and gas in those ellipticals with emission lines (Bertola et al., 1984; Ulrich et al., 1984; Bettoni, 1984; Bertola and Bettoni, 1984) is an evidence which is not consistent,in stationary conditions,with oblate spheroids.


III - Disk Galaxies

The presence of the bar in disk galaxies has suggested since long time the existence of an elongated structure in addition to the oblate disk. A twisting of the isophotes of the disk and the bulge in M 31 led B. Lindblad (1956) to propose a triaxial model for the bulge of this galaxy. Successively Stark (1977) has demonstrated that the observations are consistent with a set of triaxial bulge configurations and with a prolate one, the oblate being excluded.
From a morphological study of barred galaxies Kormendy (1979) finds evidence that the intrinsic shapes of lenses resemble moderately flattened triaxial ellipsoids with preferred intermediate to major axis ratio of $\sim 0.9$. He finds also that most outer rings are not circular, the shortest dimension being the one filled by the bar.
An interesting suggestion on the shape of the disk comes from the study of disk galaxies with polar rings (Schweizer et al., 1983). In order to explain the apparent stability of these rings Steiman-Cameron and Durisen (1982) assumes that the disk, which is seen on edge, is not truly axially symmetric but slightly triaxial. In this way a significant cross section for capture of material into orbits precessing around the major axis and leading ultimately to orbits in the plane defined by the minor and intermediate axis of the disk, is made available.

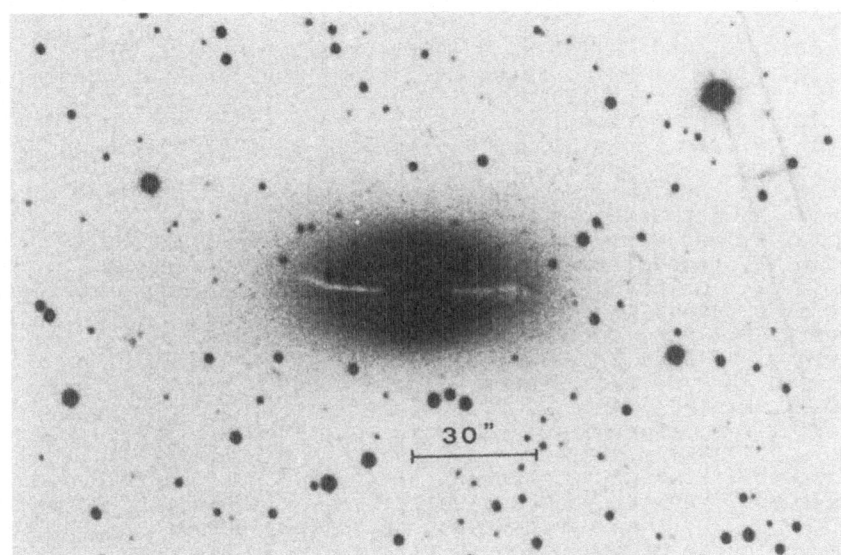

Fig. 1 - Anon 1029-459, a possible example of an elliptical galaxy with a dust lane along the major axis. The dust lane exhibits pronounced warps on both sides.
Photograph obtained on III a-J emulsion at the prime focus of the 3.9 m Anglo-Australian telescope. The image has been printed using the unsharp masking and contrast enhancement techniques by Claus Madsen at the ESO Photo Lab in Garching.

## References

Benacchio, L. and Galletta, G. 1980, M.N.R.A.S. 193, 885
Bertola, F. 1972, Proc. 15th  Meeting of Italian Astronomical Society,
          p. 199
Bertola, F. and Bettoni, D. 1984, in preparation
Bertola, F., Bettoni, D. and Capaccioli, M. 1983, IAU Symp. N° 100,
          p. 311
Bertola, F., Bettoni, D., Rusconi, L. and Sedmak, G. 1984, A.J. 89, 356
Bertola, F. and Capaccioli, M. 1975, Ap.J. 200, 439
Bertola, F. and Galletta, G. 1978, Ap.J. Letters 226, L115
Bertola, F. and Galletta, G. 1979, Astr. and Astrophys. 77, 363
Bertola, F., Galletta, G. and Zeilinger, W. 1984, in preparation
Bettoni, D. 1984, The Messenger 37, 17
Binggeli, B. 1980, Astron. and Astrophys. 82, 289
Binney, J. 1978, M.N.R.A.S. 183, 501
Binney, J. and de Vaucouleurs, G. 1981, M.N.R.A.S. 194, 679
Caldwell, N. 1984, Ap.J. 278, 96
Capaccioli, M., Fasano, G. and Lake, G. 1984, M.N.R.A.S. 209, 317
Carter, D. 1978, M.N.R.A.S. 182, 797
Contopoulos, G. 1960, Z. Astrophys. 49, 273
Davies, R.L., Efstathiou, G., Fall, S.M., Illingworth, G. and Schech-
          ter, P.L. 1983, Ap.J. 266, 516
de Vaucouleurs, G. and Olson, D.W.1981, Ap.J. 256, 346
Galletta, G. 1980, Astron. and Astrophys. 81, 179
Gott, J.R. 1975, Ap.J. 201, 296
Gott, J.R. 1977, Ann. Rev. Astron. and Astrophys. 15, 235
Hawarden, T.G., Elson, R.A.W., Longmore, A.J., Tritton, S.B. and Corwin,
          Jr. H.G. 1981, M.N.R.A.S. 196, 747
Illingworth, G. 1977, Ap.J. Letters 218, L43
Jenkins, C.R. and Scheuer, P.A.G. 1980, M.N.R.A.S. 192, 595
King, I.R. 1978, Ap.J. 222, 1
Kormendy, J. 1979, Ap.J. 227, 714
Lake, G. 1979, in, Photometry, Kinematics and Dynamics of Galaxies,
          p. 381, Ed. Evans, D.S., University of Texas, Austin
Larson, R.B. 1975, M.N.R.A.S. 173, 67
Lindblad, B. 1956, Stockholm Obs. Ann. Vol. 19, N° 2
Marchant, A.B. and Olson, D.W. 1979, Ap.J. Letters 230, L157
Merrit, D. 1982, A.J. 87, 1279
Noerdlinger, P.D. 1979, Ap.J. 234, 802
Peterson, C.J. 1978, Ap.J. 222, 84
Richstone, D.O. 1979, Ap.J. 234, 825
Schechter, P. and Gunn, J. 1979, Ap.J. 229, 472
Schweizer, F., Whitmore, B.C. and Rubin, V.C. 1983, A.J. 88, 909
Stark, A.A. 1977, Ap.J. 213, 368
Steiman-Cameron, T.Y. and Durisen, R.H. 1982, Ap.J. Letters 263, L51
Ulrich, M.H., Butcher, H. and Boksemberg, A. 1984, ESO preprint N° 327
Williams, T.B. 1981, Ap.J. 244, 458
Williams, T.B. and Schwarzschild, M. 1979a, Ap.J. 227, 56
Williams, T.B. and Schwarzschild, M. 1979b, Ap.J. Suppl. 45, 209
Wilson, C.P. 1975, A.J. 80, 175

# DYNAMICS OF ELLIPTICALS : THE CASE FOR 2-DIMENSIONAL PHOTOMETRY

R. Bacon and G. Monnet

Observatoire de Lyon

69230 Saint Genis-Laval

## I Without 2D photometry : The V/σ − ε test

Binney 1978 has introduced the so-called V/σ − ε test , where :

ε is the (constant) ellipticity of the isophotes

V/σ is the ratio of the maximum line of sight rotation velocity to the central velo-city dispersion of the stellar component.

The most simple model − an oblate distribution of mass, with isotropic velocity resi-duals (in short the Isotropic oblate model : I.O.M. ) corresponds to a predicted V/σ − ε curve, fully independent of the radial mass distribution (Binney 1978), and only weakly dependent on the (unknown) inclination of the galaxy along the line of sight (Binney 1982).

Classically, high luminosity − $M_B$ < − 20.5 − ellipticals do not fit this curve, and lie well below : Bertola and Capaccioli 1975, Schechter and Gunn 1979, Illingworth 1981, Davies 1981, Davies and Illingworth 1983. On the other hand, low luminosity ellipticals − $M_B$ > − 20.5 − fit reasonnably well (Davies et al. 1983), as well as Lenticular bulges (Dressler and Sandage 1983) and Spiral bulges (Pellet and Simien 1982, Kormendy and Illingworth 1982, Illingworth and Schechter 1982, Davies and Illingworth 1983) . See however Whitmore et al. 1984 for a dissident view.

This provides, however, for only a global test; a more powerful assessment is to compare the observed mean rotation curve V (r) and dispersion curve σ(r) along the galaxy, with the curves predicted under the I.O.M.

## II With 2D photometry : Testing locally the I.O.M.

Bacon 1985 has solved the second order hydrodynamical equations for the Isotropic oblate case and an arbitrary inclination along the line of sight.

Typical observable σ and V curves along the apparent major axis are given in figure 2 for a galaxy obeying the classical $r^{1/4}$ de Vaucouleurs density law, with a constant axial ratio q = 0.4, and an apparent axial ratio $q_a$ = 0.6 The effect of finite reso-lution of the telescope − spectrograph combination − modelled by a gaussian blurring

function of width d - is shown by the dashed curves in fig. 2. Fig. 1 gives the corresponding distorsions of the luminosity curve m(r) and the axial ratio curve $q_a$ (r).
One sees the large effect on the mean stellar rotation curve V(r) at small radii, which
is already apparent for r ∿3d, while the dispersion curve σ(r) is only moderately
affected.

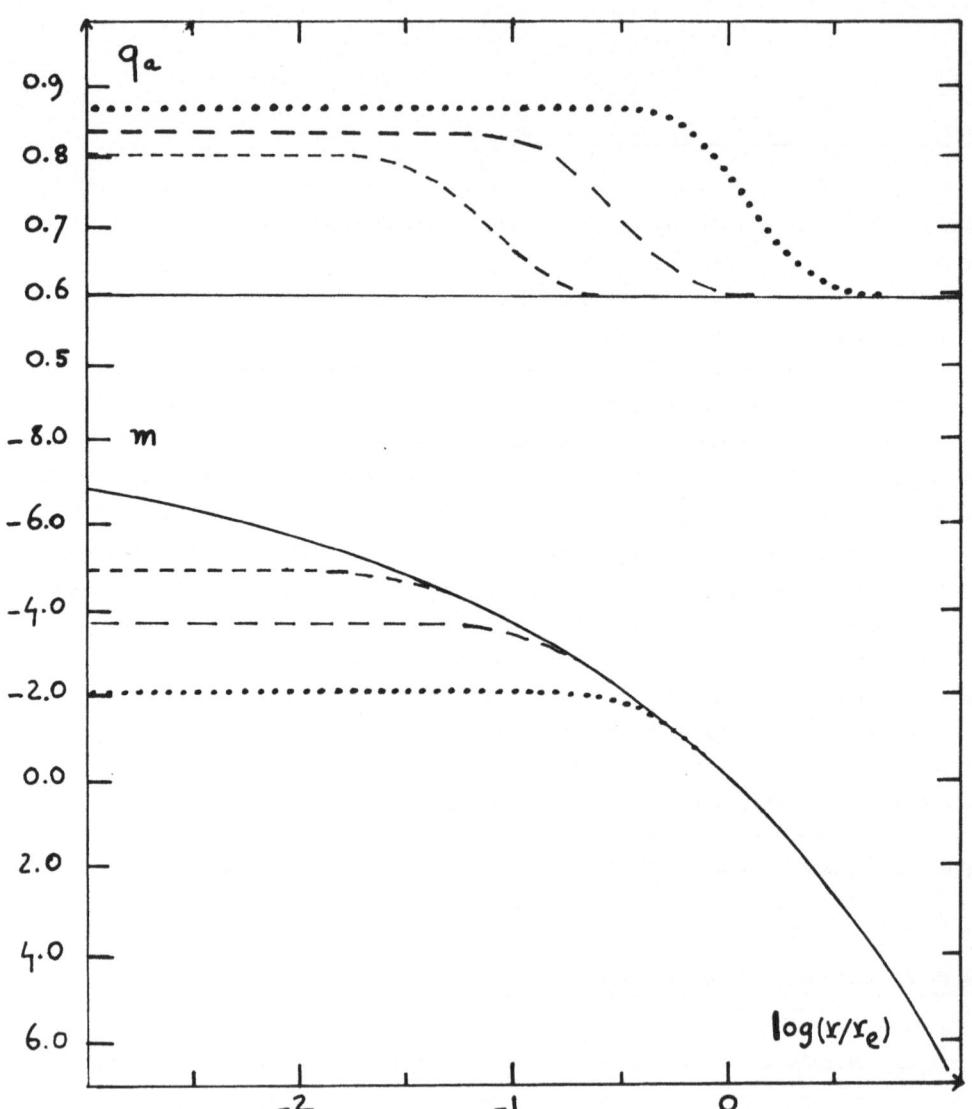

Figure 1 : $\underline{r}^{1/4}$ Oblate model - apparent axial ratio $q_a$ = 0.6
Convolved by Gaussians of r.m.s; : $d/r_e$ = 0, 0.02, 0.08, 0.3 (from Simien, private
communication)

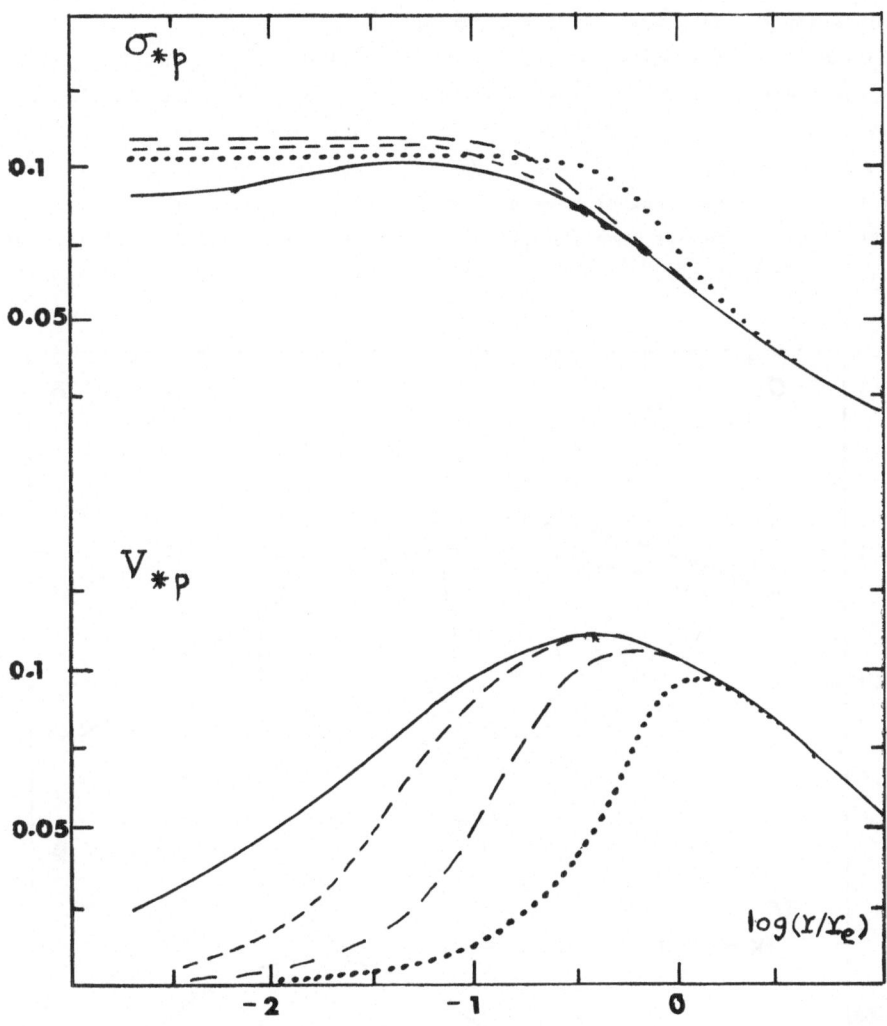

Figure 2 : Same model as in Figure 1
Predicted line of sight isotropic $\sigma$ and V curves

V and $\sigma$ are respectively given by the convolution of the theoretical line of sight first and second order velocity momenta by the transfer function.

- Application to the giant elliptical NGC 4889

Its luminosity distribution obeys a $r^{1/4}$ de Vaucouleurs law (Young 1976). Fig 3 a,b gives the theoretical V - σ curves - for the values of q and $q_a$ compatible with the observed axial ratios -, and the Davies et al. 1983 experimental points. In view of the large equivalent radius of the galaxy ($r_e$ = 25 arc.sec.), the finite spatial resolution of the data cannot have a significant effect on the theoretical curves, and the classical result from the global V/σ - ε test, i.e. incompatibility with the I.O.M., is confirmed

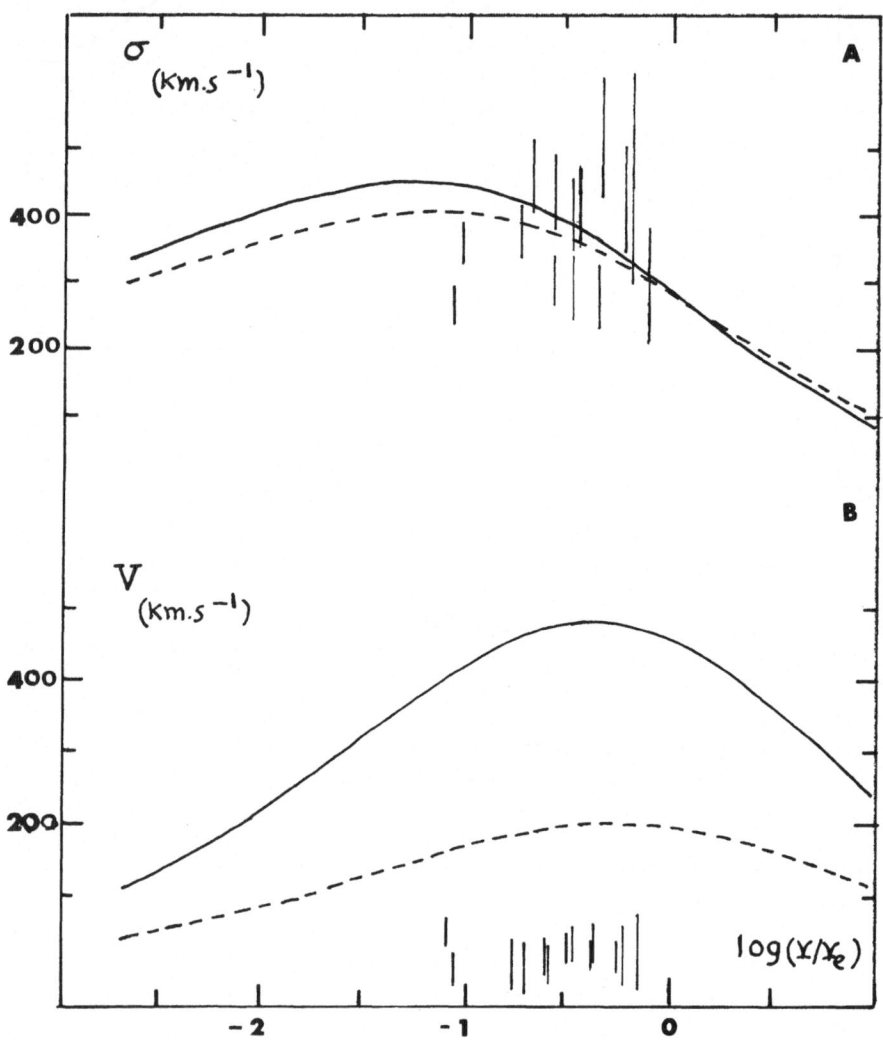

Figure 3 A, B : NGC 4889  Isotropic σ and V curves

- Application to the compact elliptical NGC 4387

Its luminosity distribution obeys a de Vaucouleurs law (Watanabe 1983). The comparison is given in fig. 4 a, b; and again the experimental points seem incompatible with the I.O.M. However, convolution with a gaussian of 5" F.W.H.M. gives a near perfect agreement, and isotropy seems quite possible : A more definite test awaits high spatial resolution spectrographic data.

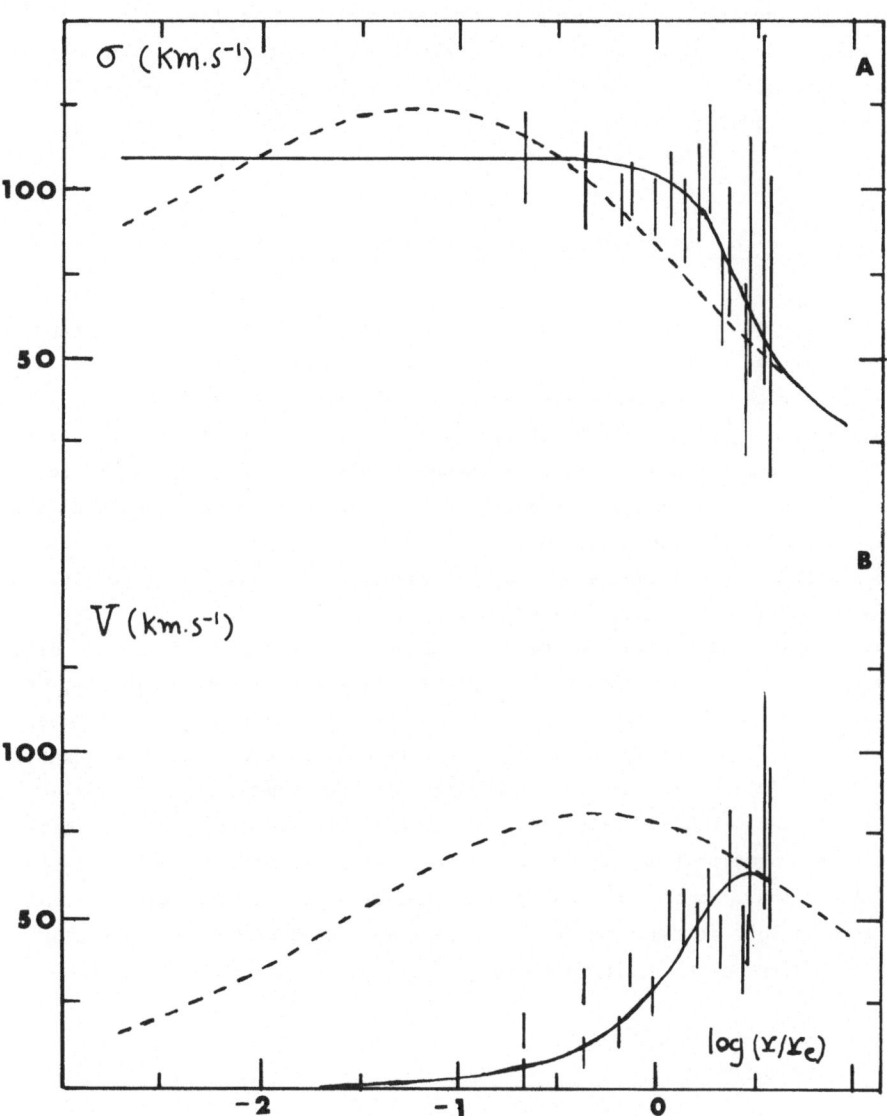

Figure 4 A, B : NGC 4387   Isotropic $\sigma$ and V curves

Such detailed analysis, on eight ellipticals, by Bacon 1985 give the following conclusions :

a) Giant Ellipticals, of the cD type, are indeed incompatible with the I.O.M.

b) Normal Ellipticals, i.e. not of the cD type, are quite compatible *when* the (usually coarse) spatial resolution along the spectrograph slit is taken into account.

It must be stressed that this strong influence of finite spatial resolution is due to the large luminosity gradient exhibited by ellipticals – at least of the $r^{1/4}$ type (de Vaucouleurs and Capaccioli 1979, Schweizer 1979) and often even more pronounced, with the presence of a nucleus (e.g. M 87 Young et al 1978, and M32 Bendinelli et al. 1977).

III The Need for High Resolution 2D Photometry : Measuring the Anisotropy

For non isotropic cD's – like NGC 4889 – the next step is to determine the kind of anisotropy generated in the galaxy (via the cD evolution by galactic cannibalism or through tidal forces ?), but assuming that it is oblate.

From Bacon 1985, we give in Fig. 5 a,b, the $\mu = (\sigma^2 + v^2)^{1/2}$ predicted anisotropy curves, respectively for a semi-isotropy (Binney's 1978 $\beta = 1 - \sigma_\theta^2/\sigma_r^2$ equal to zero), and for a large excess of radial motions ($\beta = 0.75 \ r/0.01 \ r_e + r$), as well as the Davies et al. 1983 experimental values.

As can be easily seen, no conclusions can be drawn, as this would require data at least down to radii $\sim 10^{-2} \ r_e$ (ie. $\sim 0".25$).

This is quite a general problem, as discrimination between different degrees of anisotropy uses the slope of the velocity dispersion of the experimental points, not their absolute values. High spatial resolution photometric *and* kinematical data are then essential, and, even more important, the attained resolution must be quantitatively assessed. As a first step, we urge long slit spectrographists to estimate the resolution of their kinematical data (e. g. from the one dimensional profile along the slit). A major long term improvement would be to obtain simultaneously true two dimensional photometric and kinematical data for instance with spatial scanning along a long slit spectrograph (as already tested at the AAT), or with spectral scanning Fabry-Perot Interferometers.

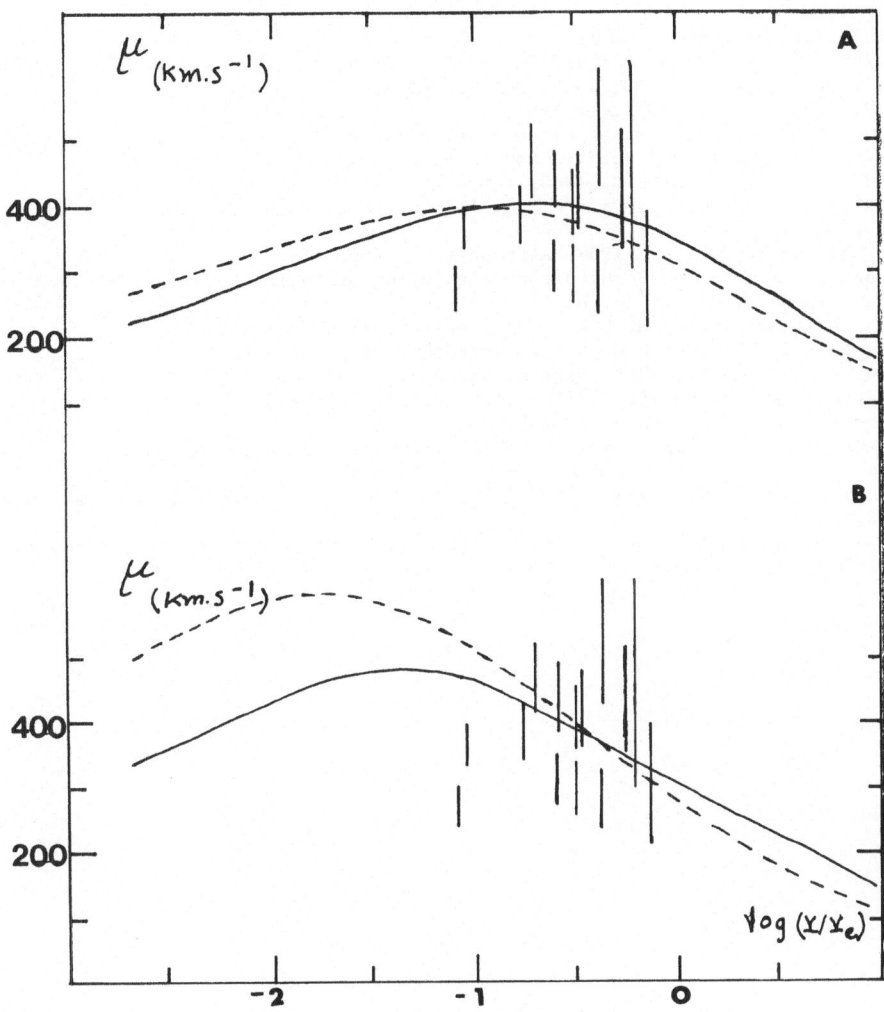

Figure 5 A,B : NGC 4889     Degree of anisotropy
upper curve  : semi-isotropy
lower curve  : quasi-radial motions
solid lines  : q = 0.4   $q_a$ = 0.6
dashed lines : q = $q_a$ = 0.8

REFERENCES

Bacon R., 1985, Astron.   and Astrophys., in press
Bendinelli O., Parmeggiani G., and Zavatti F., 1977, Mem. Soc Ast.Italiana, 48,713
Bertola F. and Capaccioli M., 1975, Astrophys. J., 200, 439
Binney J., 1978, Monthly Notices Roy. Astron. Soc. 183, 501
Binney J., 1982, Morphology and Dynamics of Galaxies, Saas Fee 1982, p. 32
Davies R.L., 1981, Monthly Notices Roy. Astron. Soc., 194,879
Davies R.L. and Illingworth G. 1983, Astrophys. J., 266,516
Davies R.L., Efstathiou G., Fall S.M., Illingworth G. and Schechter P.L., 1983, Astrophys. J. 266,41
Dressler A. and Sandage A., 1983, Astrophys. J. 265, 664
Illingworth G., 1981, The Structure and Evolution of Normal Galaxies, ed. S.M. Fall and D. Lynden-Bell, Cambridge, p. 27
Illingworth G. and Schechter P.L., 1982, Astrophys. J. 256, 481
Kormendy J. and Illingworth G., 1982, Astrophys. J., 256, 460
Pellet A. and Simien F., 1982, Astron. Astrophys. 106, 214
Schechter P.L. and Gunn J.E., 1979, Astrophys. J. 229, 472
Schweizer F., 1979, Astrophys. J. 233,23
Vaucouleurs G. de and Capaccioli M., 1979, Astrophys. J. Supp. 40, 699
Watanabe M., 1983, Annals Tokyo Astron. Obs. 19
Whitmore B.C., Rubin V.C. and Ford W.K., 1984, Space Telescope Science Institute, Preprint n° 19
Young P.J., Westphal J.A., Kristian J., Wilson C.P. and Landauer F.P., 1978, Astrophys. J. 221, 721
Young P.J., 1976, Astron. J. 81, 807

# FINE STRUCTURE IN ELLIPTICAL GALAXIES

François Schweizer and W. Kent Ford, Jr.
Department of Terrestrial Magnetism
Carnegie Institution of Washington
Washington, D.C. 20015, USA

Elliptical galaxies have long been regarded as possessing smooth light distributions. New photographic and digital techniques, however, have led to the discovery of a rich set of fine structures such as ripples, shells, streamers, dust lanes, and others (Malin 1979, Malin and Carter 1980, Schweizer 1980, Schweizer 1983, Malin and Carter 1983). For several years, we have been conducting a survey of 36 field giant ellipticals ($-19.5 > M_B > -22$) with the aim of finding fine structure in them. We have been interested especially in structures that may be signatures of merger activity, and in correlations between them and the presence of isophotal twists, color anomalies, and abnormal M/L ratios. We discuss here the analysis of the direct images obtained in our survey. First we describe a digital masking technique that is closely related to the photographic unsharp masking. Then we briefly present some statistics from the field-elliptical survey and interpret the results.

## DIGITAL MASKING

To detect fine structure in ellipticals and related merger galaxies, we use a Gould-DeAnza IP-8400 image processor attached to a VAX-11/750 computer and an enhanced version of the Tololo-Vienna Reduction System. A filtering algorithm similar to the photographic unsharp masking has proven especially useful for enhancing very faint structures. In essence, it consists of subtracting from the original image an unsharp copy obtained by digital convolution with a two-dimensional blurring function; the contrast of the difference image is then increased. Figure 1 illustrates the procedure with an image of Arp 230, a galaxy with ripples. Figure 1a shows a portion of a IIIa-J plate obtained at the CTIO 4-m telescope. The plate was scanned with a PDS microdensitometer in a 512x512 raster with 15μm = 0.″28 steps and displayed on a TV monitor without further processing. Figure 1b shows an unsharp digital mask produced from the original image as follows: first we suppressed the brightest star by interpolating the surrounding sky across it, then we cleaned all fainter stars by running a circularly shaped median filter of 6″ diameter across the whole image, and finally we smeared the image by convolution with a 2-D Gaussian weighting function with σ = 5″ and a circular boundary at r = 2σ. Figure 1c shows a masked image obtained by subtracting the

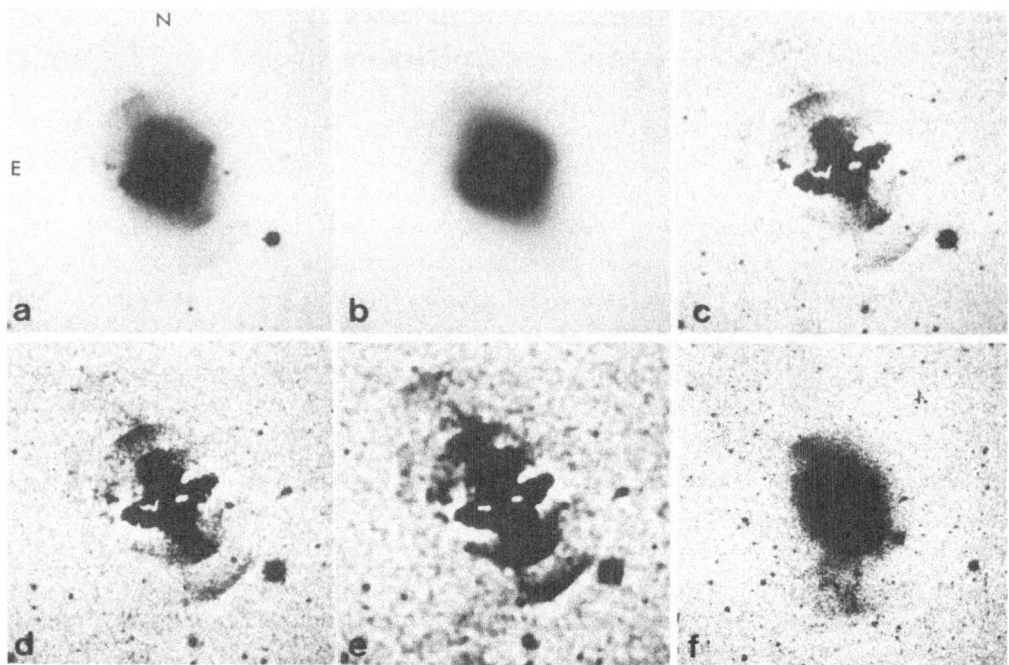

Figure 1. Steps in digital masking of Arp 230 : (a) Original scan of a 2!3x2!3 field; (b) unsharp mask; (c) masked image; (d) filtered with 3x3 median and increased contrast; (e) filtered with 11-pixel diameter median; (f) scan of 4!6x4!6 field at extremely high contrast. Note the tail extending south.

unsharp mask, multiplied by 0.8, from the original image and increasing the contrast of the display by a factor of four. Clearly, the difference image shows the ripples, inner ring, and dust lanes of the galaxy much better, at the price of increased visibility of the photographic plate noise. (This masked image is comparable to Arp's [1966] photograph, which was enhanced by electronic dodging.) The visibility of fine structure can be enhanced further by median filtering the masked image and increasing its contrast. Median filters have the desirable property of preserving edges while diminishing noise (e.g., Pratt 1978), which is exactly what is needed to enhance ripples in a noisy image. Figures 1d and 1e show the results of passing a square 3x3 median filter and a circular, 11-pixel diameter median filter, respectively, across the image. The 3x3 filter clearly brings improved visibility, whereas little gain is achieved by the larger filter. Apparently, the human brain contains excellent noise suppression circuitry and nearly resents heavy digital filtering. Finally, Figure 1f shows a 1024x1024 scan of the same galaxy, filtered with a 7x7 median and displayed at extremely high contrast. The faint filamentary appendix extending from Arp 230 to the south is suggestive of an old tidal tail, a feature which several other galaxies with ripples do also display (see Figs. 2 and 3).

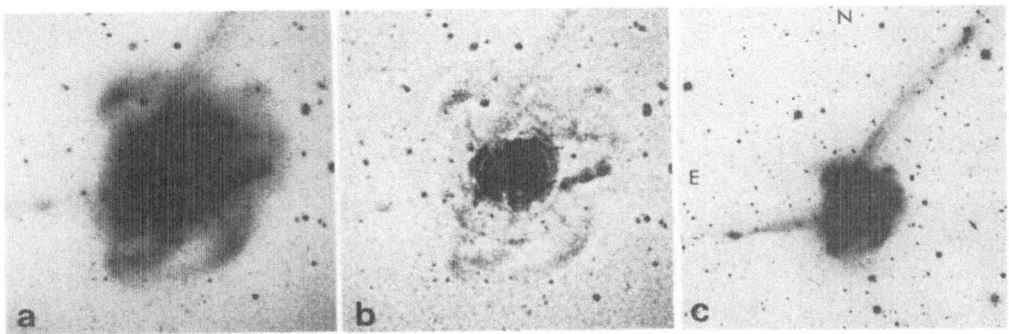

Figure 2. Fine structure in merger galaxy NGC 7252. (a) Sum of scans of three CTIO 4-m plates (3'x3' field); (b) masked image showing inner structure and ripples; (c) large-area print made from one of the plates.

When compared to photographic unsharp masking, digital masking offers significant advantages: quantitative information, strict reproducibility, elimination of stars from the mask, and better noise suppression. Also, the digital masking can be refined to yield <u>photometric</u> information by substituting a fitted model (e.g., a $r^{1/4}$ law) for the convolved mask. Disadvantages of digital masking, at least for the present, are: long computing times and limited numbers of pixels (typically $512^2$ to $1024^2$). Clearly, one of the main reasons for developing digital enhancement techniques is that astronomical images are obtained increasingly in digital form, as for example with CCD detectors; hence digital processing is a necessity if the advantages of linear detectors are not to be lost. Last, but not least, it may be easier for today's average astronomer to find a nearby image-processing facility than to find a photographic laboratory with experience in unsharp masking.

Figure 2 shows an example of fine structure in the merger galaxy NGC 7252, where digital masking has proven superior to photographic masking. From visual inspection of CTIO 4-m plates taken in 1" seeing, we have long known that there are weak ripples in this likely merger of two disk galaxies (Schweizer 1982), yet until recently we were not able to reproduce them satisfactorily on a photographic print. Problems arose from the small size of the galaxy image on the plate, the grain noise, and the large density gradients in the image. Digital enhancement of the ripples, on the other hand, has proven straightforward. Figure 2a shows an image of the main body of NGC 7252, formed by coadding scans (512x512 pixels, steps of 20μm = 0."37) of three different plates, thus reducing the noise by a factor $\sqrt{3}$. Figure 2b shows a masked image obtained by subtracting from the sum image a mask produced through convolution with a σ = 2."3 Gaussian and multiplied by 0.9. The masked image reveals three pronounced ripples in the western part of the bright inner body and two weak ripples further out to the south. Given the two long tails that extend from the main body (Fig. 2c), NGC 7252 is yet another galaxy where ripples and tails

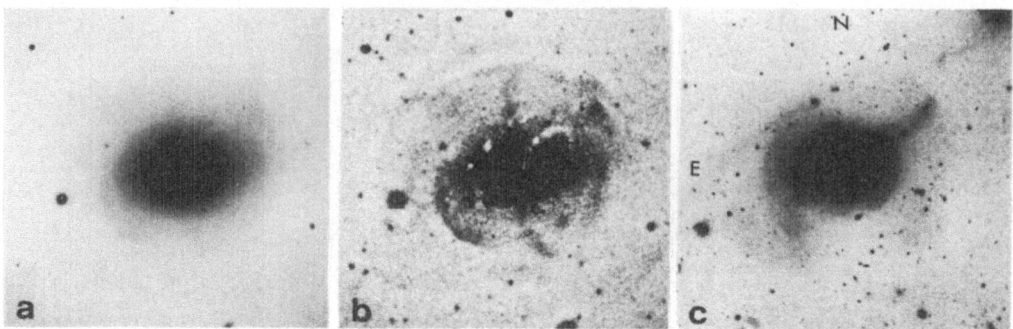

Figure 3. Fine structure in the E3 galaxy NGC 5018. (a) Scan of a IIIa-J plate (3'x3' field); (b) masked image showing ripples, dust lanes, and luminous plumes; (c) 9'x9' scan at extremely high contrast showing outer tails.

seem to be associated. In fact, it is this object which led one of us to propose that ripples may be signatures of mergers involving disk galaxies (Schweizer 1980), and we are happy to finally be able to present a crucial piece of the evidence.


SURVEY OF 36 SOUTHERN FIELD ELLIPTICALS

To study fine structure in giant ellipticals, we have selected a sample of 36 southern field E's brighter than $m_B$ = 13.0, with recession velocities less than 3300 km s$^{-1}$, and being as isolated as possible; their absolute magnitudes lie in the range $-19.5 > M_B > -22$. We have obtained deep IIIa-J plates of all of them and short exposures of most of them. Thirty-two E's were photographed with the CTIO 4-m telescope and four E's with the Palomar 48-inch Schmidt. A first search for fine structure was conducted by visual inspection of the plates and of high-contrast prints made from them. As a second step, we are now tracing the plates with a PDS microdensitometer and processing them digitally. So far, we have completed the digital unsharp masking for only a few of them. Table 1 summarizes the preliminary statistics, based on our visual inspection, of three types of fine structure: ripples, linear features, and dust patches and lanes. By "ripples" we mean the arc-shaped brightness enhancements called "shells" by Malin and Carter. These ripples occur at distances ranging from very close to the center to the distant

Table 1. Fine Structure in 36 Field Ellipticals

| Type of fine structure | # galaxies | #/36 |
|---|---|---|
| Ripples (some very weak) | 16 | 44% |
| Linear features ("plumes", "tails", ...) | 10 | 28% |
| Dust (patches or lanes) | 9 | 25% |

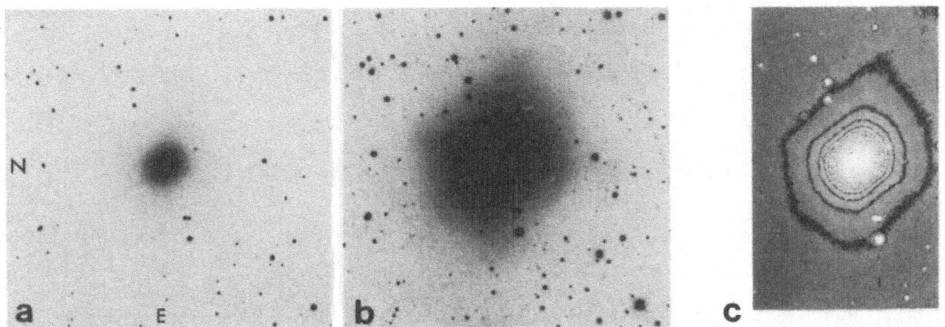

Figure 4. Crossed streamers of the E2 galaxy IC 3370. (a) Print of 10-min IIIa-J plate (4!7x4!7 field) showing apparently normal inner structure. (b) Print of 90-min IIIa-J plate showing "crossed streamers" in the outer regions. (The weak ripples are not visible in this unprocessed image.) (c) Isophotes from a CCD image obtained with the CTIO 4-m telescope by Patrick Seitzer.

outskirts of the galaxies. They are also observed in merger galaxies such as NGC 1316, 5128, and 7252. Among the ellipticals with newly discovered ripples are: NGC 596, known for its strong isophotal twists and relatively fast rotation; NGC 5018, which features also dust and two filamentary extensions reminiscent of tidal tails, as shown in Figure 3; and IC 3370, with two very unusual crossed streamers forming an X and material filling two opposite quadrants of this X (Fig. 4). Computer simulations by Quinn (1984), Toomre (unpublished; see Fig. 3 in Schweizer 1983), and Dupraz and Combes (1985) explain the ripples as the remains of disk galaxies accreted by the ellipticals. Observations of E's in spiral-elliptical pairs (e.g., NGC 474 = Arp 227) and in small goups (e.g., NGC 596) suggest that complete accretions (= mergers) need not always be invoked: partial mass transfer from a nearby disk galaxy may often suffice to create nice ripples. Some of Toomre's models support this view. If these interpretations are correct, then the statistics of Table 1 suggests that nearly one half of our survey ellipticals have accreted disks or parts of disks. If we further assume that, because of stretching and diffusion, ripples remain visible for ≤ 2 Gyr, as the Quinn and Toomre models suggest (but see Dupraz and Combes 1985), then a typical field giant elliptical must have accreted at least 2-5 (parts of) disks over the age of the Universe and, after correcting for higher interaction rates in the past, probably closer to 4-10 (parts of) disks. We suggest that it is this random accretion of luminous matter which may be the cause of the slow apparent rotation of giant ellipticals.

This work was supported in part by NSF grants AST-82 16979 and AST-83 18845.

REFERENCES

Arp, H.: 1966, Atlas of Peculiar Galaxies (Pasadena: California Inst. Technology).
Dupraz, C., and Combes, F.: 1985, this volume.

Malin, D.F.: 1979, Nature 277, 279.
Malin, D.F., and Carter, D.: 1980, Nature 285, 643.
Malin, D.F., and Carter, D.: 1983, Astrophys.J. 274, 534.
Pratt, W.K.: 1978, Digital Image Processing (New York: Wiley), pp. 330-332.
Quinn, P.: 1984, Astrophys.J. 279, 596.
Schweizer, F.: 1980, Astrophys.J. 237, 303.
Schweizer, F.: 1982, Astrophys.J. 252, 455.
Schweizer, F.: 1983, in IAU Symposium No. 100 (Dordrecht: Reidel), p. 319.

DISCUSSION

E. Sadler : What is the overlap between ripple galaxies and dust-lane galaxies in your sample, i.e., are the galaxies with ripples the same as those with dust?

F. Schweizer : Some ripple galaxies do have dust (e.g., NGC 1316, 3923, 5018, 5128, and IC 3370), others do not. Dust clearly also occurs in galaxies with no known ripples. In our sample, I cannot remember offhand the statistics of correlation or anticorrelation. I would like to point out, however, that dust seems to occur most often in small patches or lanes near the center, and I detect it in these cases only on my 1 minute and/or 10 minute IIIa-J plates of the objects, but not on the 1 - 1.5 hour plates. I quoted the dust statistics for its own interest rather than to use it in any way as evidence for ripples being signatures of disk accretions.

J.-L. Nieto : Do you have an estimate of the percentage of dull normal elliptical galaxies such as those that the theoreticians had imagined until ten years ago?

F. Schweizer : Since 44% of field ellipticals in our sample show ripples, the other 56% could be called dull and normal. However, it is my impression that with increasing signal-to-noise ratio we will find fine structure in all ellipticals.

A. Fabian : Do you actually observe a suitable population of disk galaxies around isolated ellipticals that can subsequently fall in? Presumably there must be several tens of such galaxies per elliptical.

F. Schweizer : It is difficult to find truly isolated giant ellipticals. Only 15 of our 36 E's have no companion galaxies within a projected radius of 500 kpc (for $H_o = 50$), and even these have sometimes neighbors not much further away in projection. I emphasize that both some of our photographs and the Toomre models suggest that mass transfer during disk-elliptical encounters can also create ripples. Hence, not each event need be a full-scale merger, and in sparse groups several neighboring disks may donate material, or even one disk may donate material on several occasions. So there is a suitable population of neighboring disk galaxies for present-day and future mass transfers and/or mergers, and one must assume that there were more of them in the past, some of which have merged.

R. Terlevich : What is the distribution of morphological types of shell galaxies?

F. Schweizer : Very little is known for types other than giant ellipticals. I know of a half dozen SO galaxies that have ripples and of only one Sb galaxy. The latter is NGC 3310, where I interpret Walker and Chincarini's "bow and arrow" signature (Ap.J. 147, 416, 1967) as a ripple and a likely tail, respectively. There may be a problem with ripples in E's versus SO's. Galaxy classifiers tend to shift ellipticals that have some sharp features into the SO category, and I have noticed that several galaxies with ripples have had their type changed from E to SO as better plate material has become available. Therefore, I believe that one should include SO galaxies in any survey aimed at deriving ripple statistics, and Kent Ford and I are trying to do just such an E+SO survey in the north.

F. Bertola : I would like to point out that the elliptical galaxy NGC 4125 also shows in its outer regions a cross-like structure similar to that in IC 3370.

# CAN SHELLS HELP TO DISTINGUISH PROLATE FROM OBLATE ELLIPTICAL GALAXIES ?

Dupraz Ch., Combes F., Observatoire de Meudon, 92190 Meudon, France

## I-Introduction:

Schweizer (1980,1983) noticed the frequent associations of tails and 'ripples' in mergers of galaxies, and concluded that these ripples might be the signature of a tidal encounter involving an elliptical galaxy. From an extended survey by Malin and Carter (1983), it appears now that 17% of all isolated elliptical galaxies display a system of 'shells' : numerous sharp and nearly circular optical features encircling the galaxy. In a recent article, Quinn (1984) developped the hypothesis of merging and by convincing simulations, demonstrated that the shells are formed by the mechanism of phase wrapping of the companion stars in radial orbits in the potential of the elliptical. Quinn claimed that the comparison between the model and observations (number of shells, positions..) could help to determine the mass distribution in the elliptical (presence of a dark component, ...). But he only used spherical potentials for the elliptical galaxy. Here we present 3-body simulations of a merger involving a true elliptical potential, either oblate or prolate (ellipticity in density of E3.5). We show that the shell systems formed in the two cases are quite different: shells are indeed a mean of revealing the unknown 3D shape of the ellipticals.

## II-The model:

We simulate the accretion of a small galaxy companion by a massive elliptical in a radial encounter at low initial velocity. The technique is a restricted 3-body code.

Central elliptical galaxy: The potential is that of a King model with a concentration parameter of $c = \log_{10} (r_t/r_c) = 2.25$, corresponding to the best fit to an elliptical light distribution (cf. King 1966). The chosen values for the mass and radius correspond to a typical giant elliptical $M = 10^{12}$ $M_\odot$, $r_t = 110$ Kpc ( $r_c = 0.62$ Kpc). The ellipticity is introduced in the potential, and for the sake of simplicity the axis ratio is independent of the distance from the center ((a-b)/a of the order of 0.1). By the Poisson equation we obtain a corresponding density distribution whose ellipticity is slightly increasing from the center to the outer parts. The projected density profile is compatible with the observations. The prolate or oblate galaxies constructed in this manner are of an E3.5 type (corresponding to the mean observed ellipticity).

Accreted companion galaxy: We can follow the fate of the companion material by 3000 test particles. The mass ratio between the two galaxies is 100; M(companion) = $10^{10}$ $M_\odot$. The companion can be either a spiral (disk system) or an elliptical (spherical system). In the first case, the test particles are distributed in a plane, with an exponentially decreasing surface density (characteristic radius = 15 Kpc). The potential is that of a Toomre disk (in $(r^2 + a^2)^{-1/2}$, with a = 3.83 Kpc). An initial velocity dispersion is added to the rotational velocities, which amounts to the critical dispersion necessary for axisymmetric stability (Toomre, 1964): $Q = 1$ (the central radial dispersion is $30\,Km/s$).

In the case of an elliptical companion, the particles are distributed in a sphere with a radial density distribution corresponding to the King model. They evolve in a King potential of radius $r_t = 50$ Kpc, with no rotation and a 1D central velocity dispersion of $50\,Km/s$.

The impact parameter b of the collision is very low or null. We have checked that the results are not significantly changed until $b = 20\,Kpc$ (almost radial orbits).

The geometrical parameters of the simulations are essentially:

 - $\theta$ the impact angle of the companion with respect to the axis of symmetry of the ellipsoïdal galaxy,

 - $\varphi$ the angle between the initial relative velocity and the plane of the companion, when it is of spiral type (this parameter disappears in case of a spherical companion).

We have also varied $\omega$ , the angle defining the line of nodes between the orbit plane and the companion plane, but it is not a fundamental parameter.

### III-Results:

#### 1) Comparison between oblate and prolate central ellipticals:

The companion is a disk galaxy. 24 simulations (12 for oblate, 12 for prolate) were run corresponding to 3 values of $\theta$ (33.5, 60, 80.4) and 4 values of $\varphi$ . These angles are chosen to deliminate precise range of incidence probabilities (16%, 50%, 83%).

a) for the oblate case, a clear system of shells is formed only in 5% of the initial conditions, when the companion is accreted almost in the main plane (circular section) of the elliptical. The shells system is then gathered at a low latitude around this main plane, spread around 360° (cf figure). When projected onto the sky, the shells are clearly apparent only when the galaxy is seen nearly face-on (its type is then E0).

b) for the prolate case, shells have a probability of 50% to form, i.e. when the companion arrives within 60° of the symmetry axis. They are portions of sphere aligned with the major axis of the galaxy. In projection, they disappear when the line of sight is parallel to the galaxy axis (apparent type E0), but they are clearly seen in all other directions, aligned on the apparent major axis of the galaxy.

In summary, there are two kinds of shells systems that can be observed: either multiple arcs aligned in a cône, within a certain angle from the major axis of the galaxy (like NGC 3923, Malin & Carter 1983), or portions of rings distributed at random angles around a galaxy of almost circular projection (E0) (cf NGC 474, Arp Atlas n°227). These are likely to be respectively prolate and oblate (cf figure).

#### 2) E-type companions: 6 simulations were run with the same initial conditions, but with a spherical companion. The results are quite similar to the precedent ones. The dispersion of impact angles and also in velocities ($\sigma_v \sim 50$ Km/s) does not mix out the system of shells.

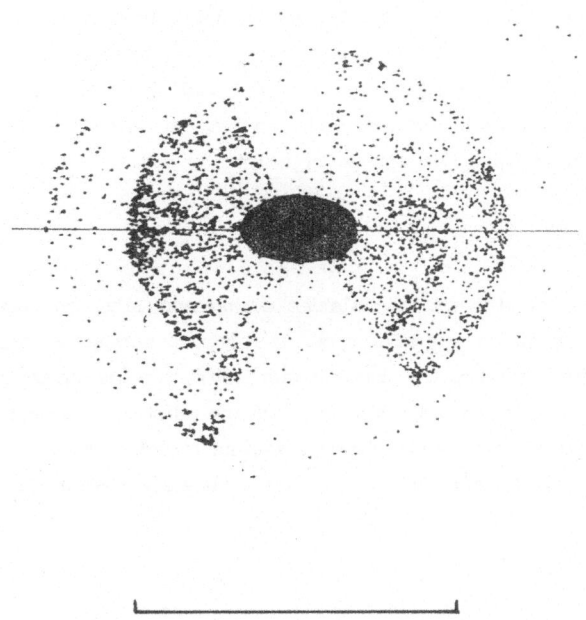

Figure: Top: prolate E-galaxy seen edge-on (i.e. E3.5); the full line represents its major-axis. Bottom: oblate E-galaxy seen face-on (E0). Scale: 100 Kpc between tick marks.

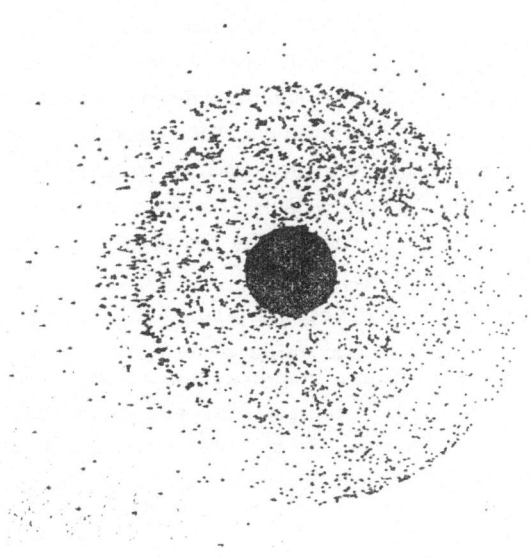

3) <u>Influence of the companion potential:</u> until now, we assumed that the bulk of the companion mass was negligible, soon after the beginning of the encounter. If we keep this potential for a longer time, and simulate the actual accretion by a realistic dynamical friction (values taken from Lin & Tremaine 1983), the mixing of the shell system is not significant. Besides, the introduction of this potential increases the ratio of distances between the first and last shells formed. This can help to reproduce the observations, since this ratio is very large in NGC 3923 for example (ratio of 10).

## IV-Conclusion:

Shells form around elliptical galaxies during and after the accretion of a small companion either of spiral or elliptical type. The impact parameter must be small (almost radial orbits). The distribution of the shells is quite different for oblate and prolate galaxies, and their observation is a clue to the 3-D shape of the elliptical. Shells are formed with 50% chance for prolate and 5% for oblate galaxies; we can therefore predict that the percentage of prolate galaxies is much higher in the galaxies with shells than in the whole sample of E galaxies.

### References:

King I.R. (1966) A. J. <u>76</u>, 64
Lin D.N.C., Tremaine S., (1983) Ap. J. <u>264</u>, 364
Malin D.F., Carter D. (1983) Ap. J. <u>274</u>, 534
Quinn P.J. (1984) Ap. J. <u>279</u>, 596
Schweizer F. (1980) Ap. J. <u>237</u>, 303
Schweizer F. (1983) I.A.U. <u>100</u>, 319
Toomre A. (1964) Ap. J. <u>139</u>, 1217

DISCUSSION

<u>F. Schweizer</u> : These are beautiful results. It seems to me you have just explained the observations of IC 3370. How long do your shells last, more than $2 \, 10^9$ years?

<u>Ch. Dupraz</u> : Intrinsically, the shells can last forever. However, the number of shells grows linearly with time, i.e. $N \sim t\sqrt{GM/R^3}$ , so that after a while the shells are so numerous and so close by, that the whole shell system is washed out. For reasonable values of M and R, the shell system can last as long as a Hubble time (we can form up to 20 shells after $10^{10}$ years).

<u>A. Fabian</u> : First, may I comment that ellipticals do contain a lot of gas (it is hot). Second, may I ask whether you can explain on a simple level how such 'cold' features as shells arise from the interaction of two 'hot' galaxies.

<u>Ch. Dupraz</u> : I want first to recall that the central galaxy just acts as a potential well and it does not matter whether it is a cold or hot stellar system. For the companion, the velocity dispersion $\Delta V$, even for an elliptical galaxy, introduces a small dispersion in the periods T of the particles in their pendulum motion in the potential well of the central galaxy. In other words, as is shown above, the dispersion in $t_{oi}$ (the initial phase associated with each particle) introduced by $\Delta V$ is small in front of $T(a_n)$, where $a_n$ is the apocentric distance of the $n^{th}$ shell.

# CONSTRAINTS ON THE POTENTIAL OF BARRED GALAXIES FROM SURFACE PHOTOMETRY AND DIRECT PHOTOGRAPHY

E. ATHANASSOULA
Observatoire de Marseille
13248 MARSEILLE CEDEX 4, France

The choice of an appropriate potential is crucial for all dynamical studies of barred galaxies. Indeed it can greatly influence the properties of the orbits, like the type, extent and stability of the families of periodic orbits, the existence and amount of ergodicity, and, through these, global properties like the stellar or gaseous responses to the bar. Some insight can be had by trying several types of barred potentials, but it would be preferable to concentrate one's efforts on potentials relevant to real galaxies. So the question arises : Which ones ?

Most studies made so far use a very simple potential. It is a Fourier series in which only the $m = 0$ (axisymmetric) and the $m = 2$ term have been retained i.e.

$$V(r,\theta) = V_o(r) + V_2(r) \cos 2\theta$$

where r and $\theta$ are the polar coordinates of a given point (e.g. Sanders and Huntley, 1976; Sanders, 1977; Contopoulos and Mertzanides, 1977; Contopoulos, 1978 and later; Contopoulos and Papayannopoulos, 1980). However this cannot be realistic for the case of a strong bar where the other harmonics of the potential and in particular the $m = 4$ are too large to be neglected. Thus another class of models, in which the bar is described by a homogeneous or nonhomogeneous ellipsoid, was introduced (Papayannopoulos and Petrou, 1983; Athanassoula et al., 1983; Pfenninger, 1984; Teuben and Sanders, 1984 etc). Although this is free of a number of the shortcomings of the previous simple model (see Athanassoula et al., 1983 and Petrou, 1984 for a discussion) it is still not necessarilly realistic. For instance if the ellipsoid ends at corotation the bar perturbation outside it is very small and cannot induce ergodicity. Whether this is also true for real galaxies and how much ergodicity should be expected is still an open question.

I will here briefly discuss two projects, whose aim is to derive some information about and constraints on the potentials that can be found in barred galaxies and to link certain properties of these potentials with morphological aspects of the galaxy. The starting point in both cases is direct photography and surface photometry.

## I. BAR POTENTIALS FROM SURFACE PHOTOMETRY

A suitable sample of plates or CCD frames of a couple of dozens barred and oval galaxies is necessary for this project. Surface photometry can give the deprojected calibrated images of these galaxies. Fourier analysis and subsequent use of an appropriate potential solver allows us to calculate the potential of the galaxy, assuming a mass to light ratio (M/L).

O. Bienayme and I have tested such procedure on NGC 1365, and have found that the m = 2 and m = 4 components of the potential are much wider and extend radially much further out than those of the homogeneous or nonhomogeneous ellipsoids ending at corotation. Thus the bar perturbation will be non negligible between corotation and outer Lindblad resonance, in contrast to what is found when the ellipsoid ends at corotation, and will incite ergodicity in this region.

Detailed analysis of the results of this and other galaxies, together with a discussion on the effects of a varying M/L and constraints from kinematics will be given elsewhere.

## II. THE SHAPE OF DUST LANES ALONG THE BARS

Prominent dust lanes can be often seen in the bars of SBb or later type galaxies. Two classes have been described by Athanassoula (1984a) as follows:

i) Straight. The best example is NGC 1300 (see fig. in Sandage, 1961). Although it has been believed that they are parallel to the bar major axis, careful inspection of large scale plates and the figures in Sandage (1984), Sandage (1961) and Sandage and Tammann (1981) gives the impression that, at least in several cases, they form an angle with it while staying parallel to each other, as is shown schematically in fig. 1a. Careful surface photometry would be needed to confirm this. In the inner parts they sometimes curl around the bulge (fig. 1b).

ii) Curved (fig. 1c), with their concave side towards the major axis. Good examples are NGC 6782 and NGC 1433.

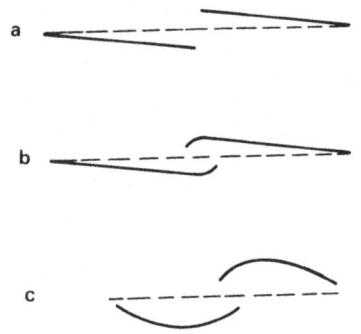

Fig. 1: Shapes of dust lanes in bars. The bar major axis is drawn by a dashed line

The link between these dust lanes and the shocks on the leading edges of the bar has been made more than twenty years ago (Prendergast, 1962) and has been well confirmed since then (Sanders and Tubbs, 1980; Roberts et al., 1979; Van Albada and Roberts, 1981 etc). It is reasonable to expect the form of the dust lane to be defined by the bar potential. For this reason I calculated the gas flow in several types of bar potentials using a gas code kindly provided by G.D. van Albada. A short description can be found in van Albada et al. (1982) and van Albada (1983). More than fifty runs were made to assess the

influence of the free parameters. We will discuss here only the influence of the
axial ratio. The other parameters were chosen so as to give dust lanes resembling
most those of figure 1 and their influence will be discussed elsewhere (Athanassoula,
1984 b).

Figure 2 shows the density response in the bar region of the first model. The bar
is a homogeneous prolate spheroid and the axisymmetric background is a Toomre disk
(Toomre, 1963) and a modified Hubble law bulge whose combination gives a flat rotation curve. The bar is at -45° with the x axis and has a 5 kpc semimajor axis (for
reference the side of the box is 16 kpc). The Lagrangian points are at 6 kpc. The
quadrupole moment is $4.5 \ 10^{10} \ M_\odot kpc^2$ and the axial ratio of the bar varies from 1.6
to 2.6 . Figure 3 shows the same result for a nonhomogeneous prolate spheroid with a
density distribution

Fig. 2: Density response in the bar region of the first model. The bar axial ratios
increase from left to right and from top to bottom (a/b = 1.6, 1.9, 2.2, 2.6)

$$\rho = \rho_o (1 - x^2/a^2 - y^2/b^2 - z^2/b^2)$$

The figures confirm the assumption that the bar potential influences the form of the dust lane. Indeed it is seen that for small axial ratios the dust lane is arced and of the form shown schematically in fig. 1c. For larger axial ratio the dust lanes are straight and parallel to each other but at an angle to the bar major axis, in good agreement with figure 1a and 1b. Thus it can be expected that the former will be preferably found in ovals or weak bars, while the latter will be found in strong bars. A cursory perusal of several plates and figures of barred galaxies conveys the same impression. Thus the shape of the dust lanes in the bar can be used to set constraints on the amplitude of the bar potential.

A more detailed discussion together with a theoretical explanation in terms of orbits and some comparisons with observations will be given elsewhere (Athanassoula, 1984b).

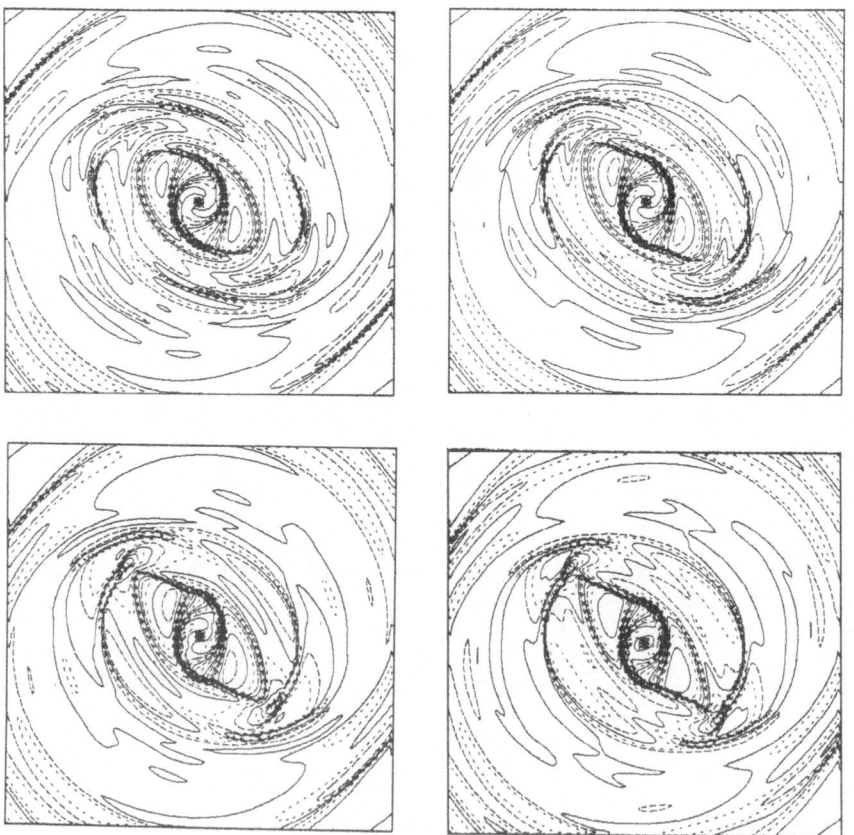

Fig. 3: Density response in the bar region of the second model. The bar axial ratios increase from left to right and from top to bottom. (a/b = 1.6, 1.9, 2.2, 2.5)

ACKNOWLEDGEMENTS

I would like to thank G.D. van Albada for providing the hydrodynamical code which I used for the results in the second part of this report and A. Sandage for allowing me to look in detail at the plates of the Shapley Ames catalogue. This work was stimulated by several discussions with A. Bosma, A. Sandage and O. Bienayme.

REFERENCES


Athanassoula, E.: 1984a, Physics reports in press
Athanassoula, E.: 1984b, in preparation
Athanassoula, E., Bienayme, O., Martinet, L., Pfenninger, D.: 1983, Astron. Astrophys. 127, 349
Contopoulos, G. and Mertzanides, C.: 1977, Astron. Astrophys. 61, 477
Contopoulos, G. and Papayannopoulos, T.: 1980, Astron. Astrophys. 92, 33
Papayannopoulos, T. and Petrou, M.: 1983, Astron. Astrophys. 119, 21
Petrou, M.: 1984, Mon. Not. R. Astron. Soc. 211, 1p
Pfenninger, D.: 1984, Astron. Astrophys. 134, 373
Prendergast, K.: 1962, in "Interstellar Matter in Galaxies", ed. L. Woltjer, W.A. Benjamin pub., and unpublished work
Roberts, W.W., Huntley, J.M. and Van Albada.: 1979, Astrophys. J. 233, 67
Sandage, A.: 1961,"The Hubble Atlas of Galaxies", Carnegie Institute of Washington
Sandage, A.: 1984, in preparation
Sandage, A. and Tammann, G.: 1981,"The Revised Shapley Ames catalogue ", Carnegie Institute of Washington
Sanders, R.H.: 1977, Astrophys. J. 217, 916
Sanders, R.H. and Huntley, J.M.: 1976, Astrophys. J. 209, 53
Sanders, R.H. and Tubbs, A.D.: 1980, Astrophys. J. 235, 803
Teuben, P. and Sanders, R.H.: 1984, preprint
Toomre, A.: 1963, Astrophys. J. 138, 385
Van Albada, G.D.: 1983, in "Internal Kinematics and Dynamics of Galaxies", ed. E. Athanassoula, p. 227
Van Albada, G.D. and Roberts, W.W.: 1981, Astrophys. J. 246, 740
Van Albada, G.D., Van Leer, B. and Roberts, W.W.: 1982, Astron. Astrophys. 198, 76


DISCUSSION

Combes F.: Does the variation of the angular velocity of the bar change significantly the shape of the dust lanes in your model ?

Athanassoula E.: I have discussed here the variation of only one parameter, the axial ratio of the bar. The other parameters were chosen so that the dust lanes resemble most those of fig. 1. The adopted angular velocity of the bar places the Lagrangian points at 6 kpc with a bar semimajor axis of 5 kpc. The shape and even the existence of the dust lanes depends heavily on the angular velocity of the bar, as can be seen from the following figure, where I compare three runs in which the Lagrangian radii are correspondingly 5, 6 and 7 kpc. It is clear that dust lanes as those shown in fig. 1 can be found only for a limited range of pattern speeds.

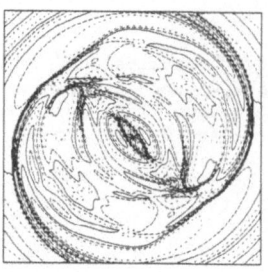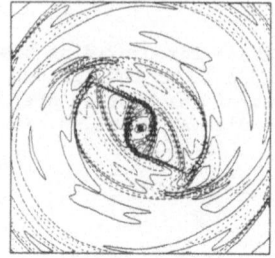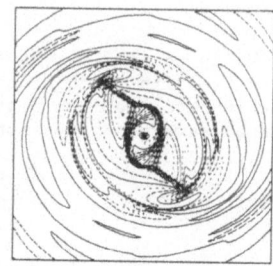

Macchetto F.: Does your model show velocity variations across the dust lanes ?

Athanassoula E.: Yes, and they are sizeable as can be seen from the figures. The length and direction of the arrows give the amplitude and direction of the velocity vectors. This model has a nonhomogeneous bar as described above, an axial ratio of a/b = 2.5 and a Lagrangian radius of 6 kpc.

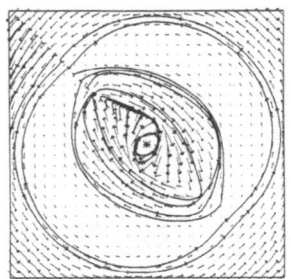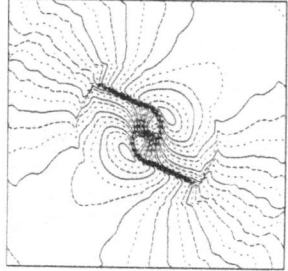

# BLUE AND NEAR-INFRARED SURFACE PHOTOMETRY OF 50 BARRED AND NONBARRED SPIRAL GALAXIES

B.G. Elmegreen and D.M. Elmegreen

IBM Thomas J. Watson Research Center

P.O. Box 218, Yorktown Heights, N.Y. 10598

## PLATE MATERIAL

Fifty spiral galaxies of various Hubble types and spiral arm classes were observed with the 1.2 m Palomar Schmidt telescope in blue (4360 Å) and near-infrared (8250 Å) passbands. The plates were digitized on a microdensitometer (Angilello et al., 1984) using an aperture size corresponding to about 1.4 arc seconds ($20\mu$ aperture in most cases) and a $512 \times 512$ pixel image size. All of the plates have calibration wedges, so relative intensities could be determined everywhere but in the overexposed nuclei.

## OBSERVATIONS OF NONBARRED GALAXIES

The spiral structures of the galaxy images were analyzed from plots of the variation of intensity with inclination-corrected azimuthal angle. Two types of spiral arms in nonbarred galaxies were found (Elmegreen and Elmegreen 1984- Paper I). "Grand design" galaxies have long and symmetric spiral arms that are nearly pure stellar density enhancements in the old stellar disks. "Flocculent" galaxies have unconnected, spiral-like pieces of arms that are very blue, and probably pure star formation.

Figure 1 shows near-infrared photographs of two galaxies, NGC 4321 and NGC 5055, which have the same Hubble type (Sbc) but different spiral arm types. Typical azimuthal profiles are also shown. Grand design galaxies, such as NGC 4321, have similar profiles in the blue and near-infrared, so the excess light in a spiral arm compared to the interarm region is almost entirely the result of an excess density of stars in the arms. The amplitude of the density pattern can be large, especially in the outer parts of the galaxies, where the arm/interarm brightness contrast is often 2 or more magnitudes, corresponding to a relative arm strength of 70%. Flocculent galaxies, such as NGC 5055, have no significant density variations in azimuth.

Individual star formation regions stand out clearly on the azimuthal scans of both NGC 4321 and NGC 5055. In the grand design galaxy, they appear as blue peaks superposed on the smooth density wave profile. The near-infrared peaks at the same locations are much fainter. NGC 5055 has similar blue-only peaks, but the background in this flocculent galaxy has no sinusoidal variations from a density wave. The star formation mechanisms may be similar in the two types of galaxies, but the overall distribution of star formation follows the density wave pattern in a grand design galaxy and is apparently random in flocculent galaxies (e.g., see Seiden and Gerola 1982).

Figure 1 - Near-infrared photographs of NGC 4321 (left) and NGC 5055, and inclination-corrected azimuthal profiles of the B and I band intensities. Blue profiles are traced by heavier lines. Each pair of profiles is for the radius indicated by the number on the left, in units of the radius at 25 magnitudes per square arc second. Marks on vertical axes indicate 2 magnitude intervals.

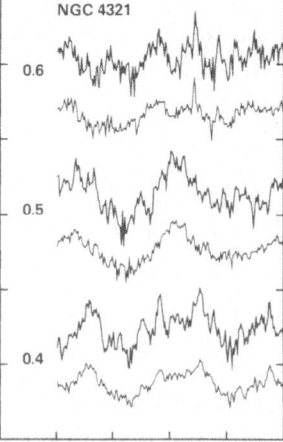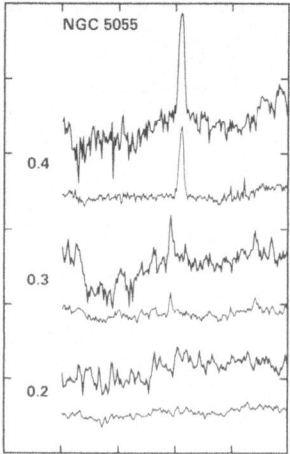

BARRED GALAXIES

Figure 2 shows two barred galaxies in the near-infrared passband, NGC 1300 and NGC 2500. The spiral arms in grand design barred galaxies (such as NGC 1300) are stellar density waves (Elmegreen and Elmegreen 1985 -- Paper II). Flocculent barred galaxies (such as NGC 2500) have either weak density waves or pure star formation arms.

The stellar nature of barred galaxy spiral arms contradicts gas dynamical models where the arms are entirely in the gas and young stars. The existence of stellar spirals may imply that bars grow slowly, and that the growth continuously excites a spiral in a not-strongly self-gravitating stellar disk (Thielheim 1980, 1981; Thielheim and Wolff 1981, Polzin and Thielheim 1981). Alternatively, a wave-mode with a bar-spiral pattern may grow if the disk is strongly self-gravitating (Toomre 1981, Bertin 1982, Haass 1982, Lin 1982). In either case, the presence of stellar arms implies that the bar-spiral pattern evolves, since the arms transfer angular momentum outward (Lynden-Bell and Kalnajs 1972). Such transfer itself may cause the bar to grow, so the bar-spiral pattern could be mutually reinforcing.

There are two types of bars in barred galaxies (Paper II). Figure 3 shows the intensity profiles of NGC 1300 and NGC 2500 along the bars and spiral arms (dots) and along the interbar and interarm regions (lines). The profiles are flat and exponential in the two cases, respectively. Flat bars generally occur in early type SB galaxies, and they tend to be larger than exponential bars, relative to both the galaxy size and the length of the rising part of the rotation curve. Figure 4 shows these correlations between relative bar size, Hubble type and bar type (squares = flat bars; triangles = exponential bars).

Flat bars are also stronger than exponential bars: the ratio of the peak amplitudes of the $m=2$ to the $m=0$ Fourier components of the bar azimuthal profiles equals 0.6 and 0.4 for the flat and exponential types, respectively. The spiral arm structures in flat and exponential barred galaxies differ as well. Flat barred galaxies tend to have long and symmetric spiral arms that decrease in amplitude with radius, while exponential bars tend to have chaotic looking arms that increase in amplitude with radius (as they do in nonbarred galaxies -- Schweizer, 1976; Paper I).

We believe that the different arm structures imply that spirals in flat-bar or early type SB galaxies are strongly driven by the bars, and that they exist between approximately the corotation radius and the outer Lindblad resonance. The spirals in exponential-bar or late type galaxies are poorly driven by the bar, and they probably lie between the inner Lindblad resonance and

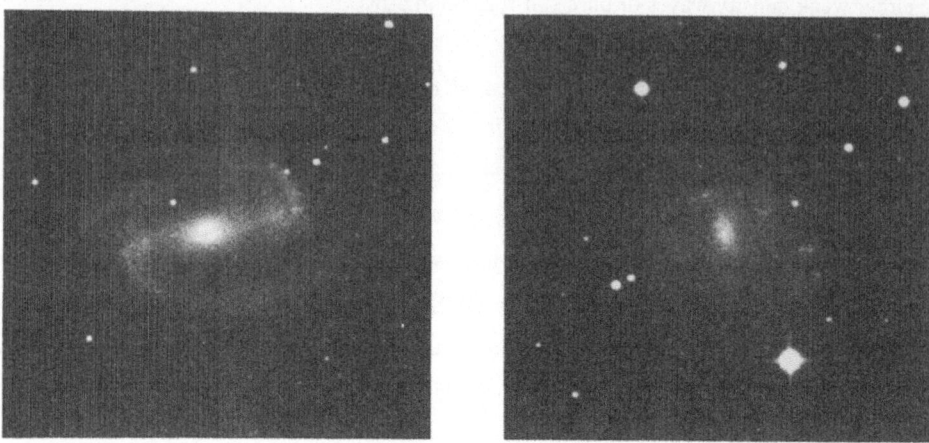

Figure 2 - NGC 1300 (left) and NGC 2500 in the near-infrared passband.

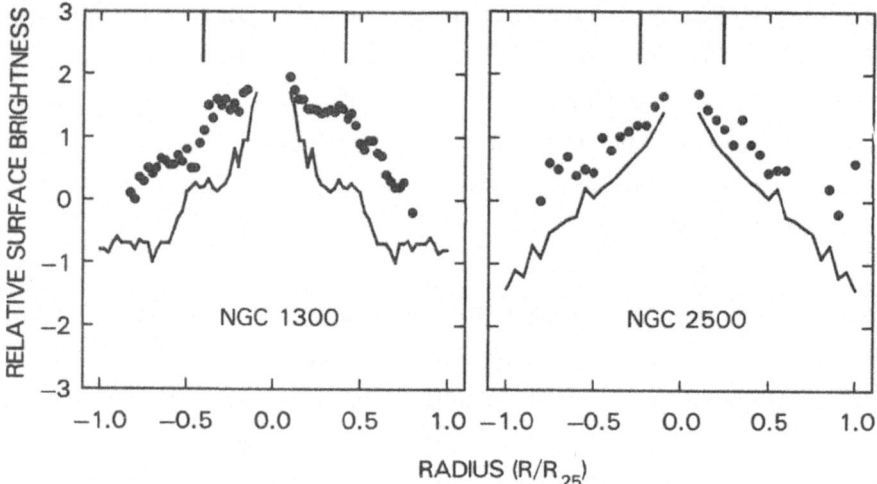

Figure 3 - Profiles of surface brightness (in magnitudes) in the near-infrared passbands of the bar and spiral arm regions (dots) and the interbar and interarm regions (lines) for NGC 1300, a typical flat-barred galaxy, and NGC 2500, a typical exponential-barred galaxy. The vertical lines at the top margin indicate the radial positions of the ends of the bars. The abcissa is radius in the galaxy, measured in units of the radius at 25 magnitudes per square arc seconds. The ordinate is relative surface brightness (arbitrary zero point) with one-magnitude intervals indicated.

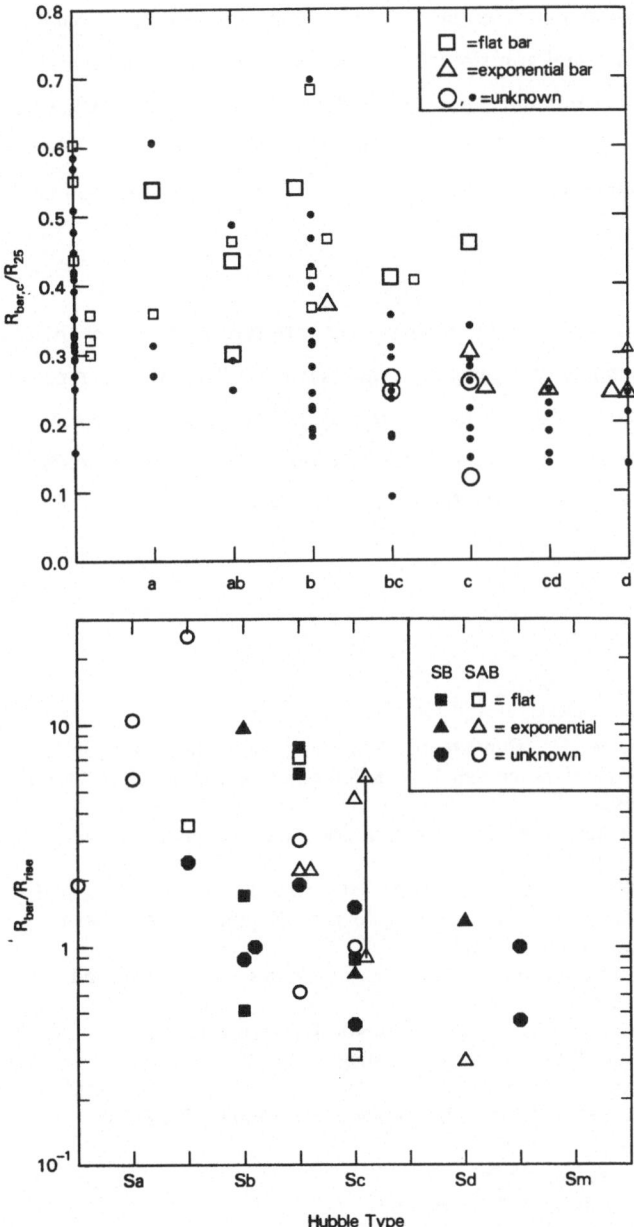

Figure 4 - The ratio of the bar length to the galaxy size (top), and the bar length to the length of the rising part of the rotation curve, are shown as functions of the galaxy Hubble type, with different symbols representing flat, exponential or unknown bar types. All galaxies with known rotation curves and surface photometry have been included. In the top diagram, large symbols represent galaxies studied in our survey, and small symbols represent galaxies whose photometry was found in the literature.

corotation, as do spirals in nonbarred galaxies. These conclusions follow from the observed radial increase or decrease in spiral arm strength for the flat and exponential types, respectively, and from the prediction that the inner or outer Lindblad resonances absorb the wave (Shu 1970; Lynden-Bell and Kalnajs 1972). Increasing spiral arms should exist between the inner Lindblad resonance and corotation and decreasing spiral arms should exist between corotation and the outer Lindblad resonance.

The relative bar sizes (Figure 4) are consistent with this placement of the orbit resonances. The exponential bars in late type galaxies extend between ~0.4 and ~1.2 times the length of the rising part of the rotation curve, so these bars may end at an inner Lindblad resonance (which usually occurs near the turnover point on the rotation curve). Also, theoretical orbits in strong bars extend out to corotation, so the flat bars, which we observe to be strong, probably extend out to corotation.

# REFERENCES

Angilello, J., Chiang, W.-H., Elmegreen, D.M., and Segmuller, A., 1984, in Astronomical Microdensitometry Conference, ed. D.A. Klinglesmith (Goddard: NASA Conference Publication 2317), p. 229.

Bertin, G. 1982, in Internal Kinematics and Dynamics of Galaxies, ed. E. Athanassoula (Dordrecht: Reidel), p. 119.

Elmegreen, B.G. and Elmegreen, D.M. 1985, Astrop.J., 287, in press, (Paper II).

Elmegreen, D.M. and Elmegreen, B.G. 1984, Astrop.J.Suppl., 54,127 (Paper I).

Haass, J. 1982, Ph.D. dissertation, MIT.

Lin, C.C. 1982, in Internal Kinematics and Dynamics of Galaxies, ed. E. Athanassoula (Dordrecht: Reidel), p. 117.

Lynden-Bell, D. and Kalnajs, A.J. 1972, Mon.Not.Roy.Astr.Soc., 157, 1.

Polzin, D. and Thielheim, K.O. 1981, Astr. Astrop., 101, 409.

Schweizer, F. 1976, Astrop.J.Suppl., 31, 313.

Seiden, P.E. and Gerola, H. 1982, Fundamentals Cosmic Phys., 7, 241.

Shu, F.H. 1970, Astrop.J., 160, 99.

Thielheim, K.O. 1980, Astrop.Sp.Sci., 73, 499.

Thielheim, K.O. 1981, Astrop.Sp.Sci., 76, 363.

Thielheim, K.O. and Wolff, H. 1981, Astrop.J., 245, 39.

Toomre, A. 1981, in Structure and Evolution of Normal Galaxies, ed. M. Fall and D. Lynden-Bell (Cambridge: U. Press Cambridge), p.111.

# BRIGHTEST MEMBER LUMINOSITIES AND DYNAMICAL TIMES OF GALAXY CLUSTERS

G. Giuricin, F. Mardirossian, M. Mezzetti

Osservatorio Astronomico

Via G.B. Tiepolo 11

I - 34131 Trieste, Italy

## INTRODUCTION

It is generally believed that galactic cannibalism plays a major role in the formation of very bright galaxies in clusters. Simple schemes for the merging of galaxies require the first-ranked galaxy in a cluster to grow at the expense of the other bright cluster members, the effect being largest in those clusters with short dynamical times. In particular, within a standard cannibalism scenario one expects to find a correlation between fundamental dynamical time scales (such as the crossing time and the relaxation time of clusters) and the magnitude difference $\Delta M_{12}$ (or $\Delta M_{13}$) between the first and second (or third) brightest cluster members (McGlynn and Ostriker, 1980).

The analysis of the data from 15 Abell clusters, carried out by these two authors, yielded moderate correlations (in the sense expected) between the two above-mentioned cluster dynamical times and $\Delta M_{12}$, which the authors regarded as support for their view. In this paper we wish to check whether the correlations claimed by McGlynn and Ostriker (1980) for galaxy clusters can be reconfirmed by an enlarged and more recent data base.

## ANALYSIS

In searching for possible correlations with cluster dynamical times we are interested in the two quantities, $t_c$ and $t_r$, taken to be proportional to the cluster crossing and relaxation times, respectively; we definite $t_c$ and $t_r$ as follows:

$$t_c = R/V \qquad (1) \qquad\qquad t_r = V \cdot R^2 \qquad (2)$$

where R is the cluster virial radius (in Mpc) and V is the cluster virial velocity dispersion (in Km/s). Searching in the literature, we have found 31 galaxy clusters with sufficiently adequate estimates of V, R, and of the magnitude differences $\Delta M_{12}$ and $\Delta M_{13}$. In a few cases we have estimated the value of R from available determinations of V and of cluster virial mass according to the virial theorem.

On the basis of a linear regression analysis of our whole data sample (and sub-sample), we wish to point out first of all that, contrary to Kashlinsky's (1983) finding, based on older data, V does not correlate with R. Besides, neither $\Delta M_{12}$ nor $\Delta M_{13}$ correlate significantly with $t_c$ or $t_r$ (the significance level is always < 90%), at variance with the marginal trends (at 95-96% significance level) claimed by McGlynn and Ostriker (1980).

Besides casting serious doubts on the validity of the standard cannibalism scenario (e.g., Ostriker's (1978) review), which involves substantial merging of galaxies occurring in the course of cluster dynamical evolution, our results can be regarded as an indication that the action of any special evolutionary processes, which are probably responsible for the development of the brightest cluster members, is not appreciably related to cluster dynamical time scales. This conclusion appears to be consistent with the results of Merritt's (1984) computer simulations of cluster evolution. In fact, his study suggests that the essential properties of the brightest cluster galaxies are determined already during the stage of cluster collapse and that little evolution is expected after cluster collapse. Other recent simulations of post-collapse cluster evolution (Richstone and Malumuth, 1983; Miller, 1983) stress the difficulties for very massive and large galaxies to grow in virialized clusters through galaxy merging processes. Their primary evolution (including merging) may occur, before cluster virialization, in subclusters (Cavaliere et al., 1984), which are gradually erased through the coalescence of lumps.

REFERENCES
Cavalière, A., et al., 1984 in "Clusters and Groups of Galaxies", p. 499.
Kashlinsky, A., 1983, Monthly Notices Roy. Astron. Soc. 202, 249.
McGlynn, T.A. and Ostriker, J.P., 1980, Astrophys. J. 241, 915.
Merritt, D., 1984, Astrophys. J. 276, 26.
Miller, G.E., 1983, Astrophys. J. 268, 495.
Ostriker, J.P., 1978, I.A.U., Symp. No 79, p. 357.
Richstone, D.O., and Malumuth  E.M., 1983, Astrophys. J. 268, 30.

# DARK HALOS AROUND LATE-TYPE GALAXIES

G. COMTE

Observatoire de Marseille

2 place Le Verrier, 13248 MARSEILLE CEDEX 4, France

In order to investigate the mass-to-light ratio in irregular Magellanic galaxies, a sample of 21 Sdm, Sm, Im objects, for which high-resolution velocity fields observed in 21 cm line are available in the literature, has been selected. Rotation curves, either taken from the authors or reconstructed from the original data and sometimes supplemented with other informations published elsewhere, are shown in Fig. 2. The homogeneous "de Vaucouleurs" distance scale was adopted.

## I. ONE-COMPONENT DYNAMICAL ANALYSIS

The galaxies were first assumed to be flat disks in pure circular rotation. The density distribution $\sigma$, projected onto the equatorial plane, was computed by Nordsieck's (1973) method (Fig. 3). The trend of $\sigma$ with radius r is found to be very generally satisfactorily fitted by an exponential law of shape:

$$\sigma(r) = \sigma(o)\ e^{-\beta r}$$

The photometric parameters of the disk are known either from surface photometry or high-quality multiaperture photoelectric photometry: in this latter case pure exponential law is supposed to hold; the effective equivalent radius $r_e^*$, which contains half the total luminosity, and $\alpha = 1.68/r_e^*$ are computed according to Olson and de Vaucouleurs (1981) from the multiaperture data.

The length scale $\alpha$ of the brightness distribution of the exponential disk:

$$I(r) = I(o)\ e^{-\alpha r}$$

is compared to the dynamical length scale $\beta$ in Fig. 1. We get in the average $1/\beta \simeq 2.7 \times 1/\alpha$, thus the "dynamical effective radius" which contains half the total mass is 2.7 times larger than $r_e^*$, or, in other words, the mass-to-light ratio apparently increases regularly with r, as:

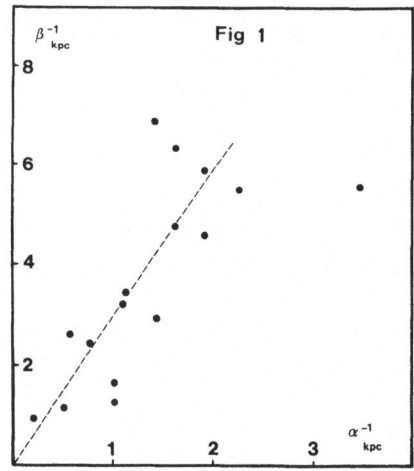

Fig 1

$$\log_{10}\ (M/L) = (M/L)_{r=o} + \gamma r$$

## II. A SIMPLE TWO-COMPONENT MASS MODEL

Let us now assume that: all the visible mass is distributed into a flat (c/a = 0.2) disk with constant M/L

- this disk is surrounded by a dark halo, supposed perfectly spherical, whose potential accounts for the apparent radial increase of M/L. At distance r from the center, $V_D$ is the rotation velocity due to the disk, $V_{obs}$ is the observed rotation velocity. The excess of potential over the disk potential felt by a test particle at r is entirely due to the halo. $M_H(r)$ being the fractional halo mass inside r, we get:

$$V_H^2 = V_{obs}^2 - V_D^2 = G\ \frac{M_H(r)}{r}$$

Differentiating: (Bahcall et al., 1982)

$$dM_H(r) = \frac{1}{G}\ (2r\ V_H\ dV_H + V_H^2\ dr)$$

Introducing the space density of halo matter:

$$\rho_H(r) = \frac{dM_H(r)}{4\pi r^2 dr} = \frac{1}{4\pi G}\ \frac{V_H^2(r)}{r^2}\ (1 + \frac{2r}{V_H(r)}\ \frac{dV_H(r)}{dr})$$

$V_D$ is computed from the disk rotation law given by Monnet and Simien (1977), with $(M/L)_{r=o}$ as constant disk mass-to-light ratio (since for r = o the contribution of the halo to the projected density is null), and with the photometric parameters $r_e^*$ and $I_e = I(r_e^*)$ as input ingredients. To avoid difficulties with the basic hypothesis, we have restricted the sample to 10 galaxies for which: i) the rotation curve is smooth enough, (objects with evidence for strong tidal deformation of the velocity field and/or non circular motions, as both Magellanic Clouds, were excluded), and ii) the photometry seems precise enough to derive $r_e^*$ with a reasonably accuracy.

Figure 4 shows the structure of the halos in terms of space density of matter; all the density laws are consistent with:

- $\rho_H(r)$ approximately constant up to a cutoff radius $r_{cH}$

- a cutoff tail where $\rho_H(r) \propto r^{-3^{+.3}_{-.2}}$

Table 1 gives, for the 10 objects of the restricted sample, some dynamical parameters of the disk and the halo, with notes referring to the sources for the HI observations and the photometry.

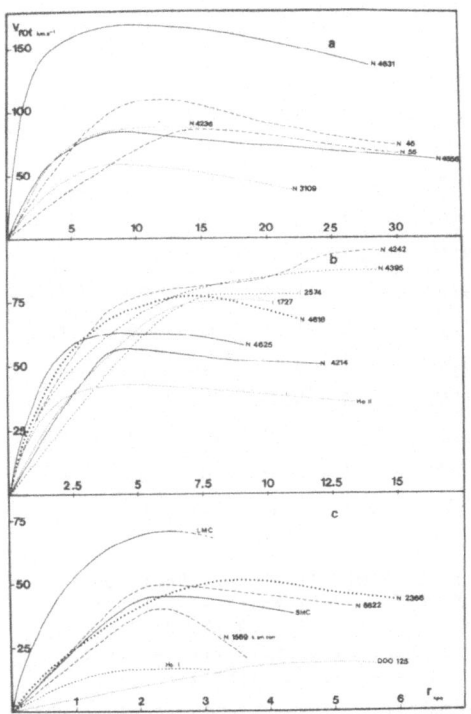

Fig. 2 : The adopted rotation curves

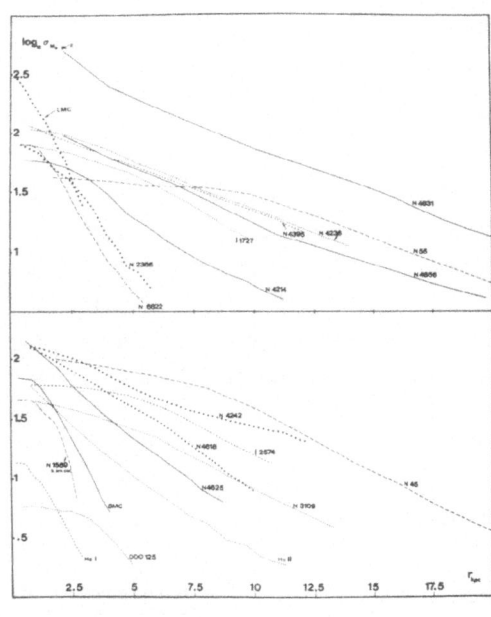

Fig. 3 (above): Projected mass density distributions

Fig. 4 (below): Density structure of the dark halos

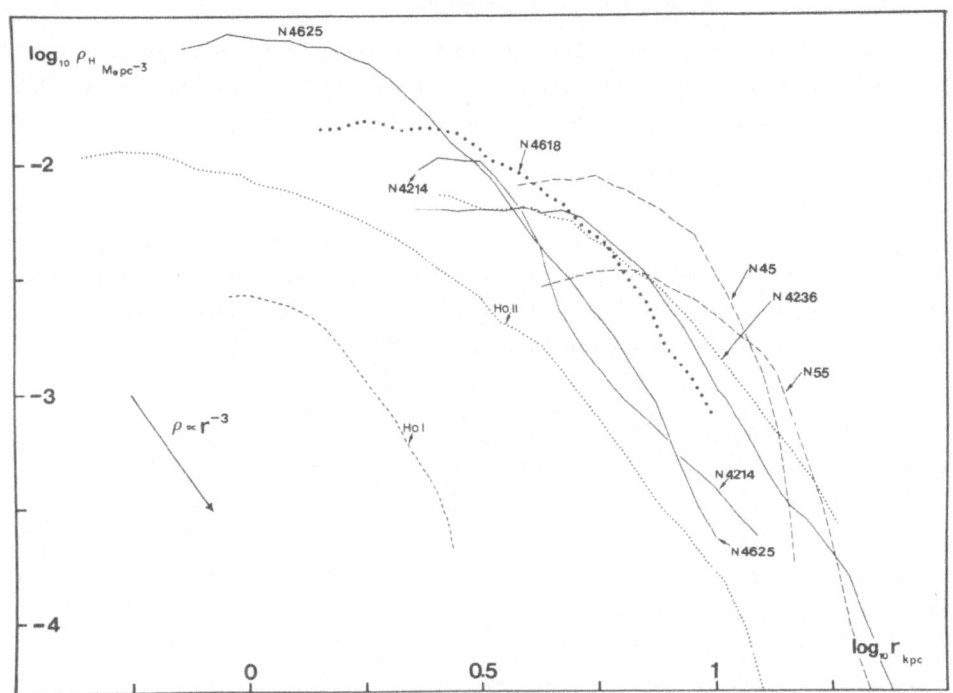

Table 1.

| (a) | (b) | (c) | (d) | (e) | (f) | (g) | (h) | (i) | (j) |
|---|---|---|---|---|---|---|---|---|---|
| NGC 45 | SA(s)dm IV-V | 1 | A | 9.32 | 1.91 | 4.56 | 39 | .10 | 1.65 |
| NGC 55 | SB(s)m : | 2,3 | B | 10.07 | 1.62: | 6.32 | 39 | .03: | 1: |
| IC 1727 | SB(s)m III-IV | 4 | C,E | 9.19 | 1.61: | 4.71 | 19 | .11 | 1.25 |
| Ho II | Im IV-V | 5,13 | D | 9.02 | 1.43 | 2.92 | 4.1 | .30 | .95 |
| Ho I | IAB(s)m V | 6 | D | 8.19 | 1.01: | 1.24 | 0.2 | 1.5 | .80 |
| NGC 4214 | IAB(s)m III-IV | 7 | E | 9.64 | 1.13 | 3.42 | 8.8 | .11 | .20 |
| NGC 4236 | SB(s)dm IV | 8,9 | E | 9.52 | 2.26 | 5.46 | 30 | .20 | 1.55 |
| NGC 4618 | SB(rs)m | 10 | F | 9.63 | 1.10 | 3.20 | 13 | .12 | .40 |
| NGC 4625 | SB(rs)m p. | 10 | F | 8.98 | .57 | 2.58 | 8 | .05 | .40 |
| NGC 4656 | SB(s)m p IV | 11,12 | G | 10.04 | 1.42 | 6.84 | 28.5 | .15 | .35 |

Notes: col. (c) gives the reference to HI work: 1: Lewis (1972), 2: Robinson, van
Damme (1966), 3: Seielstadt, Whiteoak (1965), 4: Combes et al. (1980),
5: Cottrell (1976), 6: Tully et al. (1978), 7: Allsopp (1979), 8: Shostak,
Rogstad (1973), Huchtmeier (1975), 10: Van Moorsel (1983), 11: Weliachew
et al. (1978), 12: Winter (1975), 13: Huchtmeier et al. (1981).

col. (d) gives the reference to the photometry: A: Romanishin et al. (1983),
B: de Vaucouleurs (1961), C: de Vaucouleurs, Longo (1983), D: de Vaucouleurs
et al. (1981), E: de Vaucouleurs et al. (RC2) (1976), F: Bertola (1967),
G: Stayton et al. (1983).

col. (e): $\log L^{\circ}_{T,B}$; col. (f): $1/\alpha$ in kpc; col. (g): $1/\beta$ in kpc; col. (h):
total mass in $10^9 M_\odot$; col. (i) disk/halo mass ratio; col. (j): blue M/L of
the disk alone, corrected for absorption and inclination.

## III. IMPLICATIONS FOR MASSIVE NEUTRINOS

Tremaine and Gunn (1979) and Gunn (1981) have given a simple formula to derive the
lower limit of the particle mass for massive neutrinos forming an isothermal structure of core radius a and velocity dispersion $\langle\Delta v\rangle$, when one assumes that the phase
density must not increase during the formation of the halo:

$$m_{\nu(eV)} > 101 \cdot (100 \text{ km s}^{-1}/\langle\Delta v\rangle)^{1/4} \cdot (1 \text{ kpc/a})^{1/2} \text{ g}^{-1/4}$$

where g is the statistical weight of the particle state.
Let us assume that $a \le r_{CH}$ and that $\langle\Delta v\rangle = (\sqrt{1/2}) \cdot v_{H \text{ max}}$ where $v_{H \text{ max}}$ is the maximum of rotation velocity given by the halo, and let us consider the "smallest"
galaxies in the sample:

HoI yields $m_{\nu D} > 170$ eV (for g = 1, Dirac neutrinos) or $m_{\nu M} > 142$ eV (for g = 2,
"Majorana" neutrinos).

HoII, NGC 4214, NGC 4236 yield respectively $m_{\nu D} > 133$ eV, 71 eV and 51 eV and
$m_{\nu M} > 112$ eV, 60 eV and 43 eV. All these masses seem presently excluded by particle

physicists.

However, some arguments might be stressed:

i) in the case of HoI, which has a maximum rotation velocity of only $\sim$ 25 km s$^{-1}$, a significant fraction of the disk mass may be supported by random motions, and is not accessible to the preceding analysis.

ii) an isothermal nature of the halo implies $\rho_H \propto r^{-2}$, which is not observed. Releasing this hypothesis (Basdevant, 1984), however leads to very similar results for the neutrino masses, due to the fact that an $r^{-2}$ fit to the curves in Fig. 4 remains marginally acceptable because of the sensitivity of the method to errors (in the velocity at large r as well as in the photometry of the disks).

REFERENCES

Allsopp, N.J.: 1979, M.N.R.A.S. **188**, 765

Bahcall, J.N., Schmidt, M., Soneira, R.M.: 1982, Ap. J. (Letters) **258**, L 23

Basdevant, J.L.: 1984, Astron. Astrophys. in press

Bertola, F.: 1967, Mem. Soc. Astron. Italia **38**, 417

Combes, F., Foy, F.C., Gottesman, S.T., Weliachew, L.: 1980, Astron. Astrophys. **84**, 85

Cottrell, G.A.: 1976, M.N.R.A.S. **177**, 463

Gunn, J.E.: 1981, Astrophysical Cosmology (Pontif. Acad. Sc. Scripta Varia) edited by Bruck, Loyre, Longair, p. 557

Huchtmeier, W.K.: 1975, Astron. Astrophys. **45**, 259

Huchtmeier, W.K., Seiradakis, J.H., Materne, J.: 1981, Astron. Astrophys. **102**, 134

Lewis, B.M.: 1972, Australian J. of Physics **25**, 315

Monnet, G., Simien, F.: 1977, Astron. Astrophys. **56**, 173

Nordsieck, K.H.: 1973, Ap. J. **184**, 719

Olson, D.W., Vaucouleurs, G. de: 1981, Ap. J. **249**, 68

Robinson, B.J., van Damme, K.J.: 1966, Australian J. of Physics **19**, 111

Romanishin, W., Strom, K.M., Strom, S.E.: 1983, Ap. J. Suppl. Ser. **53**, 105

Seielstadt, G.A., Whiteoak, J.B.: 1965, Ap. J. **142**, 616

Shostak, G.S., Rogstad, D.H.: 1973, Astron. Astrophys. **24**, 405

Stayton, L.C., Angione, R.J., Talbert, F.D.: 1983, A.J. **88**, 602

Tremaine, S., Gunn, J.E.: 1979, Phys. Rev. Letters **42**, 467

Tully, R.B., Bottinelli, L., Fisher, J.R., Gouguenheim, L., Sancisi, R., van Woerden, H.: 1978, Astron. Astrophys. **63**, 37

Van Moorsel, G.A.: 1983, Astron. Astrophys. Suppl. **54**, 19

Vaucouleurs, A. de, Longo, G.: 1983, The University of Texas Monographs in Astronomy No 3, Austin

Vaucouleurs, G. de: 1961, Ap. J. **133**, 405

Vaucouleurs, G. de, Vaucouleurs, A. de, Corwin, H.C.: 1976, Second Reference Catalogue of Bright Galaxies, The University of Texas Press, Austin (RC2)

Vaucouleurs, G. de, Vaucouleurs, A. de, Buta, R.: 1981, A. J. **86**, 1429

Weliachew, L., Sancisi, R., Guélin, M.: 1978, Astron. Astrophys. <u>65</u>, 37

Winter, A.J.B.: 1975, M.N.R.A.S. <u>172</u>, 1

DISCUSSION

<u>R. Terlevich</u>: 1) How the $M_D/M_H$ of your galaxies compares with the ones measured for luminous Sc's ?

2) is there any trend in $M_D/M_H$ with other intrinsic properties like luminosity or surface density/surface brightness ?

<u>G. Comte</u>: 1) $M_D/M_H$ is known for large Sc's on a reasonably sizeable sample only from Wevers'Thesis (1984, University of Groningen). On $\sim$ 30 galaxies of types Sa through Sd he finds $M_D/M_H$ from 10% to 60% but the mass disk is computed using an M/L ratio fitted to the observed mean disk color by means of a Tinsley evolutionary model: this seems questionable. Our $M_D$ are very low indeed, as well as our disk M/L's: this comes out from the extrapolation you need in the rotation curve to use Nordsieck's method: in large spiral galaxies, the plateau is quickly reached but in the smallest galaxies the maximum rotation velocity is reached only near the optical limit of the galaxy (and not far from the HI observable limit).

2) clear correlations do not appear between $M_D/M_H$ and luminosity, but the sample is statistically very poor.

# N-BODY SIMULATIONS AND GALAXY PHOTOMETRY

Ortwin E. Gerhard

Max-Planck-Institut für Astrophysik

Karl-Schwarzschild-Str. 1, 8046 Garching, FRG

Abstract. N-body models are now a standard tool to improve our understanding of galaxies. Since photometric and kinematic observations of
a galaxy do not in general very strongly constrain a dynamical model,
the main use of N-body simulations in this respect is to answer
qualitative questions about the nature of possible equilibria. This is
illustrated by presenting a model with strong radial ellipticity
gradients and one with intrinsic principal axes twists. Another
important application of N-body methods is in the study of transient
phenomena.

## 1. N-body models and galaxies.

An N-body simulation begins by specifying initial masses, positions,
and velocities for a set of N point particles. Newton's equations of
motion are then integrated to follow the evolution to some quasi-
steady state; both transient phenomena during the evolution and the
final state may be compared with galaxy observations. The information
contained in the simulation at time t is the distribution of mass
points in phase space; this may be regarded as an approximation to the
phase space distribution function of a corresponding smooth system.
Effort is made to suppress collisional relaxation, so that the
particles in the simulation may represent any form of collisionless
matter.

Some important properties required of a realistic N-body simulation of
a galaxy are that i) it has a high dynamic range, ii) it must often be

truly three-dimensional, iii) it should be collision-free over the
time-scale of interest. One commonly used program is Aarseth's direct
integrator (Aarseth 1970, Ahmad & Cohen 1973). This calculates the
force on each particle by summing predicted/corrected contributions
from all distant particles and direct Newtonian forces from near
neighbours. The presently feasible number of particles is $N \leq 10^3$. To
suppress collisions, a softening parameter $\epsilon$ is introduced; typically
$\epsilon \simeq 0.1\ r_H$ for near-spherical and $\epsilon \simeq 0.3\ r_H$ for disk/halo galaxies,
where $r_H$ is the half-mass radius of the system (e.g. White 1978,
Gerhard 1981). This then sets the spatial resolution limit except in
the outer parts where the resolution is limited by small particle
number. The one-dimensional dynamic range is $\leq 2$ decades in radius, $\leq 6$
decades in density, $\leq 3$ decades in orbital time. The small number of
particles makes it difficult to obtain, for example, reliable runs of
velocity dispersion with radius. The main application of this code to
galaxies has been the study of encounters, which is difficult for grid
simulations. A second type of N-body program uses Fast Fourier Trans-
forms to obtain the potential and force from the particle densities on
a cartesian (Miller & Prendergast 1968, Hohl & Hockney 1969) or polar
grid (Miller 1976). In the latter case, the radial grid may be adjusted
to follow the build-up of a high central density (van Albada & van
Gorkom 1977), resulting in high resolution, whereas in the cartesian
case the spatial resolution is independent of density. In these
simulations, $N \leq 10^5$ and relaxation effects are negligible over time-
scales of interest (Hohl 1973). In the collapse calculations of von
Albada (1982), the one-dimensional dynamic range is $\leq 3$ decades in
radius and $\leq 8$ decades in density. A third type of N-body code has been
introduced by Villumsen (1982). He expands the potential of each
particle in tesseral harmonics about a density center and thereby finds
the force without any gridding. The resolution of this method is at
present limited by particle number ($N \sim 10^3$) and a small softening
parameter, and the dynamic range is $\sim 2.5$ decades in r and $\sim 7$ decades in
$\rho$ with this N (Villumsen 1982). Again collisions are suppressed in the
center because of softening and further out because only the lower
order terms in the expansion are kept.


Photometric observations of galaxies measure the surface brightness of
light (IR,UV,...) on the sky. Sometimes such data are available in
several broad bands, with resolution in optical ground-based photo-

metry usually set by seing. Typically a surface brightness profile of
an elliptical galaxy may span ~3 decades in radius (Kormendy 1982).
While photometric observations provide more information than that
relevant for dynamical studies (e.g. colours), they unfortunately
contain much less information than needed to constrain the models. A
major problem in a comparison with galaxy models is that the mass-to-
light radio as a function of position in the galaxy is not known.
Velocity measurements may be used to get some idea about M/L, but the
problem is underconstrained as discussed below. Usually M/L is assumed
constant, or to vary only slowly with radius. Apart from the possible
presence of dark matter, there can be other causes why the measured
light distribution may not be representative of the mass distribution.
For example, in IC 3370 the twists in the central isophotes are much
less pronounced in R than in B, indicating that the origin of twisting
in this galaxy is dust absorption (Sadler & Gerhard 1984). Most surface
photometry of twist galaxies sofar has been done in a single band, and
a multi-colour study would clearly be useful to establish which twists
do in fact indicate triaxiality (Mihalas & Binney 1981) and which do
not.

A spherical equilibrium stellar system in the stellar-hydrodynamical
framework is described by four independent functions (Binney & Mamon
1982): luminosity density $l(r)$, mass-to-light ratio $A(r)$, radial and
tangential velocity dispersions $\sigma_r(r)$, $\sigma_\Theta(r)$. Usually, observations
determine the luminosity surface density on the sky $\Sigma(R)$ and line-of-
sight velocity dispersion $\sigma_v(R)$ (no rotation is possible in a spherical system); these are integrals over $l(r)$, $\sigma_r(r)$, $\sigma_\Theta(r)$, and an
inversion is possible. Since the spherical Jeans' equation provides one
additional relation, one free function remains. Thus only if e.g. the
mass-to-light ratio is specified as a function of radius do the
observations then completely determine a dynamical model. The situation
is much **less** determined for galaxies without such high symmetry; now
observations only provide functions of two variables on the sky plane,
whereas the dynamical quantitites sought are functions of three
variables. Thus many N-body models may in principle fit one and the
same set of observational data. The opposite phenomenon arises through
projection effects: One N-body model will give rise to an infinity of
sets of observations, depending both on aspect angle and assumed
mass-to-light ratio.

What then can be learnt from N-body models? Firstly, as to equilibrium models - to which the previous discussion applies - one may i) get an idea about what sort of dynamical configurations are possible, ii) recognize, interprete or even understand certain typical features in observations of galaxies, iii) find models consistent with a given set of observations, and iv) assess the importance of projection effects. Secondly, N-body simulations contain information about the time-evolution between initial and quasi-final state; this makes possible not only an understanding of certain transient phenomena, but also (with due caution) inferences as to the nature of the initial conditions leading to galaxies. A good example of mutual stimulation between photometric observations and N-body simulations from my own field of interest is the following. Toomre & Toomre (1972) in a study to explain "tidal bridges and tails" seen in the beautiful photographs of Arp (1966) also suggested that the orbital braking and subsequent merging in such strong collisions should lead to "nothing less than the delayed formation of some ellipticals", and they identified a list of galaxies which might be in the process of merging. Both observational (Schweizer 1982) and N-body work on the merging of disk galaxies (e.g. Gerhard 1981, Negroponte & White 1983) has thereafter shown that indeed a galaxy like NGC 7252 may be formed in this way.

## 2. Twists and Triaxiality

In the second part of this paper I would like to discuss some of my own N-body models that show twisted isophotes when projected onto the sky, similar to those observed in many elliptical galaxies. It is well-known that velocity anisotropy plays an important role in determining the shapes of elliptical galaxies. On theoretical grounds one may then argue that ellipticals will in general have triaxial figures (Binney 1978a,b). It is also known from N-body work that triaxial systems relaxed from slowly rotating aspherical initial conditions are quasi-stable, i.e. long-lived compared to a dynamical time (Aarseth & Binney 1978, Wilkinson & James 1982). A system whose density is constant on similar triaxial ellipsoids or any axisymmetric system cannot show twisting of the projected isophotes on the sky (Stark 1977), but such twists arise for triaxial galaxies whose axial ratios vary with radius (e.g. Mihalas & Binney 1981). A number of ellipticals show such isophote twists; if these are twists in the projected mass distribution this strongly suggests that those galaxies are triaxial.

Here I first describe a triaxial N-body model with shape like that of a non-rotating prolate inner bar, gradually turning into a hot oblate outer part. The model was obtained by a merger calculation, and the final configuration is stable until the end of the simulation (for 17 dynamical times). The system has no symmetry axis since the prolate bar lies in the equatorial plane of the oblate component, but it still has triaxial symmetry. One ellipticity rises slowly from $\epsilon_c = 0.35$ to $\epsilon_c = 0.4$ as one goes outwards, the other changes strongly with radius from $\epsilon_b = 0.35$ to $\epsilon_b = 0$. This is a situation in which strong twists in the projected isophotes are expected; Fig. 1 (from Gerhard 1983a) shows ellipticity profiles and position angles for a number of projections. The projection planes are indicated by their intercepts with the coordinate axes. (b) is edge-on for the oblate and along the major axis for the prolate part, (c) is edge-on/along a minor axis. Upward arrows indicate the projected half-mass radii. The general trend is an ellipticity $\epsilon \simeq 0.3$ in the center, except when the prolate part is viewed along its major axis. One conclusion is that the variety of ellipticity profiles seen in the projections of one and the same galaxy model (rising, falling, minimum, maximum, according to di Tullio 1979, Bertola & Galletta 1979) suggests that such an ellipticity profile classification may only be of limited use. Notice that some similarities with observed galaxies exist, e.g. the shape of profile (a) in Fig. 1 and NGC 3640, (b) and NGC 2300, (c) and NGC 4365 (Leach 1981), although the model does not match any observed profile exactly and the model profiles extend considerably further than the observations.

Fig. 2 (Gerhard 1983b) shows perspective projections of a second N-body model at $t = 3t_{cr}$, $t = 9t_{cr}$, and $t = 15t_{cr}$ after equilibrium was reached. In this model, the particle distribution in each of six concentric shells is characterized by the same ellipticities $\epsilon_c = 0.4$, $\epsilon_b = 0.2$. However, the overall model is no longer triaxially symmetric; the directions of the major axes determined for all particles in each of the six shells change outwards in a complex way that is perhaps best described by imagining a coordinate system with one axis parallel to

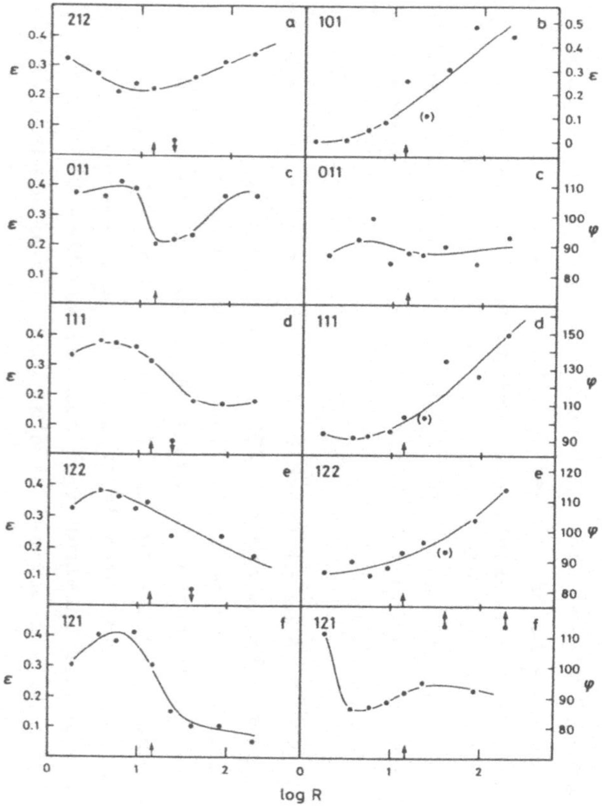

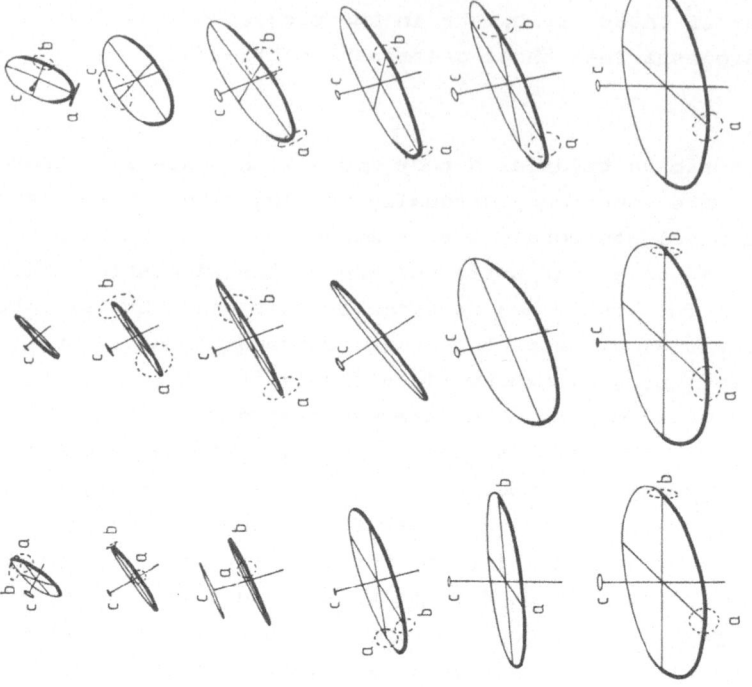

Fig. 2: Perspective projections of the principal axes in the second model, for three snapshots at times $t=3t_{cr}$(a), $t=9t_{cr}$(b) and $t=15t_{cr}$(c) after equilibrium was reached. From Mon. Not. R. astr. Soc. 203, 19P.

Fig. 1: Ellipticity profiles and position angles for a number of projections of the first model as described in the text. From Mon. Not. R. astr. Soc. 202, 1159.

and moving along a curved path while the other two rotate slowly about this axis. In Figure 2 the large ellipse in each frame represents the major plane and has the measured axial ratio when deprojected. The small circles correspond to the uncertainties in the directions of the respective axes. Details of the determination of the axes directions and errors are given in Gerhard (1983a). Thus one sees that this model has significant intrinsic principal axis twists as a function of radius, and these twists last for the entire time of the simulation although some evolution in detail is present.

In the magnitude of their ellipticity gradients and projected isophote twists both these models admittedly resemble the more extreme cases among the distribution of elliptical galaxies (Galletta 1980, Leach 1981). However, they do serve to conclude that both triaxial systems with strongly radially varying ellipticities and systems with large principal axes twists may dynamically exist for significant astronomical time-scales.

## References

Aarseth, S.J., 1970. In The Gravitational N-body Problem, p. 373, ed. M. Lecar, Reidel, Dordrecht.
Aarseth, S.J. & Binney, J., 1978. Mon. Not. R. astr. Soc. 185, 227.
Ahmad, A. & Cohen, L., 1973. J. Comp. Phys. 12, 389.
Arp. H., 1966. Atlas of Peculiar Galaxies, California Institute of Technology, Pasadena.
Bertola, F. & Galletta, G., 1979. Astr. Astrophys. 77, 363.
Binney, J., 1978a. Mon. Not. R. astr. Soc. 183, 501.
Binney, J., 1978b. Comments Ap. 8, 27.
Binney, J. & Mamon, G.A., 1982. Mon. Not. R. astr. Soc. 200, 361.
di Tullio, G.A., 1979. Astr. Astrophys. Suppl. 37, 591.
Galletta, G., 1980. Astr. Astrophys. 81, 179.
Gerhard, O.E., 1981. Mon. Not. R. astr. Soc. 197, 179.
Gerhard, O.E., 1983a. Mon. Not. R. astr. Soc. 202, 1159.
Gerhard, O.E., 1983b. Mon. Not. R. astr. Soc. 203, 19P.
Hohl, F., 1973. Astrophys. J. 184, 353.
Hohl, F. & Hockney, R.W., 1969. J. Comp. Phys. 4, 306.
Kormendy, J., 1982. In Morphology and Dynamics of Galaxies, p. 115, eds. L. Martinet & M. Mayor, Geneva Observatory, Geneva.
Leach, R., 1981. Astrophys. J. 248, 485.
Mihalas, D. & Binney, J., 1981. Galactic Astronomy, Freeman, San Francisco.
Miller, R.H., 1976. J. Comp. Phys. 21, 400.
Miller, R.H. & Prendergast, K.H., 1968. Astrophys. J. 151, 699.

Negroponte, J. & White, S.D.M., 1983. Mon. Not. R. astr. Soc. <u>205,</u>
    1009.
Sadler, E.M. & Gerhard, O.E., 1984. Mon. Not. R. astr. Soc., in press.
Schweizer, F., 1982. Astrophys. J. <u>252</u>, 455.
Stark, A.A., 1977. Astrophys. J., <u>213</u>, 368.
Toomre, A. & Toomre, J., 1972. Astrophys. J. <u>178,</u> 623.
van Albada, T.S., 1982. Mon. Not. R. astr. Soc. <u>201</u>, 939.
van Albada, T.S. & van Gorkom, J.H., 1977. Astr. Astrophys. <u>54</u>, 121.
Villumsen, J.V., 1982. Mon. Not. R. astr. Soc. <u>199</u>, 493.
White, S.D.M., 1978. Mon. Not. R. astr. Soc. <u>184</u>, 185.
Wilkinson, A. & James, R.A., 1982. Mon. Not. R. astr. Soc. <u>199</u>, 171.

DISCUSSION

**M. Capaccioli**: Although your analysis of the stability of complex triaxial or multistructured systems is very important, nontheless I believe that one should not stress too much the agreement between model calculations and observed isophotal twisting. It is not at all true that elliptical galaxies have elliptical isophotes; therefore observed twisting may in some cases reflect just distortions!

**O. Gerhard**: My impression was that in several galaxies where twists were measured the isophotes are well-described by ellipses (Carter, D., 1978, MNRAS <u>182</u>, 797). In deriving ellipticities and position angles for the projected N-body systems I have assumed that the mass density is constant on ellipses, whereas in the outer parts of the models discreteness effects make it difficult to decide whether this assumption is in fact correct. In this sense one therefore compares like with like even for "twist-galaxies" with non-elliptical isophotes.

A little dust in elliptical galaxies!

AFTERNOON SESSION IV :


COSMOLOGY AND EVOLUTIONARY
ASPECTS OF GALAXY PHOTOMETRY

# GALAXY PHOTOMETRY AND COSMOLOGY

Gustavo Bruzual A.


Centro de Investigaciones de Astronomia (C.I.D.A.)
Apartado Postal 264, Merida 5101-A
Venezuela

Abstract

A brief review is presented of the current understanding of galaxy photometry as related to cosmology. Attention is paid to the conclusions about spectral evolution of galaxies that can be derived from the deepest faint galaxy samples currently available. Some typical results are discussed.

Introduction

Galaxies are the most conspicuous and numerous detectable constituent of the universe. Despite the increasing evidence for the existence of large amounts of matter in forms that escape detection with conventional astronomical instruments, galaxies are still assumed to provide an unbiased tracer of the distribution of matter in the universe.

The study of the space distribution of relatively nearby galaxies has already shown that the universe contains sheet-like structures where galaxies are preferentially found, and large voids where galaxies are scarce. This realization has increased considerably the minimum scale in which the universe can be considered homogeneous and isotropic.

Considerable effort and telescope time are being invested in the determination of the photometric properties and space (z) distribution of faint galaxy samples. The hope is that a self consistent picture will emerge that will allow us to answer some of these basic questions of observational cosmology :

- How old and dense is the universe ?,
- How old are the galaxies we observe near us ?,
- Are galaxies coeval ?,
- How do the observed properties of galaxies evolve in time ?

Photometry plays an important role in this approach to cosmology. In the recent past it has become possible to obtain multicolor photometry for large samples of faint galaxies. Even though in some instances galaxy magnitudes have been derived from images obtained on CCD frames (Djorgovski, Spinrad and Marr 1985), the photographic plate still is one of the most efficient means of collecting large amounts of data (Koo 1981 ; 1985).

The photometric data *per se* do not contain a *priori* any clue about the size of the volume of space that is being sampled. Carefully compiled photometric catalogues, combined with redshift determinations of a large number of galaxies in these samples, have revealed a *posteriori* that at B = 23 a sizeable fraction of the universe is being observed.

From these results it can be appreciated that photometry has a fundamental role to play in cosmology. Once the behavior of the photometric properties of galaxies of different morphological (or color) classes as a function of redshift has been established, it then becomes possible to use pure photometry to probe the universe.

In practice there are several factors that make the situation quite more complicated than it would seem otherwise. With present instruments galaxies cannot be observed at arbitrarily faint light levels. Thus it is not possible to establish the loci in the color-color plane occupied by galaxies inside different redshift shells from empirical grounds only.

One can partially solve this problem by using the observed spectral energy distributions of nearby galaxies of different types to predict the photometric properties of identical galaxies as a function of redshift. However, this solution is not ideal because the evolution in time of the spectra is neglected. It then becomes necessary to model the effects of stellar evolution (or any other kind of evolution that is thought to be important) on galaxy spectra.

The data, originally meant to be used to provide information about cosmology, must also be used to obtain information about the correctness of our assumptions about galaxy evolution. Even though we are far from understanding spectral evolution completely, some clues are beginning to emerge that make the problem more tractable than a few years ago (Bruzual 1983).

Once spectral evolution has been understood, more realistic predictions of the photometric properties of distant galaxies become possible, and the original problem of obtaining cosmological information out of photometry can be approached.

See Koo (1985) for an excellent treatment of this subject.


## Models for the spectral evolution of galaxies

The models for the spectral evolution of galaxies constructed by the author
(Bruzual 1981, 1983a,b,c,d; Koo 1985) have been used by several authors as a guide
in the interpretation of the photometric data of distant galaxies. In the rest of
this paper these models will be used with the same purpose. For reasons of space
no details are given here about the models. The reader should consult the
references mentioned above


## Faint galaxy samples

In a general sense the existing faint galaxy samples can be grouped in three
different categories according to the criteria under which they have been
selected.

> Blind deep samples. These are magnitude limited faint galaxy
> samples chosen without any information about the nature of the
> objects included.

> Cluster galaxy samples. These samples are chosen by systematically
> studying several galaxies belonging to a cluster. The membership of
> the galaxies must be established. Cosmological information is
> derived by choosing clusters at different redshifts.

> Peculiar galaxy samples. These samples contain galaxies that are
> chosen according to a criterium that singles them out. For example,
> the optical counterparts of 3C radiosources, or the first-ranked
> galaxy in a set of rich clusters of galaxies. There is some danger
> in applying the conclusions derived from peculiar galaxies to the
> general population of galaxies.


## Results and interpretations

In the rest of this paper the main results obtained by authors working with some
of the different samples mentioned above will be summarized. For reasons of time
and space, only the results and interpretations of the data based on the models of
the author will be mentioned. These results are presented in pictorial and tabular
form in the indicated papers. The reader should consult the original papers for
details.


## Blind deep samples

Kron, Koo and Windhorst (1985) and Koo (1985) have done the most careful work with a complete magnitude limited sample. From their data and a comparison with the author's model the following conclusions emerge :

1) Spectral evolution of galaxies is a slow process. At least to z = 0.7, a large fraction of galaxies is consistent with no or very slow spectral evolution. The observed colors of galaxies fall in the range expected for slowly evolving stellar populations.

2) To the limit of their sample, the early stages of galaxy formation have not been detected. If the opposite were the case, an absence of red galaxies should be apparent at their highest z values. Red galaxies, instead, are well represented at all z's.

3) Up to their limit of z = 0.7, galaxies of all color classes seem to coexist at different epochs. The most natural interpretation of their color vs. z diagrams in terms of Bruzual's model is that the reddest galaxies at any z will evolve into galaxies similar to present day elliptical galaxies. The bluest galaxies at any z seem to be the progenitors of present day spiral and irregular galaxies. There is no evidence in their data to support the view that these blue systems will become as red as nearby ellipticals.

Peculiar galaxy samples

It has been claimed by several authors (e;g; Djorgovski, Spinrad, and Marr in this volume ; Lilly 1983) that they can detect spectral evolution in the 3C sources beyond z of 1 and up to the limit of their current data (z of about 2). Their data show indeed that the distant galaxies are markedly brighter in the ultraviolet than nearby elliptical galaxies. However, these galaxies are known to have a non-thermal continuum. Before these colors are taken as evidence of spectral evolution it should be proved that the observed colors reflect the colors of the stellar population and are not affected by this emission.

One point in favor of their interpretation is the fact that the observed points cluster remarkably well around the model prediction for a softly evolving stellar population. If the non-thermal emission were dominant effect in these galaxies, there is no a priori reason to expect that this emission will evolve in time as to mimic the emission from an evolving stellar population, and not in a more or less random fashion.

Thuan et al. (1984) find that the strong 3CR radio galaxies and the optically selected radio-quiet field galaxies cannot be distinguished by their infrared and optical-infrared colors. They conclude that the colors are probably due to a stellar population and also claim that they see in their sample the effects of a mildly evolving population.

## Conclusions

The quality of the data on faint galaxy photometry has improved by a large factor in the recent past. However, the questions asked at the beginning of this paper are still unanswered. At most, the data provide some information on the relevance of spectral evolution for the problems of interest.

Much more data are needed before the effects of evolution can be accounted for properly and the cosmological problem approached in a safe ground.

Even though the 3C radio galaxies seem to show a detectable amount of spectral evolution, the selection effects and peculiarities of this sample have not been fully evaluated. Thus it seems dangerous to extend this result to the general population of galaxies.

The deep galaxy samples have shown clearly that different galaxy types coexist at a given redshift, and that evolution, if it occurs, is much slower than previously thought. The early stages of galaxy evolution occur long before the epoch that is being sampled.

The particular cosmological model used in the interpretation of the observations is of minor importance for the conclusions. The evolutionary effects are much greater than the differences introduced by different cosmologies.

## References

Bruzual A., G. 1981, Ph. D. thesis, University of California, Berkeley.

Bruzual A., G. 1983a, Ap. J., 273, 105.

Bruzual A., G. 1983b, Rev. Mexicana Astron. Astrofis., 8, 29.

Bruzual A., G. 1983c, Ap. J. Suppl., 53, 497.

Bruzual A., G. 1983d, Rev. Mexicana Astron. Astrofis, 8, 63.

Djorgovski, S., Spinrad, H., and Marr, J. 1985 (this volume).

Koo, D.C. 1981, Ph. D. thesis, University of California, Berkeley.

Koo, D.C. 1985, A.J., (in press).

Kron, R.G., Koo, D.C., and Windhorst, R.A. 1985, Astr. Ap., (in press).

Lilly, S.J. 1983, Ph. D. thesis, University of Edimburg, Scotland.

Thuan, T.X., Windhorst, R.A., Puschell, J.J., Isaacman, R.B., and Owen, F.N. 1984, Ap. J., (in press).

## Discussion

### M. Capaccioli :

IUE observations of nearby ellipticals show a rather large scatter in the amount of the UV excess in these galaxies. Just with these data and no evolution we can predict : 1) a significant spread of colors at large z, and 2) that local first ranked ellipticals if seen at high z, would appear as blue but fainter than Sb galaxies.

### G. Bruzual :

I can only agree with your statement. However, I see no astrophysical reason by which you can suppress the effects of stellar evolution on the resulting spectrum of a galaxy. I think that in the scheme that you are proposing, you have to invoke a statistical distribution of the source or sources of the UV excess that by some reason does not change over cosmological time scales. This is an *ad hoc* assumption, without an astrophysical basis. In some of Spinrad's and Koo, Kron, and Windhorst's data, there are distant galaxies that are bluer than what you would predict from nearby IUE-bright ellipticals.

### T. Jaakkola :

a) Just in a preceding paper Laurikainen and I showed that some previous data, including yours, can be used to show that evolution can be interpreted simply by the familiar K-term. Why are observers so anxious to fit their data into fashionable ideas such as cosmic evolution ?

b) Isn't redshift just the effect which you need to bring distant blue galaxies into your telescope (thanks to the K-term)?

### G. Bruzual :

a) I think that you have to differentiate two things. You may not believe in an evolving universe and invoke some sort of steady state cosmology. However, and I think that all steady state cosmologists agree on that, stellar evolution is a proven fact that this or any theory cannot ignore. I think that stellar evolution is the most clear example of cosmic evolution.When I speak of spectral evolution I am only using the results of the theory of stellar structure to follow the effects of stellar evolution on the spectrum of an evolving stellar population. You may dislike the evidence that I showed in favor of the detection of spectral evolution, but I do not think that that is an argument against cosmic evolution.

b) When one does evolutionary calculations and predict the colors of a distant galaxy, the K-term is automatically included when the spectrum is shifted. The evolutionary-term is small for low z, but it can be larger than the K-term for high enough z. I do not think, as I said before, that you can explain everything with the K-term. The universe is much more complicated than the K-term or than my simple-minded evolutionary models.

# OBSERVING THE GALAXY EVOLUTION AT HIGH REDSHIFTS

S. Djorgovski, H. Spinrad, and J. Marr
Astronomy Department
University of California
Berkeley, CA 94720, USA.

ABSTRACT:    Magnitudes and colors for a sample of  distant 3CR radio-galaxies, with redshifts up to z=1.82,  show very dramatic evidence for luminosity and color evolution of these objects.  Cosmological differences  are small compared to the detected evolutionary effects.  Elongated shapes  and very extended  (both in space and velocity)  [O II] emission lines  suggest violent merging activity  at the early epochs, and possibly mark the beginnings of activity of powerful radio-galaxies.

During the last few years, we have pursued spectrophotometry of faint 3CR radio-galaxies (Spinrad & Djorgovski 1982; 1984a; 1984b; Spinrad, Djorgovski & Marr 1983). We have obtained  a substantial number of  high redshifts,  with some 15-20 galaxies beyond z=1, and up to z=1.819 (for 3C256).  This means that we can now observe galaxies over  look-back times  of the order of ~70% of the age of the Universe,  for a "sensible" range of Friedman models.  One may expect  to find prominent evolutionary effects at such high redshifts.

One simple way of addressing this question  is through the use of  magnitude vs. redshift (Hubble) diagrams,  as shown in Figure 1.  The low-z end of the Hubble diagrams is  now fairly  well determined,  primarly through the efforts  of Mt. Palomar groups  (Gunn & Oke 1975;  Sandage et al. 1976; Kristian et al. 1978; Hoessel et al. 1980; Schneider et al. 1983).  Here we make the assumption that, by analogy with the relatively nearby cases,  the powerful radio-galaxies  which we observe at high redshifts are the same kind of objects  as the first-ranked cluster galaxies,  normally used as the "standard candles" for this purpose.  Because of such assumptions, various selection effects, aperture and seeing corrections (Djorgovski 1983), etc.,  the Hubble diagrams  can be very treacherous -- but they still  can be used  as a "first look" tool.

Good CCD magnitudes  are available  for some of the galaxies  in our sample, but for a majority we had to use  the synthesized magnitudes from our spectrophotometry. From a variety  of internal  and external  tests and  comparisons,  we estimate  the errors to be ~$0^m.5$ .  No aperture, seeing, color system, etc., corrections have been applied at this time, so that the results presented here are very preliminary.

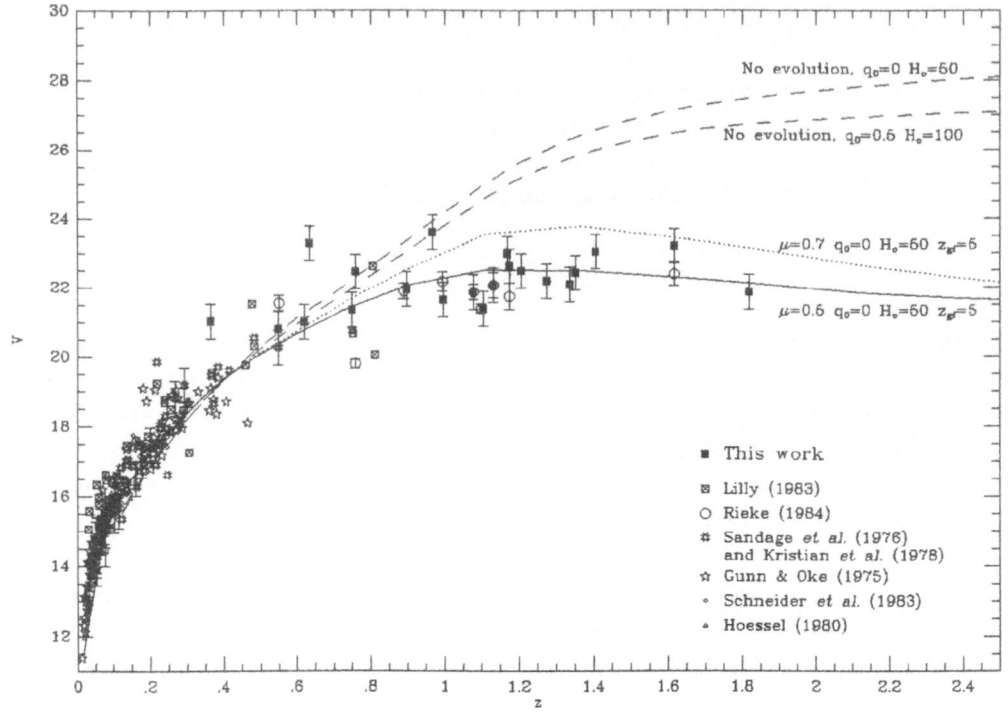

Figure 1.    The V band   Hubble diagram  for the brightest  cluster members  and the
powerful radio-galaxies.  Two Bruzual models are shown for comparison.  The "no evolution" lines were computed by  redshifting an old  elliptical galaxy spectrum,  and
assuming the particular cosmologies.

We compare our results with  the evolutionary models of Bruzual (1981; 1983; and
this volume).  Simple  exponential SFR models  with e-folding times  of the order of
1.5 Gyr (Bruzual $\mu=0.5$), Salpeter IMF,  and constant galaxy mass,  fit the data very
well.  This is somewhat surprising,  since some mass infall and merging is to be expected, and is  probably detected (see below).  Perhaps we are lucky,  and the vaga-
ries of individual galaxy evolution  statistically average into a good approximation
of an exponential SFR,  for the sample as a whole.  It is worth noting that the same
models fit the data in all the bandpasses used in this study  (B,V,R vs. z, and B-V,
V-R vs. z), and are consistent with the similar studies in the infrared (Lilly 1983;
Rieke 1984).  Moreover,  most of the observed spectra  show a fairly low ionization,
and allow only  a small contribution of  a non-thermal continuum.  Together with the
consistency of the fitted models, this is,  perhaps, evidence that we are indeed observing the light which is mostly of a stellar origin, and that the observed effects
are reflecting evolution of stellar populations at high redshifts.

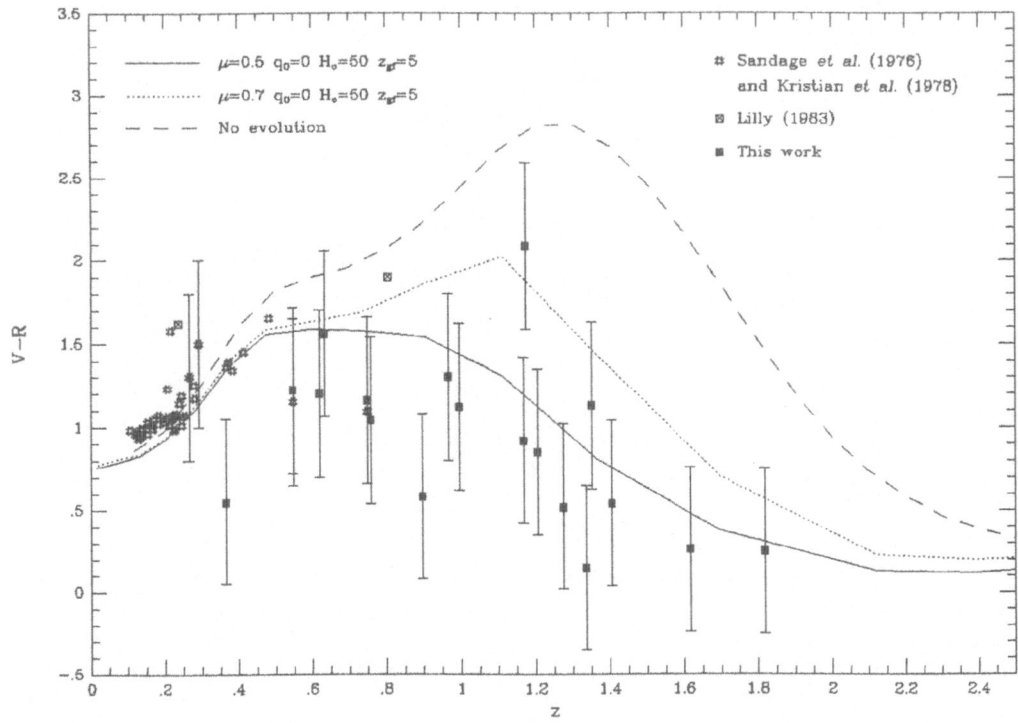

Figure 2.    The $(V-R)_{obs}$ color evolution for the radio-galaxies in our sample.  The
same two Bruzual models as in Figure 1 are shown for comparison.  The "no evolution"
line was computed by redshifting an old elliptical galaxy spectrum.

It is remarkable that on the Hubble diagram shown in Figure 1, the data are some
$5^m$-$6^m$ away  from the K-corrected  "no evolution" curves,  in the regime which we are
now exploring.  This is even more pronounced in the B band (but the data are scarce)
and somewhat less in the R.  The evolutionary effects  are now very dramatic,  a far
cry from  the lower redshift regimes  (z < 0.6-0.8),  to which most  of the previous
studies were confined,  and which yielded  no persuasive evidence for the evolution.
It is also notable that the evolutionary effects  (and their uncertain ties!) exceed
by far the differences introduced by the different cosmologies;  in other words, the
value of $q_0$  is not very relevant for such a diagram.  This means  (rather unambiguously) that the Hubble diagrams in the visible regime are practically worthless as a
tool for measuring the $q_0$.  This conclusion is certainly  less drastic in the infrared, but probably still valid.  Perhaps the estimates of evolutionary effects in the
visible  can be used to make the cosmological use of the infrared ($\lambda=3\mu$, and beyond)
Hubble diagrams  viable.  In any case,  as evidenced here,  the Hubble diagrams  may

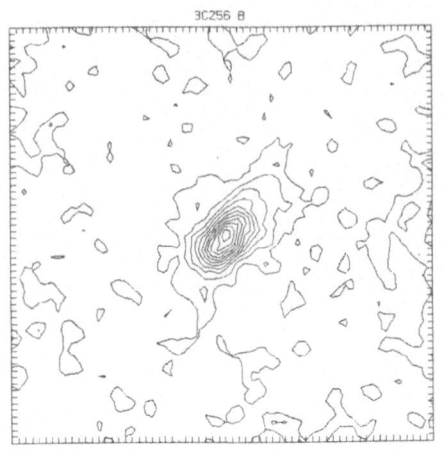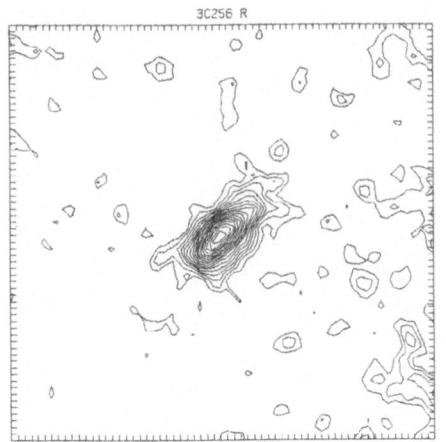

Figure 3.    The contour maps of the radio-galaxy 3C256 (z=1.819), in B (left) and R (right). The fields are approximately 19 arcsec square. Strong gaseous velocity fields are often associated with the major axes of such elongated distant radio-galaxies.

---

teach us something about the galaxy evolution. The only cosmological inference which we can make from fitting the evolutionary models to our data is that the longer cosmological time scale seems prefered, but this is a very tenuous conclusion at the moment.

Because of the possible apparent magnitude selection effects, it may be better to examine the color vs. redshift diagrams, as shown in Figure 2. Here we see that the data are slightly bluer, but still consistent with the proposed $\mu=0.5$ evolutionary model. Note also that the data are up to $2^m$ bluer than the K-corrected "no evolution" line. This is completely independent of the assumed cosmology, and presents unambiguous evidence for more vigorous star formation, and the presence of more early-type stars in the past.

One has to bear in mind that the giant radio-galaxies are not necessarly evolving in the same way as the more "normal", lower luminosity or field galaxies. We are currently pursuing a search for tentative fainter cluster members around these objects. Evidence for a cluster is known in at least one case, 3C324, with z=1.206 (Spinrad & Djorgovski 1984a), and possibly in some others. Multi-slit spectroscopy may yield some definite results in the near future.

Imaging of those distant objects yielded one unexpected result: their shapes are often very elongated and extended, quite unlike the nearby powerful radio-galaxies (Figure 3). The shapes of isophotes are same in all bandpasses used (B,V,R), indicating that their origin is from the starlight, rather than the extended line emission. However, such objects invariably have extended strong [O II] 3727 emission, covering some tens of kpc, and having velocity field amplitudes exceeding 1000 km/s in the galaxy rest-frame (Spinrad & Djorgovski 1984a; 1984b). This collisionally excited line emission may be due to violent cooling flows at that epoch (Fabian, Nulsen & Canizares 1984), but we think that the high-velocity collisions of gas-rich neighbors with the giant radio-galaxy provide a more likely cause. In some cases (e.g., 3C267, or 3C368) the oxygen line emission and the optical images are clearly multimodal. One speculative possibility is that we are witnessing the epoch of "turn-on" of the powerful radio-galaxies, which some models predict to occur near z~2 (Windhorst 1984).

A more detailed and more careful discussion of these effects will be presented in future papers. This work was supported in part by the NSF grant AST81-16125 to H.S., and the University of California Moore Fellowship and the AAS travel grant to S.D. This work is largely based on the data obtained at the Kitt Peak National Observatory.

REFERENCES:

Bruzual, G. 1981, Ph. D. thesis, University of California at Berkeley, USA.
Bruzual, G. 1983, Astrophys. J. <u>273</u>, 105.
Djorgovski, S. 1983, J. Astrophys. Astron. <u>4</u>, 271.
Fabian, A.C., Nulsen, P.E.J., and Canizares, C.R. 1984, Nature <u>310</u>, 733.
Gunn, J.E., and Oke, J.B. 1975, Astrophys. J. <u>195</u>, 255.
Hoessel, J.G., Gunn, J.E., and Thuan, T.X. 1980, Astrophys. J. <u>241</u>, 486.
Kristian, J., Sandage, A., and Westphal, J.A. 1978, Astrophys. J. <u>221</u>, 383.
Lilly, S.J. 1983, Ph. D. thesis, University of Edinburgh, UK.
Rieke, M. 1984, Private communication.
Sandage, A., Kristian, J., and Westphal, J.A. 1976, Astrophys. J. <u>205</u>, 688.
Schneider, D.P., Gunn, J.E., and Hoessel, J.G. 1983, Astrophys. J. <u>264</u>, 337.
Spinrad, H., and Djorgovski, S. 1982, Bull. Am. Astron. Soc. <u>14</u>, 959.
Spinrad, H., and Djorgovski, S. 1984a, Astrophys. J. Lett. <u>280</u>, L9.
Spinrad, H., and Djorgovski, S. 1984b, Astrophys. J. Lett. <u>285</u>, L000 (October 15).
Spinrad, H., Djorgovski, S., and Marr, J. 1983, Bull. Am. Astron. Soc. <u>15</u>, 932.
Windhorst, R.A. 1984, Ph. D. thesis, University of Leiden, Netherlands.

DISCUSSION:

B. Rocca-Volmerange : Most of your fits are obtained with $H_o$=50 km/s/Mpc. Do you think that you can obtain any good fit by adopting $H_o$=100 km/s/Mpc ?

S. Djorgovski : Yes, there is some freedom in adjusting the cosmological time-scale and the galaxy evolution time-scale, but it may be necessary to do some fine tunning of the parameters.

R. Terlevich : I would like to be sure that the colours you derive for high red-shift galaxies are measuring the colors of the stellar population, and not the co-lours of the non-thermal underl ying continuum. This aspect is crucial, because you are comparing with stellar evolution models. Moreover, there is a colour selection effect in that it will be easy to measure redshifts for those objects with stronger lines, and therefore bluer (due to a larger contribution of the non-thermal conti-nuum).

S. Djorgovski : We think that the contributions of possible non-thermal continua are probably not very important. Namely, the observed spectra are generally of a very "soft", low-ionization variety, and require relatively little ionizing conti-nuum. We should know more about this when the spectra are analysed in a greater detail. It is possible that some individual objects are afflicted. For your second comment, I do not think that there is any color preselection in measuring the red-shifts.

# FAR-UV TO INFRARED PHOTOMETRIC STAR FORMATION TRACERS

B. Rocca-Volmerange

Institut d'Astrophysique, 98 bis, bld Arago

F-75014 PARIS

## Abstract

A short review of quantitative photometric tracers of star formation and their limits is given. We separate distance-dependent indicators of current star formation rates (SFR) from those (colors and equivalent widths) which, distance independent, yield access to the present/past SFR ratio. Extinction and Initial Mass Function (IMF) are considered. A comparison of SFR values from various sources shows agreement at a local (per unit mass) but not global scale while no discrepancy appears for present/past SFR ratio.

## Introduction

First significant rates of star formation in external galaxies were approached from integrated optical broad-band colors by comparison with theoretical evolutionary models (Searle et al., 1973, Larson-Tinsley, 1978, Huchra, 1977). If their conclusions about stellar bursts were most interesting, more accurate information about energetic and spatial distribution of star formation can be presently obtained from far-UV, far-IR (down to radio) fluxes.

According to a now classical parametrization, SFR by mass unit and IMF normalized to unity are the two fundamental functions to determine. They can be approached with a relative accuracy for massive stars. Due to their short lifetimes, separation between the two functions SFR and IMF is still possible. Their direct effective UV emissivity, counts in HR diagrams or their indirect and no less effective far IR (or radio) fluxes are a priori good indicators of massive star formation. The problem is more delicate for less massive stars : estimates of the SFR implicate assumptions about the IMF (continuity, time variation, etc...) and conversely the determination of the IMF from luminosity functions strongly depends on the present/past SFR ratio (see review of Lequeux, 1980 and Scalo, 1984). In section 1, we propose to separate i) SFR estimators (absolute dereddened fluxes or star counts in HR diagrams) which are dependent on distance effect, absolute absorption correction, dark mass factor, emitting surface, IMF determination,... and ii) colors or equivalent widths, well correlated to present/past SFR ratio (Larson and Tinsley, 1978, Rocca-Volmerange et al., 1981, Kennicutt, 1983), which are independent on distance and dark mass fraction values, and for which only differential extinction intervenes.

Section 2 and 3 present some recent ideas about dereddening and IMF and last section summarizes present and past SFR estimates from various sources and consequences about gas consumption time-scale.

## 1.1 Distance Dependent SFR estimators
### 1.1.1 Far-UV fluxes

From satellites D2B (Maucherat-Joubert et al., 1980) or O.A.O. (Code and Welch, 1982), integrated fluxes in Magellanic Clouds and more generally irregular or spiral galaxies (Mochkovitch and Rocca-Volmerange, 1984 and references therein, Donas and Deharveng, 1984) as well as spectrophotometry with I.U.E. satellite from HII regions and blue compact galaxies (Israël, 1980 ; Lequeux et al., 1981, Viallefond & Thuan, 1983) or from various galactic inner bulges (Bertola et al., 1982, Ellis et al., 1982, Rocca-Volmerange & Balkowski, 1984) gave possible estimates of SFR for massive stars M>2 M⊙.

$$SFR_{M\odot} \ yr^{-1} = \left(4\pi \ d^2 \ f_{uv}^{obs} \cdot 10^{\frac{A_{uv}}{2.5}} - F_{uv}^{ev}\right) \cdot \frac{1}{\xi} \cdot \frac{\int_{Mi}^{Mu} \phi(m) dm}{\int_{Mi}^{Mu} \phi(m) f_{uv}(m) dm}$$

- $f_{uv}^{obs}$   observed flux in erg $s^{-1}$ $Å^{-1}$ $cm^{-2}$ at any far-UV wavelength
- d   is the distance in $cm^2$
- $A_{uv}$   is the extinction. More details about it will be given in the next section.
- $\phi(m)$   is the initial mass function, supposed to be time independent. See also next section.
- $\xi$   is the visible fraction of the stellar mass in formation.
- $f_{uv}(m)$ gives, at any wavelength UV, the flux emitted by a star of initial mass m. It is one of the most uncertain parameters since it depends on stellar evolution and atmosphere models. Moreover stellar mass loss, metallicity, helium abundance would have to be taken into account in a not well determined way.

. $F_{uv}^{ev}$   is the contribution of the evolved hot stellar population which could be not negligible in galactic nuclei according to the UV excesses observed in early type galaxies and may be reckoned from synthetic population model.

Nebular emission and starlight reflected by dust particles can be neglected (Israël & Koornneef, 1979) but, if generally adopted, uncertain is such a faintness of the scattered light. Contribution of a non-thermal emission could be possible in some cases.

In HII regions, required parameters are about identical but :
. Time scale is shorter because evolution of massive stars is <$10^{7}$ years.
. Contribution of evolved stars becomes negligeable.
. Reddening is spatially inhomogeneous and varies on a very large range ≃ 2-3 mag.
High spatial resolution should allow a best information on such local SFR process.This is waited for from Space Telescope Observatory.


### 1.1.2 Far infrared ($\lambda > 10$ µm) fluxes

The far infrared emission is essentially due to reemission by grains of the Lyman continuum photons absorbed in UV light. Consequently it is a good indicator of star formation rate for ionizing stars only if a value of the conversion factor f is given. According to Boissé et al, 1981, the factor f depends on absorption by 1) the intercloud medium, 2) the own cloud assumed to be optically thin, 3) the dark clouds. Puget, 1984 gives variations of the resulting factor with stellar mass : it increases from 0.15 for 1 M⊙ stars to 0.50 for 0 and B stars. Then the present SFR may be estimated from the observed $\ell_{IR}^{obs}$

$$SFR_{M\odot} \ yr^{-1} = \left(4\pi \ d^2 \ \ell_{IR}^{obs} - L_{IR}^{ev}\right) \frac{\int_{Mi}^{Mu} \phi(m) dm}{\int_{Mi}^{Mu} \phi(m) L_{TOT}(m) f(m) dm} \cdot \frac{1}{\xi}$$

. As for estimates from UV, dependence on distance d and stellar evolutionary models ($L_{TOT}(m)$) still exists but we note the absence of the extinction parameter which is now negligible.

$.L_{IR}^{ev}$     The contribution of evolved population (essentially old red
giants) can reach ≈ 30 % in the giant complexes. It can be
estimated from a stellar population model.

From such a synthetic model, conversion from total to infrared
luminosities can be reckoned and compared to the input f factor as did
Guiderdoni & Rocca-Volmerange, 1982 in their analysis of the Galactic
disk.

Some problems arise at 10 μm due to a possible strong absorption
feature of silicates (Genzel et al., 1982). Moreover in some cases the
10 μm emission is not related to star formation. A non-thermal process
is needed to explain emission line ratios (Laurence et al., 1984).

Radio fluxes are frequently used as estimators of massive star
(M≥10 M⊙) in HII regions (Smith et al., 1978 ; Lequeux et al., 1981).
We should go beyond the fixed limits of this paper to develop this
subject about which a good review is given by Lequeux, 1980. Let us
only note that the main difficulty is due to the uncertainties about
the respective thermal and non-thermal radio flux contributions.
According to Israël and Van der Hulst 1983, Gioia et al., 1982, the
total radio flux density at 6.3 cm (≈5 GHz) is essentially non
thermal. The same is ascertained by Kennicutt, 1983, even at 10 GHz.
Excellent correlations appear between radio continuum density and the
two infrared (60 μm and 100 μm) total emissivity from irregular or
spiral galaxies (Klein et al., 1984). Authors conclude in favor of
the fact that the main sources of relativistic electrons would be due
to the star formation process and by means of supernova explosions.
This idea is not in agreement with the explanation of   radio
non-thermal emission from Sancisi and Van der Kruit (1981), neither
with an only thermal radio emission, frequently observed in HII
regions (Viallefond et al., 1983). Better information about that will
be given with high spatial resolution map, correlated with absolute
photometry, as realized in this last paper.

## 1.2 Distance independent SFR estimators

Due to uncertainties tied to distance parameter, it appears more
fruitful to relate tracers of present SFR to other indicators also
depending on $d^2$ such as other fluxes. We then obtain colors or
equivalent widths, witnesses of present/past SFR ratio.

An estimate of such a ratio is most useful :
- For galactic evolution : associated to a present SFR value, it

gives the past SFR. Mochkovitch & Rocca-Volmerange, 1984, showed that integrated SFR determination is faintly depending on reddening.

- For IMF of low mass stars, the determination of which generally implies a constant rate for star formation history (Miller & Scalo, 1979). Fig. 17 of Scalo, 1984, evidences the effect of such a ratio on the slope determination.

## 1.2.1. Colors

It is now well known that a color is a good indicator of the present/past SFR ratio : this was verified from synthetic models in visible (Larson and Tinsley, 1978) and confirmed in UV light (Rocca-Volmerange et al., 1981)

The relation $m_\lambda - m_{\lambda'}$ $\sim \dfrac{\text{present } M_*}{\text{ever formed } M_*}$ $\sim \dfrac{\text{present SFR}}{\text{T.past SFR}}$, if we assume

that $M_*$ is the stellar mass, is true if the IMF does not vary with time.

If distance effects and IMF dependance are ruled out, two important problems are still present :
- The dereddening which essentially affects the UV light
- The stellar evolutionary tracks and other input stellar data
  Another difficulty is aperture corrections if fluxes are obtained
  by different telescopes.

Are different colors equivalent to estimate the present/past SFR ratio ? A priori most extreme UV and near infrared (1 $\mu m < \lambda < 5$ $\mu m$) fluxes would have to be best tracers of respectively young and old populations. In most cases, that is true. However sometimes both of them have noticeable disavantages : dereddening in UV light can be very difficult to be estimated due to a patchy structure. Otherwise near infrared light (K and H wavelength bands) can be partly due to red supergiants of intense star formation sites. In such cases witnesses of old population become optical fluxes and V-K or B-H colors are well correlated to the giants/red supergiants number ratio. This effect explains the strong increase of near-infrared flux (2.4 $\mu m$) in our Galactic disk emitted from the intense star formation zones (Guiderdoni & Rocca-Volmerange, 1982). At last, it must be notified that in optical bands (U,B,V and R) the respective contribution of young and old population differs so much from one band to another that the most extreme colors U-R or B-R appear to be good estimators.

## 1.2.2. Equivalent widths

Equivalent widths of ionized gas emission lines (Kennicutt, 1983) are also good indicators of the ionizing stars (M >10 M$\odot$);

$$EW_{H\alpha} = \frac{\Phi (H\alpha)}{\Phi \text{ (red continuum)}} \simeq \frac{\text{present SFR}}{\text{past SFR}}$$

A correction to take into account the ionizing star/all star number ratio has to be done. As colors, they are strongly dependent on dereddening and stellar evolutionary tracks. A possible contribution of underlying stellar H$\alpha$ absorption or [N II] emission can exist. Moreover photoionization models are needed. It is interesting to note that Kennicutt, who tried such estimates, found such large numbers of ionizing stars that he finally calculated photoionization statistically by computing the ultraviolet luminosities.

It is worthy to note here the relation of the H$\alpha$ equivalent width of a galaxy with its HI neutral gas surface density (Guiderdoni and Rocca-Volmerange, 1984). Galaxy samples from field and Virgo Cluster correspond to a large dynamic range for the two variables. Such a relation implies the high sensitivity of the present SFR to the neutral gas abundance, whatever the molecular hydrogen component. More details about that will be given in a next paper.


## 2. Initial Mass Function parameter

According to the classical parametrisation $d^2 N(m,t) = \phi(m) \; \tau_*(t) \; dm \; dt$, the IMF $\phi(m)$ is assumed stationary in time. Moreover it is also supposed to be homogeneous in space. In fact, if we except very massive stars M>60-100 M$\odot$, IMF appears quite similar in Solar Neighborhood and the Magellanic Clouds (Vangioni-Flam et al., 1981). A recent paper about correlation between radio and 60 $\mu$m or 100 $\mu$m data from spiral galaxies observed by the I.R.A.S. satellite (Klein et al., 1984) confirms the idea of an universal I.M.F. in galactic disks.

The mass limits are presently fixed to $M_{up} \approx 100-200$ M$\odot$ and $M_{inf} \approx 0.01$ M$\odot$ (Scalo, 1984). The contribution of very massive stars to ionizing fluxes is sufficiently high to make evaluations of such fluxes very sensitive to the upper limit of IMF. About the slope for massive stars, more recent results are from Garmany et al., 1982 with a Salpeter index varying from -1.36 to -1.65. For less massive stars, Scalo 1984 presents current results with a dispersion due to various present/past SFR ratio values.

But, to summarize, all these results concern :
. only massive or intermediate stars M⩾2 M☉
. a disk stellar population
. a normal (or lightly deficient) metallicity
. a present estimate of IMF
and very little is known otherwise.


## 3. Extinction correction

This correction is crucial in assessing the intrinsic photometry properties. It requires the estimates of two parameters : the internal color excess $E_i(B-V)$ and the extinction at any wavelength relative to visible : $A_\lambda - A_V$. We only consider here questions arising from internal extinction.


### 3.1. Internal color excess

It strongly depends on the uniformity of the dust layers. Several approaches were proposed : the "cosec|b|" models (de Vaucouleurs et al., 1983) available if the patchiness of the interstellar dust is not too serious and the method depending on the gas density N(HI + HII) which assumes that the gas/dust ratio is constant in the galaxy. We recently showed that this ratio varies with metallicity by comparing our Galaxy to the metal-deficient Small Magellanic Cloud (Lequeux et al., 1984). Moreover, according to the important abundance gradient observed in our Galaxy (Shaver et al., 1983) the $N_{gas}/E_{B-V}$ ratio would also vary with the galactic radius.


### 3.2 Relative extinction

Extinction in UV or IR light relative to visible are deduced from an extinction law $(A_\lambda/E_{B-V})$, preliminary established from the pair method. However, for some cases in far-UV light, such a dereddening estimate appears to be largely insufficient : several of the stars observed in 30 Doradus by the I.U.E. satellite are systematically much redder, if compared to the rest of the nebula (Fitzpatrick & Savage, 1983). In our analysis of spirals from O.A.O. Satellite (Code & Welch, 1982), the observed galaxies are much redder that waited for from synthetic models though agreement in optical light is excellent. (Mochkovitch & Rocca-Volmerange , 1984). Variations of the extinction law with metallicity (Prevot et al., 1984 and reference therein) or anomalies of the main features detected in several points of our

| (1)<br>Galaxy | (2)<br>$^{\tau}M\odot$ kpc$^{-2}$ yr$^{-1}$ $(10^{-3})$ | (3)<br>$^{\tau}M\odot$ Gyr$^{-1}$ per mass unit $(10^{-2})$ | (4)<br>$^{\tau}M\odot$ yr$^{-1}$ per galaxy ($H_o$=100 km s$^{-1}$ Mpc$^{-1}$) | (5)<br>Source |
|---|---|---|---|---|
| SN | 3 | | | von Hoerner 1975 |
| SN<br>5 kpc ring | 5<br>12 | 3.5<br>2.6 | 5 | Smith et al.1978<br>Mezger 1979<br>radio |
| SN | 3/7 | | | Miller & Scalo 1979 |
| Galaxy<br>SMC<br>LMC<br>Mean value on 11 late type galaxies | <br>885<br>1135<br><br>23±20 | 2.7<br>4<br>4<br><br>2.4 ± 1.5 | 3.5<br>0.08<br>0.40 | Israël 1980 radio |
| SN | 3 | 8 | | Tinsley 1980 (B,V,R) |
| SN<br><br>5 kpc ring<br><br>SMC<br>LMC | 3.2<br><br>10<br><br> | 2.3<br><br>2.5<br><br>3<br>6 | <br><br><br><br>0.04<br>0.40 | Rocca-Volmerange et al., 1981 (UV→R band) and Guiderdoni & Rocca-Volmerange 1982(IR, Radio) |
| M33 | 5.2 | 7 | | Berkhuijsen,1982 |
| 170 spiral galaxies | | | 5 | Kennicutt,1983 (Hα) |
| 40 irregular and spiral galaxies | | | 0.3/0.8 | Donas & Deharveng 1984 (UV) |

Table 1. Comparative estimates of SFR ($\tau$) in our Galaxy (SN = Solar Neighborhood, 5 kpc ring = 5 kpc galactocentric distance zone), Magellanic Clouds (SMC and LMC), and other late-type galaxies.

Galaxy (Massa et al., 1982) cannot explain such a large reddening discrepancy.

Then a common interpretation is proposed : massive hot stars assumed to be born in dusty interstellar medium and possibly still embedded in molecular clouds are more reddened than the other stars. This effect is stronger in UV light. As a numerical exemple, we can deredden spiral galaxies by assigning a complementary color excess $\Delta E_{B-V}$ varying with the stellar mass. A consequence of such an analysis is that the mean color excess of a stellar population will depend on the massive star sub-population and that could strongly affect irregular or blue compact galaxies (Mochkovitch & Rocca-Volmerange, 1984)

## 4 Comparative results

An analysis of the SFR estimates from various sources (UV, far-IR, H$\alpha$, radio, etc..) will allow to verify if any consistency exists. We summarize in the following table : SFR per unit surface area (column 2), SFR per mass unit (col. 3) and SFR integrated on the whole galaxy (col. 4). The most striking features are :

- the Star Formation Rate per unit mass is about constant in a galactic disk.
- the mean value calculated from Our Galactic Disk (solar neighbourhood and 5 kpc galactocentric ring) and 11 late type galaxies is : $\tau* \simeq 3 \ 10^{-2}$ M$\odot$ Gyr$^{-1}$ per mass unit

This result was already obtained from our previous analysis (Guiderdoni & Rocca-Volmerange, 1982) of the Galactic 5 kpc ring compared to the solar neighbourhood for far-infrared emissivity. A better statistics is presently obtained.

What is the meaning of such an apparent uniformity ? SFR per unit mass can be related to SFR per gas mass unit by using the gas mass density. An estimate of the gas mass density in our Galactic disk may be deduced from the structure model of Smith et al., 1978 : $\sigma_{gas}$ = $M_{gas}/M_{TOT} \simeq 5 \ 10^{-2}$ from 4 to 11 kpc galactocentric distance. Vangioni-Flam et al., 1980 give $\sigma_{gas}$ = 0.13 and 0.46 respectively for SMC and LMC and Donas & Deharveng (1984) give $\langle M_{gas}/M_{indicative}\rangle$ = 0.05 calculated on 35 spiral galaxies.

A preliminary deduction of such results would be that the uniformity of the SFR per unit mass might be associated to the uniformity of the $M_{gas}/M_{TOT}$ ratio in late-type and spiral galaxies to conclude that SFR is well correlated to gas mass unit. This conclusion would need more work to be interpretated in terms of star formation process.

About present/past SFR ratio, estimates from UV light (Rocca-Volmerange et al., 1981) in the Magellanic Clouds, from EW(Hα) (Kennicutt, 1983) in spiral galaxy sample are about concordant. This ratio is about equal to unity : this means that the star formation rate was constant in time. According to our recent paper about spiral galaxies observed from the O.A.O. Satellite in far UV light (Code & Welch, 1982), this ratio could decrease with more evolved spirals in which the bulge component becomes more important. So we conclude that a constant ratio is roughly available for a disk population.

Last but not least problem is the time scale gas consumption. With such estimates of SFR, this time is about 4 to 5 Gyr! Different solutions were proposed by Larson et al.,1978, Kennicutt, 1983 but none appears as evident.

Next study will be to correlate statistical results on SFR by unit mass, gas density and the mystery of the gas consuming time scale as a function of the morphologic type of galaxies.

References

. Berkhuijsen, E.M., 1982, Astron. Astrophys., 112, 369
. Bertola, G., Capaccioli, M., Oke, J.B., 1982, Ap.J., 254,454
. Boissé, P., Gispert, R., Coron, N., Wijnbergen, J., Serra, G., Ryter, C., Puget, J.L., 1981, Astron. Astrophys., 94, 265
. Burstein, D., Heiles, C., 1984, Ap.J. Suppl. Series, 54, 33
. Code, A.D., Welch,G.A., 1982, Ap.J., 256, 1
. de Vaucouleurs, G., de Vaucouleurs, A., Buta, R., 1983,A.J., 88,764
. Donas, J., Deharveng,J.M., preprint
. Ellis, R.S., Gondhalekar, P.MK, Efstathiou, G., 1982, M.N.R.A.S., 201, 223
. Fitzpatrick, E.L., Savage, B.D., 1983, Ap.J., 267, 93
. Garmany, C.D., Conti, P.S., 1982, Ap.J., 263, 777
. Genzel, R., Becklin, E.E., Wynn-Williams, G.G., Moran, J.M., Reid, M.J., Jaffe, D.T., Downes, D., 1982, Ap.J., 255, 527
. Gioia, J.M., Gregorini, L., Klein, U., 1982, A.& A., 116,164
. Guiderdoni, B., Rocca-Volmerange, B.,1982, Astron.Astrophys, 109,355
. Guiderdoni, B., Rocca-Volmerange, B., 1984, Proceedings of the ESO Worshop on the "Virgo Cluster of Galaxies", ESO Garching

. Hoerner, S. von, 1975, in HII regions and related topics. Ed
T.L.Wilson and D. Downes, Springer Verlag, Heidelberg p.53
. Huchra, J., 1977, Ap.J., 217, 928
. Israël, F.P., 1980, Astron. Astrophys., 90, 246
. Israël, F.P., Koornneef, J., 1979, Ap.J., 230, 390
. Israël, F.P., Van der Hulst, M.A., 1983, A.J., 88, 1736
. Kennicutt, R.C. Jr., 1983, Ap.J., 272, 54
. Klein, U., de Jong, T., Wielebinski, R., Wunderlich, E., preprint.
. Larson, R., Tinsley, B.M., Caldwell, C., Nelson, 1980,Ap.J., 237,692
. Larson, R., Tinsley, B.M., 1978, Ap.J., 219, 46
. Laurence, A., Ward, M., Elvis, M., Fabbiano, G., Willner, S.,
Carleton, S., Carleton, N., Longmore, A., preprint
. Lequeux, J., Maurice E., Prévot, L., Prévot-Burnichon, M.L.,
Rocca-Volmerange, B., "Structure and Evolution of the Magellanic
Clouds", p.405, Ed. Van den Bergh and de Boer, Tübingen, 1984
. Lequeux, J., Maucherat-Joubert, M., Deharveng, J.M., Kunth, D.,
1981, Astron. Astrophys., 103, 305
. Lequeux, J., 1980, 10th Advanced Course of Swiss Society of
Astronomy and Astrophysics, Saas-Fee,
. Massa, D., Savage, B.D., Fitzpatrick, E.L., 1983, Ap.J., 266, 662
. Maucherat-Joubert, M., Lequeux, J., Rocca-Volmerange, B., 1980,
Astron. Astrophys., 86, 299
. Miller, G.E., Scalo, J.M., 1979, Ap.J. Suppl. Series, 41, 513
. Mochkovitch, R., Rocca-Volmerange, B., 1984, A. & A., 137, 298
. Prevot, M.L., Lequeux, J., Maurice E., Prévot, L., Rocca-Volmerange,
B, 1984, Astron. Astrophys., 132, 389
. Puget, J.L., 1984, preprint, Les Houches
. Rocca-Volmerange, B., Lequeux, J., Maucherat-Joubert, M., 1981,
Astron. Astrophys., 104, 177
. Rocca-Volmerange, B., Balkowski, Ch., 1984, Proc. 4th European IUE
conference, Roma, p.69
. Sancisi R., Van der Kruit, P.C., 1981, in "Origin of Cosmic Rays",
IAU Symp. n°94, Ed. G. Setti, G. Spada, A.W. Wolfendale, D. Reidel
Dordrecht p.209
. Scalo, J.M., 1984, preprint to be published in Fundamentals of
Cosmic Physics.
. Searle, L., Sargent, W.L.W., Bagnolo, W.G., 1973, Ap.J., 179, 42
. Shaver, P.A., Mc Gee, R.X., Newton, L.M., Danks, A.C., Pottasch,
S.R., 1983, Mon.Not.Roy.Ast.Soc., 204, 53
. Smith, L.F., Biermann, P., Mezger, P.G., 1978, A. & A., 66, 65
. Vangioni-Flam, E., Lequeux, J., Maucherat-Joubert, M.,
Rocca-Volmerange, B., 1980, Astron. Astrophys., 90, 73

. Viallefond, F., Thuan, T.X., 1983, Ap.J., <u>269</u>, 444
. Viallefond, F., Donas, J., Goss, W.M., 1983, A. & A. , <u>119</u>, 185.

## Discussion

<u>F. Israël</u>    This is just a comment. It is very unfortunate that we really do not have a reliable sample of UV fluxes of external galaxies. IUE has too small an aperture to be very useful, ANS produced an extremely heterogeneous set of observations that is difficult to use statistically and worst of all perhaps, OAO2 had filters with sometimes appreciable red leaks, so that one has to apply colour dependent corrections which is extremely unfortunate for applications as the ones just discussed.

<u>B. Rocca-Volmerange</u>  I quite agree with you about the poorness of the integrated UV photometric data. We actually need large field observations such as from balloons (SCAP 2000, Marseille) or from a future UV Schmidt Telescope. But presently for lack of better data, we can take into account possible red leaks of UV filters by simultaneously analyzing continuous UV to red energetic spectra through the filter transmission profiles.

<u>T. Jaakkola</u>    I want to emphasize once more that one observe at large Z as the brightest objects automatically, almost by definition, blue galaxies and not red as close to us and this is due to the redshift itself (K-term) and not due to the evolution. I don't remember but I expect that the K-correction at $Z \approx 1.5$ is for blue spirals of the order of 2 mag smaller than that for red ellipticals two mag is a figure pointed out by Djorgovski who labelled it as evolution.

<u>B. Rocca-Volmerange</u> The problem of isolating evolutionary effects from redshift effects is still open and it is somewhat premature to be affirmative in this domain.

# EVOLUTION OF SPIRAL GALAXIES IN A CLUSTER ENVIRONMENT

B.Guiderdoni

Institut d'Astrophysique, 98 bis, bld Arago

F-75014 PARIS

Abstract

The HI deficiency and the integrated colors of spiral galaxies in the Virgo Cluster are studied by means of a model of photometric evolution.

## I. Introduction

The spiral galaxies in the Virgo Cluster (hereafter VC), as well as in other nearby clusters (Coma, A 1367, A 262, A 2147), seem to exhibit a very poor HI content with respect to "field" counterparts of the same morphological type and luminosity or optical surface (Davies and Lewis, 1973, Sullivan and Johnson, 1978, Giovanelli et al., 1981, 1982). Despite the fact that this HI deficiency is still controverted (Bothun et al., 1982, Tully and Shaya, 1984), various dynamical mechanisms have been invoked : collision between disks (Spitzer and Baade, 1951), ram-pressure stripping (Gunn and Gott, 1972) and evaporation (Cowie and Songaila, 1977) in the hot intergalactic medium (IGM), or tidal disruption of an extended gaseous halo that might be the reservoir for star formation in the disk (Larson et al., 1980).

Such a deficiency must obviously have some influence on the integrated colors of the galaxy. As it is expected, the average $(B-V)_T^0$ color is slightly redder for spirals in the VC than for "field" counterparts (Holmberg, 1958, Kennicutt, 1983). Nevertheless, problems of internal extinction and uncertainties about the galactic absorption towards the North Galactic Pole (the center of the VC is at $b = 75°$) have forbidden any precise interpretation. The publication of a catalog of galactic absorptions by Burstein and Heiles, 1984, partly improved the situation.

## II. An analysis of the properties of VC spirals

We have analyzed the correlation between gas content and color properties of 107 spiral galaxies in the VC, from a compilation of HI data and Hα, UBV and λ2421 Å photometry (Guiderdoni and Rocca-Volme-

-range, 1984,paper I). Most HI deficient objects are located in Virgo
I, the VC core (average deficiency of a factor 2.3), in association
with the IGM around M87-M86 and M49 shown by the X-ray map from Jones
and Forman, 1984. Early-type spirals are more frequently HI deficient
than late-type ones. Their Hα equivalent widths, which characterize
the formation rate of massive stars, are consistent with their HI
content. The most deficient early-type spirals exhibit the anemic
appearance described by Van den Bergh, 1976. The VC spirals have
redder average colors $<\Delta(B-V)_T^\circ>\approx 0.05$ and $<\Delta(U-B)_T^\circ>\approx 0.07$ which
roughly correlate with the amount of deficiency. Far-UV color
$(m_{2421}-B)^\circ$ very well correlates with the deficiency, but we lack a
reference sample for any quantitative comparison.

Such photometric properties of member spirals do not seem to be
very common in most other clusters (Bothun et al., 1982, 1984,
Kennicutt et al., 1984). Nevertheless, severe selection effects,
especially including the difficulty of a precise morphological
classification of early-type disk galaxies which leads these authors
to an a priori rejection of SO-like systems, make us believe that the
problem certainly needs further examination.


III. Photometric evolution of VC spirals

We have tested the scenario of a total gas removal followed by gas
replenishment from stellar rejections. The influence on the color
evolution of a typical Sb galaxy has been calculated by means of a
model of photometric evolution (Guiderdoni and Rocca-Volmerange, in
preparation, paper II). Fig. 1 and 2 respectively display some
preliminary results in the log $\sigma_H$ versus $(B-V)_T^\circ$ and $(m_{2421}-B)^\circ$
diagrams for VC spirals. $\sigma_H$ is the HI mass divided by the optical
surface of the disk. Very HI deficient and mildly deficient objects
are denoted by dots and small open circles, normal objects by large
open circles and HI overabundant objects by large open circles with a
cross. Underlined symbols denote upper values of log $\sigma_H$. The star
shows the location of a 12 Gyr old model Sb galaxy in the
log $\sigma_{gas}$-extinction-free B-V and $m_{2421}$-B diagrams. Model and observations can be compared after shifts of the model points, accounting
for face-on internal extinction (reasonable estimates from paper II
are $E_i(B-V)\approx 0.06$ and $E_i(m_{2421}-B)\approx 0.40$) and He, $H_2$ components
(Y=0.30 and $M_{H_2}/M_H\approx 1$), which are respectively denoted by
horizontal and vertical arrows. Smaller shifts are probably more
suitable to gas poor galaxies. The colors after complete gas removal

at age >$t_1$ are denoted by short arrows. The dashed curve is the locus
of the 12 Gyr old model Sb galaxy after complete stripping during 0.5
Gyr (crossing time of the inner degrees of Virgo I) at age $t_1$(written
beside the ticks). The cross shows the effect of two complete
strippings during 1 Gyr, at ages $t_1$=7 Gyr and $t_1$=9 Gyr.

A complete gas removal (or several short strippings) from age $t_1$<5
Gyr would lead to too red colors. Moreover, the surface brightness
would be much too low, as in the case of a single early stripping (At
$t_1 \approx 2$ Gyr, half the mass of the disk disappears). It is clearly ruled

Fig.1

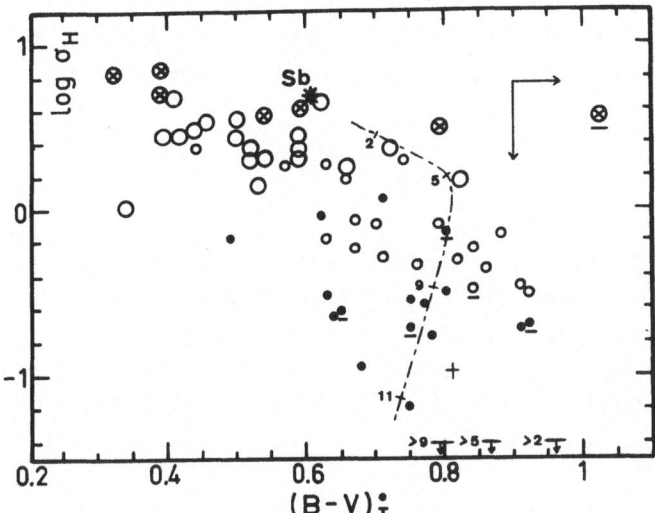

Fig.2

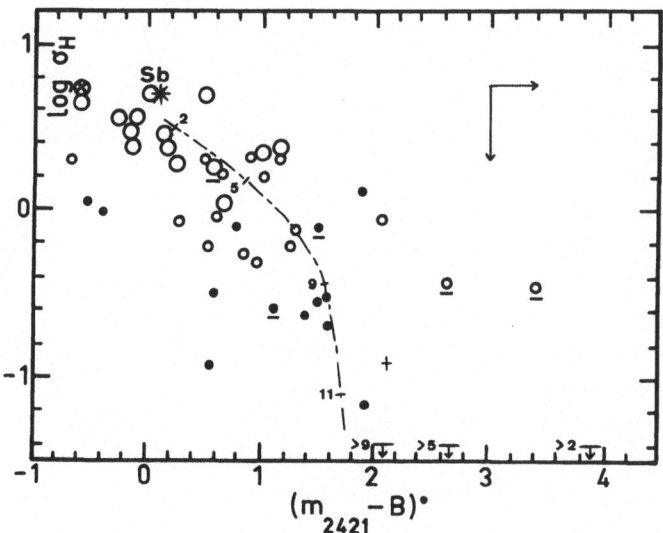

out for our sample of bright VC spirals. On the contrary, the properties of the most deficient galaxies are consistent with one or two recent strippings ($t_1 > 7$ Gyr). Dynamical models (Rivolo and Yahil, 1983, Tully and Shaya, 1984) suggest that most VC spirals have recently entered the VC core, after infall from the outer regions.

Finally, it is seen that, in such a dynamical scenario, recent strippings (or a single stripping with the short time scale associated with the crossing of the core) cannot have formed a present SO from a Sb. Very early total gas removal would lead to a disk galaxy with present bulge/disk ratio, gas content and colors of a SO ($<(B-V)_T^o> = 0.87$ and $<m_{2421} - B)^o> = 2.95$ in the VC), but with low surface brightness of the disk and low total luminosity in the B-band. At least bright SO's can difficultly originate from gas stripping of a typical spiral (see also Sandage, 1983).

## References

Bothun, G.D., Schommer, R.A., Sullivan, W.T., 1982, An.J., 87, 731

Bothun, G.D., Schommer, R.A., Sullivan, W.T., 1984, An.J., 89, 466

Burstein, D., Heiles, C., 1984, Ap.J.Suppl., 54, 33

Cowie, L.L., Songaila, A., 1977, Nature, 266, 501

Davies, R.D., Lewis, B.M., 1973, M.N.R.A.S., 165, 231

Giovanelli, R., Chincarini, G.L., Haynes, M.P., 1981, Ap.J., 247, 383

Giovanelli, R., Haynes, M.P., Chincarini, G.L., 1982, Ap.J., 262, 442

Guiderdoni, B., Rocca-Volmerange, B., 1984, submitted

Guiderdoni, B., Rocca-Volmerange, B., in preparation

Gunn, J., Gott, J., 1972, Ap.J., 176, 1

Holmberg, E., 1958, Medd. Lund. Ser. II, No136

Jones, C., Forman, W., 1984, in Clusters and Groups of Galaxies, F. Mardirossian et al., D. Reidel, p. 319

Kennicutt, R.C., 1983, An.J., 88, 483

Kennicutt, R.C., Bothun, G.D., Schommer, R.A., 1984, An.J., 89, 1279

Larson, R.B., Tinsley, B.M., Caldwell, C.N., 1980, Ap.J., 237, 692

Rivolo, A.R., Yahil, A., 1983, Ap.J., 274, 474

Sandage, A., 1983, in Internal Kinematics and Dynamics of Galaxies, IAU symp. No100, p 367.

Spitzer, L., Baade, W., 1951, Ap.J., 113, 413

Sullivan, W.T., Johnson, P.E., 1978, Ap.J., 225, 751

Tully, R.B., Shaya, E.J., 1984, Ap.J., 281, 31

# GALAXY PHOTOMETRY AT FAINT LIGHT LEVELS - INTERACTION WITH THE ENVIRONMENT

D. Carter,
Mount Stromlo and Siding Spring Observatories,
Private Bag, Woden, A.C.T. 2606, AUSTRALIA.

## I - Introduction

Galaxies interact with their environment in many ways, and many of these interactions produce features and effects at low surface brightnesses which can be diag-
nostics for those interactions. Techniques for detecting and measuring the brightness
of faint features are discussed elsewhere in this volume, and recent advances in
photographic, electronographic and CCD technology have made such measurements much
more precise.

When galaxies interact with each other they produce features such as tidal
tails, loops, shells, ripples and dust lanes, and photometric investigation of all
of these features is important. More subtle effects such as tidal distension of
envelopes also occur. Galaxies also interact with the intergalactic or interstellar
medium, X-ray evidence suggests that particularly in galaxies in the centres of rich
clusters, but also in a number of isolated elliptical galaxies, hot gas is cooling
and being accreted by the galaxies, the form that this material eventually takes is
still not determined. Photometric and spectroscopic investigations of galaxies with
cooling accretion flows are very important.

## II - Interactions of two disc galaxies

Numerical models of the interaction of pairs of disc galaxies (Toomre and
Toomre 1972; Toomre 1974; White 1979; Quinn 1982) show that a bound interacting
pair will merge quite rapidly, and give rise to a merger remnant which looks something like an elliptical galaxy. Schweizer (1982) looked for photometric evidence of
this process in the disturbed system NGC 7252. He found that while the gas kinematics
show two velocity systems, indicating a recent merger, the azimuthally averaged
radial light profile is of the $EXP\{r^{\frac{1}{4}}\}$ form (deVaucouleurs 1959) characteristic of
many elliptical galaxies. This suggests that violent relaxation in the stellar
component of a merging system operates on a shorter timescale than dissipation in
the gaseous component.

## III - Interactions between disc and elliptical galaxies

A very comprehensive study of the giant elliptical galaxy NGC 1316 was made by

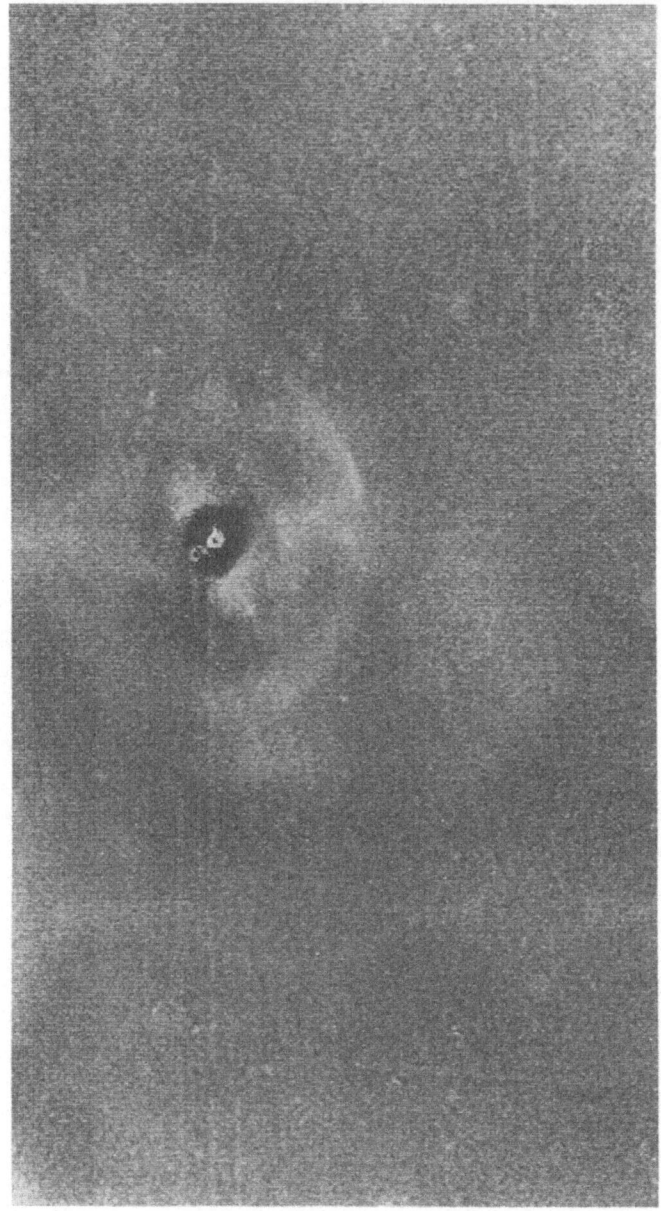

Figure 1 - The shell system of the elliptical galaxy NGC 2865 after subtraction of an EXP$\{r^{\frac{1}{4}}\}$ law model for the surface brightess distribution of the galaxy. North is at the top and East to the right, the image is approximately 4.4 x 7.0 arcmin in size. The passband is approximately V.

Schweizer(1980). Schweizer assembled evidence that NGC 1316 has accreted one or more gas rich companion galaxies in the past $2 \times 10^9$ years. He identified an inclined disc of gas in the inner regions, and gas from an infalling companion is probably responsible for fuelling the radio outburst (NGC 1316 is identified with the radio source Fornax A). The photometric profile of the envelope resembles that of a D type galaxy, but superimposed on this there are a number of "ripples", which Schweizer interprets as the remnant of the stellar component of the companion.

Similar features in the envelopes of a number of normal elliptical galaxies were noted by Malin and Carter (1980, 1983) who interpreted them as shells seen in projection. These features have been explained variously as remnants of encounters between low mass disc galaxies and more massive ellipticals (Quinn 1984; Dupraz and Combes 1984) or as stars formed in shock or blast waves in an outflowing interstellar medium (Fabian et. al. 1980; Williams and Christiansen 1984). As the features are very faint quantitative measurements are difficult, efforts have been confined to attempts to measure broadband colours in order to determine what the shells consist of. Carter, Allen and Malin (1982) measured optical and infrared colours of a shell near the normal elliptical galaxy NGC 1344, and found it to be somewhat bluer than the main body of the elliptical. They interpreted this result as supporting the collision models for the origin of the shells, but as the initial mass function of the stars formed in the blast wave models is unknown this conclusion is tentative. Recent CCD optical photometry in four passbands by Fort et. al. (1985), who measure the colours of the shells around three galaxies after subtraction of an $EXP\{r^{\frac{1}{4}}\}$ law model from their images, shows that the colours of the shells are bluer in those shell systems whose morphology suggests a younger dynamical age. Again this evidence that the dynamical ages correlate well with the ages of the stars in the shell systems supports the dynamical models for their origin.

## IV - Tidal distension of the envelopes of galaxies

Kormendy (1977), fitting deVaucouleurs (1959) $EXP\{r^{\frac{1}{4}}\}$ empirical fitting function to the surface brightness distribution of a number of elliptical galaxies observed by King (1978) noted that those galaxies with close massive neighbours tend to have envelopes which are more distended than this empirical law would predict. Moreover he found the same result was obtained if he used instead the empirical fitting function of Hubble (1930) or the models of King (1966). He interpreted this result as evidence for tidal distension of the envelopes of the galaxies by the massive neighbours.

## V - cD galaxies

Clusters of galaxies often contain one or occasionally more very luminous galaxies at their dynamical centres. These galaxies are termed cD galaxies (Matthews, Morgan and Schmidt 1964). The origin of cD galaxies is a controversial subject, it has been suggested that a large part of the envelopes of such galaxies form from debris from tidal encounters in the outer parts of the cluster (Richstone 1976) or from cannibalism of entire galaxies by the cD (Hausman and Ostriker 1978; McGlynn and Ostriker 1980). Photometric studies of cD galaxies (Oemler 1976; Carter 1977; Carter and Dixon 1978; Thuan and Romanishin 1981) have shown that their structure is quite different from that of normal ellipticals; their photometric profiles do not fit at all well to deVaucouleurs' law, but rather fall off as a power law with an index of around 1.6. Thuan and Romanishin (1981) show that the structure of brightest cluster galaxies is environment dependent; cD galaxies in poor clusters do not have such extended power law envelopes as those in rich clusters, indeed they resemble normal ellipticals much more closely.

Measurements of colour gradients in the envelopes of cD galaxies will place important constraints upon theories of their origin. If the envelopes are composed of debris from tidal encounters between smaller galaxies, or of whole cannibalised galaxies then one might expect them to have colours characteristic of these galaxies, which tend to be bluer than bright ellipticals. Boronson et. al. (1983), using a scanning CCD, have measured a strong colour gradient in the halo of M87, the galaxy is bluer by 0.12 magnitudes in (B-V) at 80 arcsec radius than near the nucleus. The results of Valentijn and Moorwood (1984) are more surprising, they performed optical (electronographic) and infrared photometry of the cD galaxy in Abell 496. They found that while the (B-V) colour gets steadily bluer with increasing radius the optical to infrared colours get steadily redder. They show that a stellar population domina- ted by dwarf stars at the centre, with a contribution from giant stars which increases radially from the centre, as does the upper mass cutoff of the main sequence, fits the observed colour gradients adequately. X-ray observations (e.g. Fabian 1984) show that accretion of the hot intracluster medium is an important environmental effect upon galaxies in the centres of rich clusters, and models of star formation in accretion flows (e.g. Fabian et. al. 1982) suggest that low mass stars might be the final form of much of this accreted material. Thus the unusual stellar population required by Valentijn and Moorwood to explain their photometric results might just result if most of the stars in the galaxy were formed out of a cooling accretion flow. Further photometric and spectroscopic tests of their result are required.

VI - The halos of spiral galaxies

One of the most interesting questions in extragalactic astronomy concerns the existence, and the nature if they do exist, of dark halos around spiral galaxies. Several attempts have been made to measure light from the halos of spiral galaxies using various techniques: Heygi and Gerber (1977) used an ingenious annular scanning photometer; Davis, Feigelson and Latham (1980) used "grid photography"; Spinrad et. al. (1978), Jensen and Thuan (1982) and others have used a combination of photograp- hic and photoelectric techniques. The edge-on Sb galaxy NGC 4565 has been observed by many groups, and the data are summarised by Jensen and Thuan. There is a "corona"

component to this galaxy which is probably quite well fitted by deVaucouleurs' (1959)
empirical light distribution for elliptical galaxies, however this radial light
distribution is much too steep to give rise to the flat rotation curve in NGC 4565,
so this component is not a tracer of the dark halo. The search for the dark halo
might best be conducted in the infrared, where low mass stars, one possible constituent of such a halo would be more readily detected. Infrared searches are being
carried out, but no results are yet available.

## VII - Isophotometry

Studies of the shapes of isophotes have been carried out by a number of authors
(Evans 1952; Carter 1978; Williams and Schwarzschild 1979; King 1978; Bertola and
Galetta 1978; Barbon et. al. 1980; di Tullio 1979; Leach 1981). For elliptical galaxies these studies have revealed that the position angles of the isophotes often
twist as a function of radius, indicating that the basic shape of these galaxies is
triaxial. At low light levels isophotometry can also reveal the effects of tidal
interactions or mergers, which might cause isophotes not to be concentric, or to be
twisted, or to deviate from elliptical shape.

## VIII - Interacting spiral galaxies

An example of the use of surface photometry to investigate the effects of
interactions on individual galaxies is provided by a study of the pair of interacting spirals NGC 5426/5427 by Blackman (1982). Blackman finds that the adjacent
halves of these two galaxies have been dimmed with respect to the rest of the galaxies, he attributes this to star formation being retarded by the interaction, although
I prefer the alternative explanation that they are obscured by dust.

## IX - Dust lanes

A very substantial proportion of elliptical galaxies possess dust lanes, often
near the nucleus, probably as a result of the quite recent accretion of a gas and
dust rich dwarf companion. The accretion of a substantial amount of gas and dust can
have the effect of fuelling a radio outburst or causing a burst of star formation.
Dufour et. al. (1979) used photographic broadband surface photometry to investigate
the nearest and most spectacular dust lane elliptical NGC 5128 (Centaurus A). From
their digitised images they constructed maps in (B-V), (U-B), and the index $Q =$
(U-B) - 0.72(B-V), which is supposed to provide a "reddening free" index which will
indicate regions of recent star formation. They conclude that vigorous star formation
is occurring throughout a disc which is largely masked by the dust lane, and that the
ages of the blue stellar associations in this disc are in the range $2x10^6 - 4x10^7$
years. This burst of star formation has presumably been triggered by the same interaction which created the dust lane.

Application of similar techniques to more recent CCD images, which have a
higher dynamic range than photographic images, will prove a powerful tool in the

investigation of dust lane galaxies. An example is provided by Carter et. al. (1983), who use (B-R) and (B-I) colour maps to investigate the morphology of the dust lanes in NGC 1316 (Fornax A) and NGC 1052.

## X - The environment of radio galaxies

Lilly, McLean and Longair (1984) have considered the photometric structure of powerful radio galaxies, and their relationship to brightest cluster galaxies. They consider a dimensionless structure parameter, $\alpha = d(\log L)/d(\log r)$, where L is the luminosity within radius r and the parameter is defined at a rather arbitrary radius of 19.2 $h^{-1}$ kpc. This structure parameter was introduced by Gunn and Oke (1975), and has been studied in brightest cluster galaxies by Hoessel (1980) and Schneider, Gunn and Hoessel (1983). There is some evidence that $\alpha$ is a probe of the environment of the galaxy, and particularly that galaxies separate in an $M_V - \alpha$ diagram according to how many mergers they have undergone. Lilly et al. find differences between the powerful radio galaxies and the brightest cluster galaxies which indicate that the radio galaxies have undergone much less merging than the brightest cluster members.

## XI - Future work

Photometry, particularly panoramic surface photometry, can provide important evidence on the effects of mergers, accretion and tides on galaxies. With the application of panoramic infrared detectors in the future it will become possible to make colour maps over a wide baseline in wavelength, and to obtain much more information on the effects of environment on stellar populations in galaxies. The space telescope will enable such studies to be extended into the ultraviolet, where dust has a much greater effect and hot gas emits its strongest emission lines. More sensitive X-ray telescopes will enable us to observe the accretion by galaxies of the hot intergalactic medium, perhaps the most important environmental effect on galaxies. Finally improvements in technology for ground based telescopes, such as the development of larger and more uniform CCDs and more sensitive photographic emulsions will provide more reliable quantitative measurements on the remnants left by collisions between galaxies in the distant past.

## References

Barbon, R., Benacchio, L., Capaccioli, M., De Biase, G., Santin, P., and Sedmak, G., 1980. Proc. S.P.I.E., 264, 250.
Bertola, F., and Galetta, G., 1978. Ap. J., 226, L115.
Blackman, C.P., 1982. M.N.R.A.S., 200, 407.
Boronson, T.A., Thompson, I.B., and Shechtman, S.A., 1983. Astron. J., 88, 1707.
Carter, D., 1977. M.N.R.A.S., 178, 137.
Carter, D., 1978. M.N.R.A.S., 182, 797.
Carter, D., Allen, D.A., and Malin, D.F., 1982. Nature, 295, 126.
Carter, D., and Dixon, K.L., 1978. Astron. J., 83, 574.
Carter, D., Jorden, P.R., Thorne, D.J., Wall, J.V., and Straede, J.O., 1983. M.N.R.A.S., 205, 377.
Davis, M., Feigelson, E., and Latham, D.W., 1980. Astron. J., 85, 131.
Dufour, R.J., et. al., 1979. Astron. J., 84, 284.

Dupraz, Ch., and Combes, F., 1984. This volume.
Evans, D.S., 1952. M.N.R.A.S., 112, 606.
Fabian, A.C., Nulsen, P.E.J., and Canizares, C.R., 1982. M.N.R.A.S., 201, 933.
Fabian, A.C., Nulsen, P.E.J., and Stewart, G.C., 1980. Nature, 287, 613.
Fabian, A.C., 1984. This volume.
Fort, B.P., Prieur, J.-L., Carter, D., Meatheringham, S.J., and Vigroux, L.,
    1985. In preparation.
Gunn, J.E., and Oke, J.B., 1975. Ap. J., 195, 255.
Hausman, M.A., and Ostriker, J.P., 1978. Ap. J., 224, 320.
Heygi, D.J., and Gerber, G., 1977. Ap. J., 218, L7.
Hoessel, J.G., 1980. Ap. J., 241, 493.
Hubble, E., 1930. Ap. J., 71, 231.
Jensen, E.B., and Thuan, T.X., 1982. Ap. J. suppl., 50, 421.
King, I.R., 1966. Astron. J., 71, 64.
King, I.R., 1978. Ap. J., 222, 1.
Kormendy, J., 1977. Ap. J., 218, 333.
Leach, R., 1981. Ap. J., 248, 485.
McGlynn, T.A., and Ostriker, J.P., 1980. Ap. J., 241, 915.
Malin, D.F., and Carter, D., 1980. Nature, 285, 643.
Malin, D.F., and Carter, D., 1983. Ap. J., 274, 534.
Matthews, T.A., Morgan, W.W., and Schmidt, M., 1964. Ap. J., 140, 35.
Oemler, A., 1976. Ap. J., 209, 693.
Quinn, P.J., 1982. Thesis, Australian National University.
Quinn, P.J., 1984. Ap. J., 279, 596.
Richstone, D.O., 1976. Ap. J., 204, 642.
Schneider, D.P., Gunn, J.E., and Hoessel, J.G., 1983. Ap. J., 268, 476.
Schweizer, F., 1980. Ap. J., 237, 303.
Schweizer, F., 1982. Ap. J., 252, 455.
Spinrad, H., Ostriker, J.P., Stone, R.P.W., Chiu, L.T.G., and Bruzual, G., 1978.
     Ap. J., 225, 56.
Thuan, T.X., and Romanishin, W., 1981. Ap. J., 248, 439.
di Tullio, G., 1979. Astron. Astrophys. suppl., 37, 591.
Toomre, A., 1974. I.A.U. symposium no. 58, p347.
Toomre, A., and Toomre, J., 1972. Ap. J., 178, 623.
Valentijn, E.A., and Moorwood, A.F.M., 1984. E.S.O. preprint no. 334.
deVaucouleurs, G., 1959. Handbuch der Physik, 53, 275.
White, S.D.M., 1979. M.N.R.A.S., 189, 831.
Williams, R.B., and Christiansen, W.A., 1984. E.S.O. preprint no. 322.
Williams, T.B., and Schwarzschild, M., 1979. Ap. J., 227, 56.

DISCUSSION

F. Schweizer : From the shell colours in NGC 3923, 5018, and 2865, can you say whether in any of these cases the infalling galaxy could have been an elliptical?

D. Carter : In NGC 3923 the infalling galaxy could definitely have been an elliptical, the colours of the shells are similar to the colours of the outer parts of NGC 3923. In NGC 2865 and 5018 the colours may be too blue.

S. Djorgovski : Can you estimate a typical total luminosity of the material in a shell?

D. Carter : In NGC 1344 we estimate it is $3 \times 10^8$ solar in the outer shell.

J.-L. Nieto : In NGC 3379 there seems to be a change in the geometry and a change
in the colour behaviour of the galaxy at 25 arcsec from the nucleus, where there is
a kink in the luminosity profile. This might be a signature of a gravitational
interaction with NGC 3384. In M87 we see exactly the same kind of correlation
between profile, geometry, and colour in your data (Carter and Dixon 1978), and
this may also be attributed to an interaction with the environment (Nieto and Vidal,
Astron. Astrophys. 135, 190).

M. Capaccioli : Concerning Kormendy's profile types T1, T2, and T3, how much does
the result depend upon the way you fit the deVaucouleurs law to the observations?

D. Carter : Kormendy claims that you get similar results if you fit a Hubble law
or a King model.

A.C. Fabian : The shells are blue. Using a standard IMF (whatever that is) can any
of the stars be younger than the dynamical ages inferred for the shells?

D. Carter : I would need to check the transformations between the CCD magnitudes
and Johnson magnitudes before I could answer that question.

S. Djorgovski : You subtract a uniform ellipticity, uniform position angle model
from your data. If there are any ellipticity gradients or isophotal twists in the
real galaxy, you would create artificial ripples.

D. Carter : There is a substantial isophote twist in the central regions of NGC
2865 which contributes to the poor fit of the model in the central regions of this
galaxy. NGC 3923 and 5018 appear not to have large isophote twists. The ellipticity
and position angle we use in the model are determined in the region of the galaxy
near the radius of the shells we are measuring.

## MALMQUIST BIAS IN "TULLY-FISHER" RELATION

L. Bottinelli[1], L. Gouguenheim[1], G. Paturel[2], P. Teerikorpi[3],

(1) Observatoire de Paris, section de Meudon, département
de Radioastronomie F-92195 Meudon Cedex, France
and Université Paris Sud, Centre Orsay
F-91405 Orsay Cedex, France

(2) Observatoire de Lyon F-69230 Saint-Genis-Laval, France

(3) Observatoire de Paris, section de Meudon, Département
de Radioastronomie F-92195 Meudon Cedex, France
on leave from Turku University Observatory, Tuorla, Finland

The Tully-Fisher (1977) relation is a tight correlation between the absolute magnitude
of a galaxy, M, and its maximum rotational velocity, $V_m$, measured from the width of
the 21-cm line :

$$M = a \log V_m + b \qquad (1)$$

The present treatment of the Malmquist bias in relation (1) is based on the discussion
given by Teerikorpi (1975, 1984). In a sample of galaxies with absolute magnitude $M_0$,
taken from a catalogue with sharp limiting magnitude $m_l$ and uniformly distributed, the
bias $[\Delta M]_d$ at a given distance d, is an increasing function of d, through the quantity

$$M(d) - M_0 \qquad \text{with} \qquad M(d) = m_1 - 5 \log d - 25$$

Each curve is characterized by $M_0$, $m_1$ and the intrinsic scatter of relation (1), the
errors on $\log V_m$ and the luminosity function of the galaxies. The unbiased range,
corresponding to the plateau in figure 1, goes up to a limiting distance $d_1$ which
depends on $M_0$.

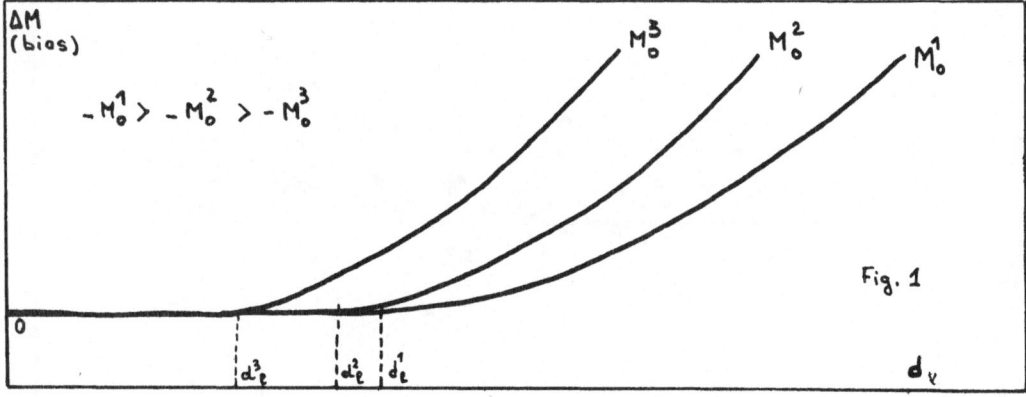

Fig. 1

The bias is not easy to put in evidence for two main reasons. First, the effect vani-
shes when plotting $\Delta M$ (or $H_0$) vs. biased distances. Second, in a sample of galaxies
with different $M_0$, the different curves are mixed.

Previous studies have tried either to avoid the bias by using restricted samples
(Aaronson et al., 1982 ; Bottinelli et al.,1983; Tully and Fisher,1977) or to correct
it from the formula $\Delta M = -1.382\,\sigma^2$ (Bothun et al., 1984). However, these restrictions
or corrections appear to be still insufficient.

The method used here relies on the two main concepts of sosies and kinematic
distances. We have first identified the Malmquist effect and second obtained an unbiased estimate of $H_o$.

Sosies: The effect of the adopted value of the slope of TF relation and a possible
type dependence on the value of $H_o$ has been much discussed. In order to bypass these
difficulties, Paturel (1984) has introduced the concept of "sosies". Two sosies are
identical; they are identified as sosies from their parameters which are not distance
dependent (type, log $V_m$, axis ratio) and it is then concluded that they have the same
absolute magnitude, which gives their relative distances. In particular, all the sosies
of a given calibrator are expected to be affected by the same bias, which corresponds
to a given curve in Figure 1.

Kinematic distances: Our work relies on a strong hypothesis. We assume that the relative distances of galaxies, $d_v$, in units of Virgo Cluster distance, may be calculated
with sufficient accuracy from Peebles (1976) symmetric differential expansion model.
We have adopted the following parameters: mean velocity of Virgo Cluster, with respect
to LG: $V_o = 980$ km s$^{-1}$ ; infall velocity of LG towards Virgo Cluster: $v_{LG} = 350$ km s$^{-1}$.

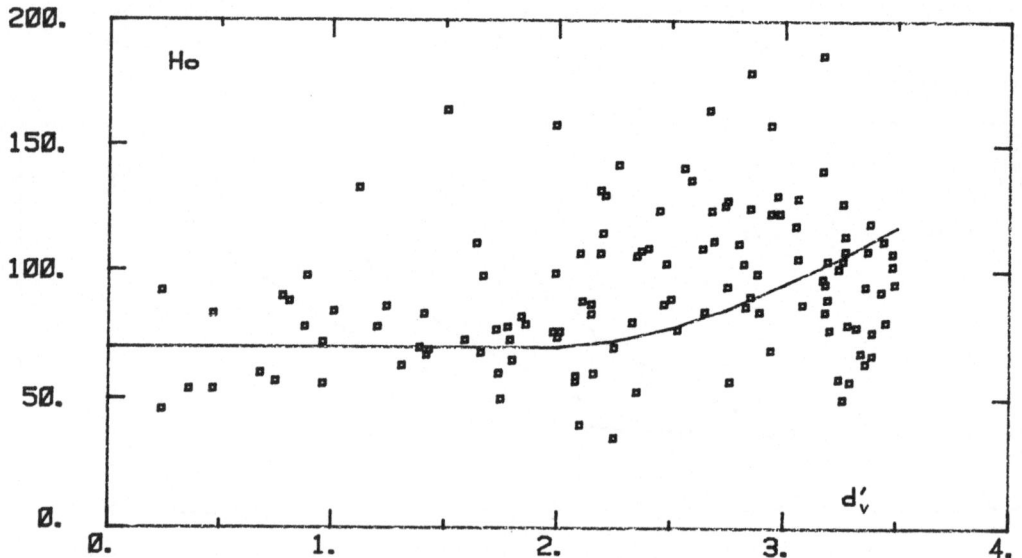

Figure 2

The plot of $H_o$ vs. $d_v$ for different groups of sosies shows the trend expected from Figure 1. In order to go further and to separate the galaxies situated within their own unbiased distance range, we have introduced the concept of normalized distance $d'_v$ for which $M(d'_v) - M_o^1$ has the same value for different $M_o^1$ : all the curves in Fig.1 are reduced to a unique curve. When applying this normalization to the sosies, we obtain an $H_o$ vs. $d'_v$ diagram (72 points) in which there is no more segregation of the different classes of sosies; it remains a clear increase of $H_o$ with $d'_v$ and the scatter is considerably reduced. The number of sosies in the unbiased range is too small for a reliable determination of $H_o$ by this method, though the diagram points at $H_o$ around 70-80.

In order to increase the amount of data, we have then used the large sample of galaxies with distance determined from TF relation (Bottinelli et al., 1984a,b), restricted to the more accurate data ( $\sim$ 400 galaxies). From the 41 galaxies in the unbiased region (Fig.2) we find the mean logarithmic value of $H_o$ : $H_o = 72 \pm 3$ km s$^{-1}$Mpc$^{-1}$.

## CONCLUSIONS

(1) From a plausible local velocity field model, the TF relation suggests the value $H_o = 72 \pm 3$, when the local distance scale is due to de Vaucouleurs.

(2) This result relies strongly on the assumption that Peebles' model gives a plausible evaluation of an unbiased distance scale, but not on the accurate values of the model parameters.

(3) The Malmquist effect is much more subtle than the simple $M = - 1.382 \sigma^2$ formula.

(4) It is not sure that the use of galaxies within clusters prevents against the bias, the incompleteness of the samples being larger at low luminosities.

(5) We have not tried to fit to the observed $H_o$ vs. $d'_v$ points a theoretical curve: such a curve relies on a number of badly known parameters and on several assumptions. It appears safer to identify the unbiased range.

(6) There are several arguments favouring the conclusion that our value of $H_o$ is global and not local. One of them concerns the rather large radial velocity range encountered in the unbiased region from about 200 to 2500 km s$^{-1}$.

## REFERENCES

Aaronson M.,Huchra J., Mould J.,Schechter P.L.,Tully R.B.,1982 Astrophys.J. 258, 64
Bothun G.D.,Aaronson M.,SchommerB.,Huchra J.,Mould J., 1984 Astrophys.J. 278, 475
Bottinelli L.,Gouguenheim L., Paturel G.,de Vaucouleurs G.,1983 Astron.Astrophys.118,4
Bottinelli L.,Gouguenheim L.,Paturel G., de Vaucouleurs G.,1984a Astron.Astrophys.
        Suppl.Ser. 56, 381
Bottinelli L.,Gouguenheim L.,Paturel G., de Vaucouleurs G.,1984b Astron.Astrophys.
        Suppl.Ser. in the press
Paturel G.,1984 Astrophys.J. 282, 382
Peebles P.J.,1976 Astrophys.J. 205, 318
Teerikorpi P.,1975 Astron.Astrophys. 45, 117

Teerikorpi P.,1984 Astron.Astrophys. in the press
Tully R.B.,Fischer J.R.,1977 Astron.Astrophys. 54, 661

DISCUSSION

G.Bruzual: What value of $H_o$ should we believe ?

L.Gouguenheim: With primary calibration in de Vaucouleurs scale, $H_o$ = 72 ± 3; with the old Sandage and Tammann scale, $H_o$ = 56 ± 3; with the new ST scale, $H_o$ = 63 ± 3.

M. Capaccioli: If not the exact value, can you give us a lower limit of $H_o$ ?

L.Gouguenheim: Some improvements of the primary calibration can be expected from Madore et al. IR study of cepheids. The first results go in the sense of a decrease of the distances of the local calibrators, and thus an increase of $H_o$. It seems thus that $H_o$ = 70 is the lower limit.

# PHOTOMETRY OF QUASAR HOST GALAXIES AND COSMOLOGICAL IMPLICATIONS

Thomas Gehren

Institut für Astronomie und Astrophysik der Universität München
Scheinerstr. 1, 8000 München 80
Federal Republic of Germany

## SUMMARY

Surface photometry of sky-limited red photographic and CCD observations, corrected for galactic extinction, the K term and seeing image degradation, reveals the
decomposition of low-redshift quasar images into a central point source and an extended underlying nebulosity. The investigation of the statistical properties of a
well-resolved subsample of these nebulosities shows that

(a) Quasar redshifts cannot be explained by "tired light" assumptions in a
static universe or a local hypothesis. The observed central surface brightness is,
however, roughly compatible with the prediction of an expanding FRIEDMANN universe.
Taking into account the integrated magnitudes and the physical association of the
nuclei and the extended component, we conclude that the underlying nebulosities are
in fact the host galaxies of quasar nuclei.

(b) Correlations with redshift of the integrated magnitudes, the isophotal
diameters and the surface brightness at constant angular distances from the center
indicate, that the host galaxies of low-redshift quasars ($z < 0.6$) strongly evolve in
luminosity. 5 to 8 Gyr ago, quasar host galaxies were on average 5 times brighter
than they are today.

(c) From the concentration of galaxies in the immediate neighborhood of quasars
($< 200$ kpc) there is strong evidence that quasars are born in and evolve with galaxy
clusters. On an even smaller scale ($< 50$ kpc projected distance), quasar host galaxies in many cases appear to be heavily distorted by interaction with faint companion
galaxies.

## I. QUASARS AND COSMOLOGY

If quasar redshifts originate from the expansion of our universe,
quasars are probably the most important tracers of cosmic evolution.
Their extreme luminosities combined with their blue non-thermal energy
distributions make them observable out to distances which correspond to
an appreciable fraction of the HUBBLE time. Thus, quasars are cosmological lighthouses which point to a region of the sphere where we have
to look for evidence of the evolution of our universe, if we knew what
to look for. At present there do not seem to exist crucial differences
between the spectra of low- and high-redshift quasars, and the information about their central engines' total luminosities cannot yet be

interpreted unambiguously in terms of cosmic evolution.

The situation is somewhat relaxed for the low-redshift quasars, which have recently been resolved into a central point source and an extended underlying nebulosity, the latter showing some of the common features of bright galaxies (Wyckoff et al. 1981, Hutchings et al. 1982, Gehren et al. 1984, Malkan et al. 1984, Green and Yee 1984). Spectroscopic observations have revealed that the redshifts of quasar nuclei and their underlying nebulosities are the same to within reasonably narrow limits whenever emission or absorption lines could be detected. While different investigators have found diverging evidence concerning the morphological types of the nebulosities, they agree that virtually all quasars with redshifts < 0.4 are resolved. The fraction of resolved quasars decreases strongly with redshift and, as would be expected for FRIEDMANN models, no quasars with z > 0.7 have been resolved at a surface brightness level of 26 mag arcsec$^{-2}$.

Although the photometric observations favour the interpretation of these nebulosities being quasar host galaxies, as yet the only direct evidence that their light is of stellar origin has been published by Boroson and Oke (1982), who detected Balmer absorption lines in the nebulosity of 3C 48, and Balick and Heckmann (1983), who found absorption lines in a few other quasar nebulosities. Whereas a number of independent methods for the determination of extragalactic distances (cf. Sandage and Tammann 1982) leaves no doubt that the HUBBLE relation between apparent magnitude and redshift is in fact a distance law for the brightest members of galaxy clusters, such an outstanding confirmation of general relativity theory is not possible for quasars at present, since independent determinations of quasar distances do not exist. Moreover, the distribution of quasar apparent magnitudes with redshift shows quite clearly that quasars are far from being "standard candles". The cosmological redshift hypothesis for quasars is, however, not undisputed. Arp and his co-workers (see Arp 1983, for a review) have observed a number of quasar associations including objects of different redshifts which appear to be arranged along a straight line. A number of quasars with comparatively high redshifts have also been detected in the immediate vicinity of nearby bright galaxies (Arp et al. 1975, Burbidge 1979). In both cases the non-cosmological origin of quasar redshifts (local hypothesis) is claimed on the basis of statistical arguments which appear to support the physical association of objects with different redshifts. It is therefore of primary importance to obtain independent observational support for either the cosmological or the local hypothesis.

## II. THE ORIGIN OF QUASAR REDSHIFTS

The propagation of photons in different world models has been investigated by Hubble and Tolman (1935) who first compared expanding models of the FRIEDMANN type with predictions of the static EINSTEIN universe. Since the latter gives no explanation for the redshift, they adopted the ad-hoc hypothesis of "tired light", which vaguely describes the interaction of photons with other particles on their way to the observer. Let $I_O$ denote the central intensity of an extragalactic object in its co-moving frame at redshift z, and $i_O$ the observed intensity. The surface brightness theorem states that in expanding world models

$$i_O = I_O (1+z)^{-4} ,$$  (1)

whereas in a static world model

$$i_O = I_O (1+z)^{-1} .$$  (2)

Unless the redshift is explained by a relative motion of object and observer, Eq.(2) is also valid for the local hypothesis. In either model, one factor of (1+z) results from the energy loss of the redshifted photons, irrespective of the redshift origin. In the expanding world model, another factor of (1+z) is due to time dilatation between object and observer, and a factor $(1+z)^2$ to the increase in angular area of the object resulting from the expansion of the universe. The importance of the theorem has been emphasized in a number of publications (Sandage 1961, Kristian and Sachs 1966, Gudehus 1975), yet its application is strongly complicated because of the necessary corrections for the K term and atmospheric seeing. Phillipps' (1982) attempt to prove that quasar redshifts originate from the expansion of the universe failed because he used the central surface brightness data of QSO nebulosities published by Wyckoff et al. (1981), which were only tentative estimates resulting from extrapolation. Improved surface photometry based on the decomposition of quasar images is now obtained in a quantitative and repeatable way (Gehren et al. 1983), and the present results are based on a subsample of 26 well-resolved low-redshift quasars and BL Lac objects.

While the details will be published in a forthcoming paper, we note here that the observed surface brightness data have been corrected for galactic absorption and the K term as described by Gehren et al. (1984). The seeing corrections, however, depend on the assumption of the two-dimensional intensity distribution of the quasar nebulosities. Though individual objects of our subsample may be forced into an exponential profile fit, statistical evidence favours the approximation by a power law of the HUBBLE type,

$$I(r) = I_o \ (1+r/R)^{-2} \ , \tag{3}$$

where R denotes the scale height. Seeing corrections thus were obtained by determining the scale height from an empirical curve-of-growth relation (Sandage 1972), and convolving Eq.(3) with the observed stellar point spread function. Fig.1a demonstrates the large scatter of the corrected central surface brightness $\mu_{o,c}$ of quasar nebulosities ($\sigma$ = 0.80 mag) about the regression line,

$$\mu_{o,c} = 12.64(\pm 3.74) \ \log(1+z) + 16.29(\pm 0.34) \ . \tag{4a}$$

While the probability that such a correlation arises from a parent population defined by an expanding universe (Eq.1) is 0.51, the corresponding probability for a parent population in accordance with a static universe or the local hypothesis (Eq.2) is only 0.02. Part of the individual scatter of $\mu_o$ can be removed. Though systematic errors resulting from K correction or the modelling of the seeing cannot be ruled out, the tight correlation between individual deviations from the regression lines of $\mu_{o,c}$ (Fig.1a) and $m_c$ (Fig.2, no seeing corrections!) indicates that the scatter in $\mu_{o,c}$ mainly represents the distribution of intrinsic luminosities at constant redshift. Fig.1b shows the central surface brightness after correction for the individual scatter of the luminosities, $\mu_{o,c}^* = \mu_{o,c} - \delta m_c$. Whereas the slope of the regression line,

$$\mu_{o,c}^* = 13.31(\pm 1.97) \ \log(1+z) + 16.24(\pm 0.18) \ , \tag{4b}$$

has changed only marginally with respect to Eq.(4a), the scatter is reduced to $\sigma$ = 0.42. The probability in favour of an expanding universe

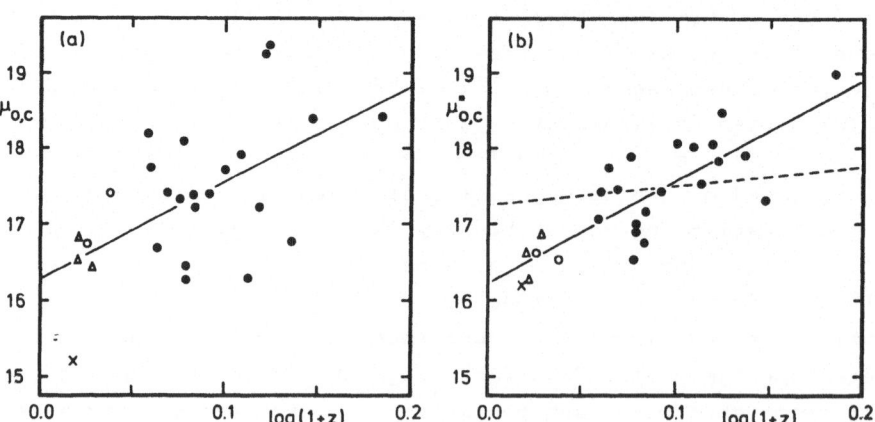

**Fig. 1:** Central surface brightness $\mu_o$ of quasar host galaxies as a function of redshift z. Open and filled circles are optical and radio quasars respectively. Triangles are BL Lac objects, and the cross is the X-ray quasar 0241+622.
(a) $\mu_{o,c}$, corrected for galactic extinction, K term, and seeing image degradation.
(b) $\mu_{o,c}^*$, including an additional correction for the intrinsic scatter of luminosities at constant redshift (see text). The solid line represents the linear regression, the dashed line shows the prediction from Eq.(2)

now would be 0.11, while the corresponding value for a static universe or the local hypothesis becomes as small as $10^{-5}$.

These results depend mainly on the assumption that the seeing corrections do not introduce strong redshift-dependent systematic errors. Although this is difficult to verify, it is possible to show that at high redshifts the present seeing corrections based on Eq.(3) are rather upper limits. In fact, using exponential profiles instead of a power law leads to a significantly steeper slope in Fig.1, while modelling with a deVAUCOULEURS law or a KING profile does not remove the discrepancy. Statistically, the observed properties of quasar nebulosities clearly show that our current notion of <u>cosmological</u> quasar redshifts is acceptable, whereas the static "tired light" universe and the local hypothesis are not. Moreover, since the physical association of the QSO nuclei and the extended nebulosities seems to be out of question and the observed integrated magnitudes of the nebulosities are of the same order as for the brightest cluster galaxies, the extended nebulosities surrounding the low-redshift quasars are almost certainly <u>galaxies</u>.

## III. THE EVOLUTION OF QUASAR HOST GALAXIES

The surface brightness theorem may depend on the cosmic luminosity evolution of quasar host galaxies, since $I_0$ could be a function of

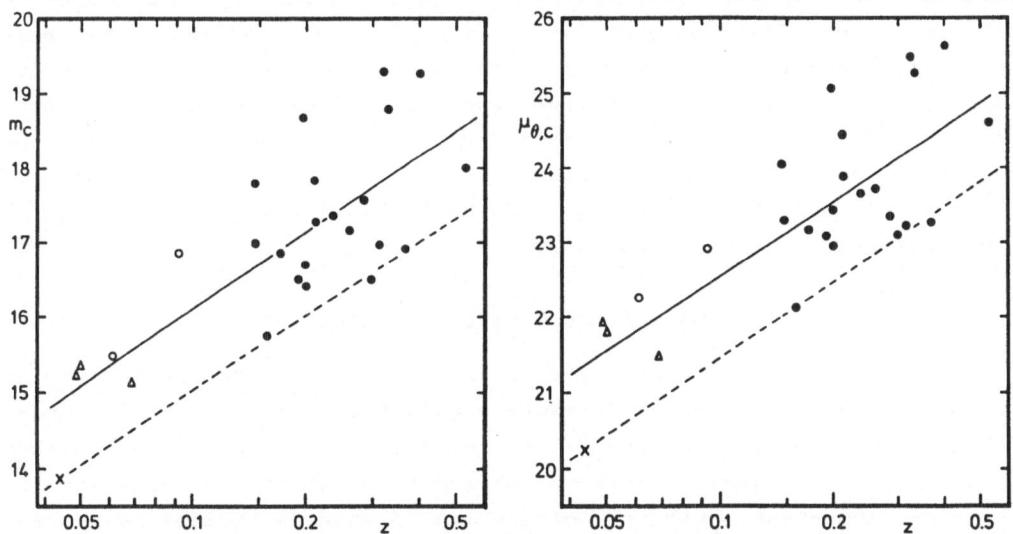

**Fig. 2:** Apparent magnitude m and surface brightness $\mu_\theta$ at constant angular distance $\theta = 5"$ from the center of quasar host galaxies as functions of redshift. The linear regression and a straight line connecting the brightest objects are indicated by solid and dashed lines, respectively. A similar diagram can be constructed for the isophotal radius $\theta_{26}$ of the host galaxies

redshift. Fig.2 indicates that the host galaxies of low-redshift quasars strongly evolve with cosmic epoch. Although the data base is
fairly small and inhomogeneous (including optical and radio-loud quasars as well as BL Lac objects), it seems relevant that the evolution
with redshift of the average QSO galaxy (continuous lines) follows the
same slope as it does for the brightest objects (dashed lines). The
fact that the host galaxies of BL Lac objects and optical quasars
appear to determine the correlations simply results from the decreasing
number of quasars with z < 0.2. Such objects are mostly classified as
N- or SEYFERT galaxies and are not included here. While the upper left
part of the diagrams may contain some extremely faint QSO host galaxies, the area below the dashed lines is definitely devoid of objects
with redshifts < 0.5.

The regression lines in Fig.2 would imply a cosmic luminosity
evolution according to
$$L \sim z^{0.64} .$$ (5)
Since, introducing an appropriate cutoff at large radii, Eq.(3) yields
$$L \sim I_0 R^2 ,$$ (6)
such an evolution may be due to $I_0$ or R or a combination thereof
evolving with redshift. If the luminosity of QSO host galaxies were
evolving only with $I_0(z)$ the resulting modification of the surface
brightness theorem would be incompatible with our observations (Fig.1).
Thus, an increase with redshift of the central intensity $I_0$ measured in
the co-moving frame of QSO host galaxies can be ruled out. The same
holds for a strong decrease of $I_0(z)$ compensated by an increase of
R(z). In conclusion, the observed luminosity evolution of quasar host
galaxies must be mainly due to a decrease with cosmic time of the
galaxy scale height parameter R(z), and it seems that at a redshift of
z = 0.6 (about 5 to 8 Gyr ago) QSO host galaxies were on average 5
times brighter than they are today.

## IV. QUASARS AND CLUSTERS OF GALAXIES

At present there is no evidence as to how the quasar phenomenon is
triggered, and the model for active galactic nuclei is still under
debate (see Rees or Shklovsky, this conference). The issue whether
quasars are isolated events or related to clusters of galaxies is
therefore important. While the clustering of quasars themselves represents an as yet unsolved problem, observations of quasar fields including spectroscopy (Stockton 1978,1980) or a statistical comparison of
the area number densities of faint galaxies in the immediate vicinity
of quasars (Gehren et al. 1984) seem to indicate that at least low-

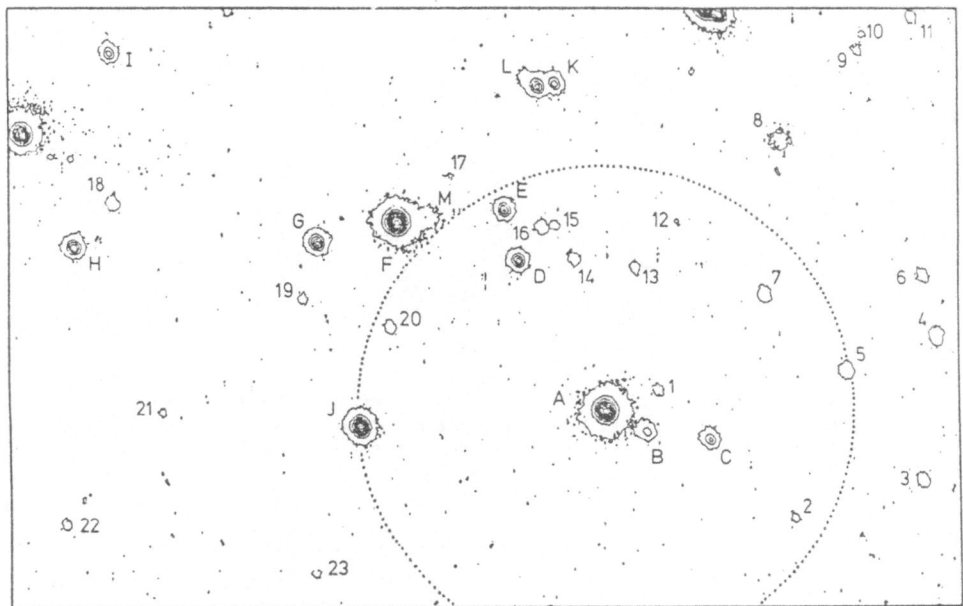

**Fig. 3:** Isophote map of CCD field (2 by 3 arcmin) around the radio quasar 2201+315 (4C 31.63, z = 0.297). A and B denote the QSO host galaxy and a compact companion. C to M are stars, numbers 1 to 23 are faint galaxies. The dotted circle marks the average core radius of galaxy clusters (150 kpc for $H_0$ = 75 km s$^{-1}$ Mpc$^{-1}$ and $q_0$ = 1)

redshift quasars are associated with groups or clusters of galaxies. The statistical approach is demonstrated in Fig.3 for the radio quasar 2201+315. The isophote map of the CCD field shows a total of at least 10 faint galaxies (m < 22) within a projected distance of 150 kpc from the QSO, corresponding to the average core radius of galaxy clusters (Bahcall 1975), whereas large-scale background counts of Tyson and Jarvis (1980) would only predict less than 1. The apparent clustering of galaxies around quasars is a common property of <u>all</u> low-redshift quasars that have been investigated. While spectroscopic observations are necessary to confirm the redshifts of these galaxies, their magnitude distribution and in particular the gap between the magnitudes of the quasar host galaxy and the second brightest galaxy in the field is in accordance with what is known about normal clusters containing a first-ranked elliptical galaxy.

Even more important for the formation and evolution of quasars is the existence of faint galaxies within a projected distance of less than 50 kpc. Fig.4 shows three isophote maps, again fairly representative of the low-redshift quasar sample. Although we will not enter here in a discussion of apparent luminous bridges and extensions, we note

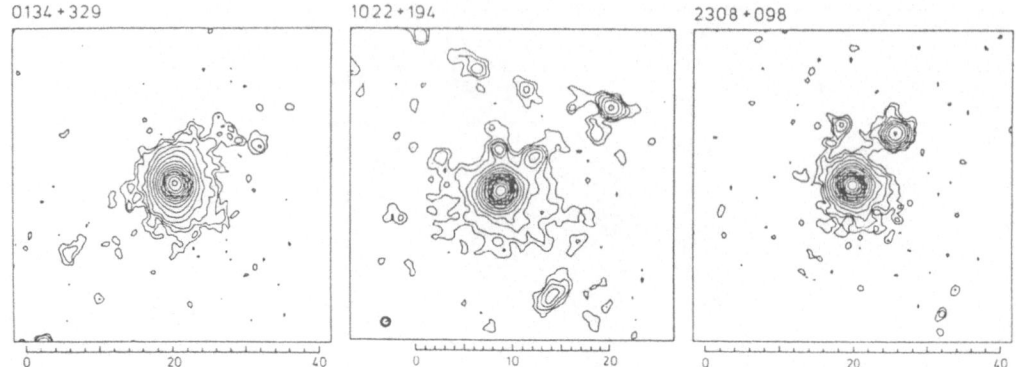

**Fig. 4:** Faint galaxies in the immediate neighborhood of 3 radio quasars. The morphology of the quasar host galaxies and the luminous extensions suggests that the QSO galaxies and their companions are in a state of tidal interaction. The vertical bars correspond to a projected distance of 50 kpc

that if these faint galaxies are physically associated with the quasar host galaxies, there is probably no way avoiding tidal interactions on scales of 10 to 50 kpc.

## V. CONCLUDING REMARKS

The current state of quasar surface photometry is significantly affected by the lack of completeness at low redshifts. Although there may be marginal differences with respect to spectra and images between the various types of active galactic nuclei, advances in our understanding of quasars require a more comprehensive view. The simple approach of reclassifying a quasar as a SEYFERT galaxy every time such a quasar is resolved adds nothing to our knowledge, since classification criteria used for active galactic nuclei have proved to depend on operationally defined quantities such as sensitivity or resolution of a detector.

At this time of writing, no quasar with a redshift larger than 0.7 has been resolved, and none of the higher redshift quasars has definitely proved to sit in a cluster of galaxies. If we accept that quasars are located at cosmological distances, this result is not surprising. The predicted diameters of QSO host galaxies as well as the integrated magnitudes of the cluster galaxies steeply fall off with redshift. Already at $z = 0.8$ the mean isophotal diameter at a surface brightness of 26 mag arcsec$^{-2}$ of a quasar host galaxy is only 6 arcsec, whereas a normal cluster galaxy of $M = -21$ will transform into an apparent magnitude of $m = 21$ to $22$, depending on the world model. Both quantities are further affected by a K dimming of 2 to 3 magnitudes. Hence, even at this redshift, the limitations of terrestrial observations become obvious.

Almost certainly observations with Space Telescope will extend the present ground-based results to higher redshifts. It will be important to obtain redshifts of host galaxies as well as faint companion or cluster galaxies in order to provide a comparison between "normal" and quasar galaxy clusters. If the existence and cosmic evolution of quasars does at all depend on tidal interaction or galaxy merging in a cluster environment, the luminosity evolution of quasar nuclei and perhaps their host galaxies should reflect the amount of matter available to supply the QSO central engine.

## REFERENCES

Arp, H. 1983, in "Quasars and Gravitational Lenses", Proc.24th Intern. Astrophys. Symp. (Liège), p.307

Arp, H., Pratt, N.M., and Sulentic, J.W. 1975, Astrophys.J. 199,565

Bahcall, N.A. 1975, Astrophys.J. 198,249

Balick, B., and Heckman, T.M. 1983, Astrophys.J.Lett. 265,L1

Boroson, T.A., and Oke, J.B. 1982, Nature 296,397

Burbidge, G. 1979, Nature 282,451

Gehren, T., Fried, J., Wehinger, P.A., and Wyckoff, S. 1983, in "Quasars and Gravitational Lenses", loc.cit. p.489

Gehren, T., Fried, J., Wehinger, P.A., and Wyckoff, S. 1984, Astrophys.J. 278,11

Green, R.F., and Yee, H.K. 1984, Astrophys.J. 280,79

Gudehus, D.H. 1975, Publ.Astron.Soc.Pacific 87,763

Hubble, E., and Tolman, R. 1935, Astrophys.J. 82,302

Hutchings, J.B., Crampton,D., Campbell, B., Gower, A.C., and Morris, S.C. 1982, Astrophys.J. 262,48

Kristian, J., and Sachs, R.K. 1966, Astrophys.J. 143,379

Malkan, M.A., Margon, B., and Chanan, G.A. 1984, Astrophys.J. 280,66

Phillipps, S. 1982, Astrophys.Lett. 22,153

Sandage, A. 1961, Astrophys.J. 133,355

Sandage, A. 1972, Astrophys.J. 173,485

Sandage, A., and Tammann, G.A. 1982, in "Astrophysical Cosmology", eds. H.A. Brück, G.V. Coyne, and M.S. Longair (Città del Vaticano: Pontif.Acad.Sci.), p.23

Stockton, A. 1978, Astrophys.J. 223,747

Stockton, A. 1980, in "Objects of High Redshift", eds. G.O. Abell and P.J.E. Peebles (Dordrecht: Reidel), p.89

Tyson, J.A., and Jarvis, J.F. 1980, in "Objects of High Redshift", loc.cit. p.1

Wyckoff, S., Wehinger, P.A., and Gehren, T. 1981, Astrophys.J. 247,750

# DISCUSSION

<u>A. Fabian</u>: Two questions: first, have you considered local QSO hypotheses in which the redshift is due to velocity, and second, is there a correlation between the luminosities of the QSO and the host galaxy?

<u>T. Gehren</u>: As yet local hypotheses explaining the quasar redshift by DOPPLER velocities have not been considered. There is a correlation between the luminosities of quasar nuclei and their host galaxies (cf. Gehren <u>et</u> <u>al</u>. 1984, Fig. 4) such that for low redshifts the luminosity ratio is about constant, however, with a large scatter.

<u>S. Djorgovski</u>: Two remarks: first, I do not believe that you can correct for the seeing as well as you assert. I did very extensive modeling of seeing effects in situations similar to the ones you are investigating, and I think that it is impossible to do it as well as you say you did. The seeing corrections are strongly systematic with redshift, and even a slight miscompensation of seeing can produce spurious correlations and parameter couplings. Second, there is a correlation of QSO nuclear luminosity; on the other hand, most QSO samples are flux-limited, and one expects more luminous objects at higher redshifts. This can leak into a artificial correlation of host galaxy luminosity with redshift.

<u>T. Gehren</u>: The seeing dependence on redshift caused by a change of the angular galaxy scale height has been accounted for using the curve-of-growth method of Sandage (1972). Tests with a variety of parameters indicate that a possible systematic error over the whole redshift range ($z < 0.6$) is smaller than 0.5 mag, which does not affect my conclusions. Your second remark implies that our QSO sample misses some faint quasars at higher redshifts, which is certainly true. However, the same holds for very low redshifts as well, and the mean relation may not be as unreliable. Note that all the <u>brightest</u> objects with $z < 0.5$ are included, and we may replace the mean relation by that of the brightest quasar host galaxies.

# POSSIBLE EVIDENCE FOR EVOLUTION IN THE PHOTOMETRIC PROPERTIES
## OF NEARBY GALAXIES

R. Brent Tully
University of Hawaii

Abstract :

Given the four observables--optical magnitudes, infrared magnitudes, optical dimensions, and H I line profile widths--then 6 pairwize combinations of these observables can be constructed that are distance dependant and 4 triplet combinations can be constructed that are distance independant. Tight correlations are found between all possible combinations of these parameters with samples of galaxies drawn from the Virgo and Ursa Major clusters. However, when large samples of mostly non-cluster galaxies are considered, the scatter in the same relationships increases. The scatter in certain of the distance--independant plots is <u>not symmetrical</u> with respect to the lay of the cluster data. If the data are good, then the interpretation could be that some galaxies in our large sample are <u>redder</u> than normal for their mass. Such galaxies are expected to exist if spirals evolve to become gas-depleted lenticulars.

MORNING SESSIONS

# ON THE STRUCTURE AND KINEMATICS OF THE SMALL MAGELLANIC CLOUD

E. Maurice, N. Martin, L. Prévot and E. Rebeirot
Observatoire de Marseille
2, Place Le Verrier
13248 MARSEILLE CEDEX 04

The first attempt to a kinematical study of the Small Magellanic Cloud (SMC) was performed by the Radcliffe astronomers (Feast et al., 1960, 1961) on the basis of the radial velocities of 40 stars and 13 HII regions. Objective-prism observations by Florsch (1972a) resulted in about 100 stellar radial-velocities and a compilation by Maurice (1979) of all then known slit-spectrograph radial velocities gave data for 80 supergiants, 35 HII regions and 12 planetary nebulae.

In January 1981 a copy of the radial-velocity spectrometer CORAVEL, already in use for several years in Haute Provence Observatory, was installed at the focus of the Danish 1.54 m telescope at La Silla, Chile (ESO). It is operated jointly by ESO and Copenhagen, Geneva and Marseille Observatories. In three years, observations with this instrument provided velocities of 195 SMC member-stars of late spectral type, that is almost the double of the previously existing velocity data. We intend to give here, in advance of a more complete paper (Maurice et al., 1985), some results of these observations.

## INSTRUMENTATION AND OBSERVATIONS

A complete description of the operation and performance of CORAVEL is given by Baranne et al. (1979) and Mayor et al. (1983). Among the stars observed in the SMC are 16 F-K5 type stars, 175 K5-M type stars and four Cepheids; these latter stars were observed regularly and a mean of 50 measurements were made per star. In addition we have measured velocities of 33 foreground stars and of 4 stars probably situated in space between the two Magellanic Clouds or between them and the Galaxy.

The F-K5 stars were taken from lists of already known SMC members (Feast et al., 1960; Sanduleak, 1968, 1969; Azzopardi and Vigneau, 1979, 1982). The K5-M stars were taken from the recently published catalogue by Prévot et al. (1983).

The magnitude of the fainter stars observed was of the order of $V = 14.5$. For this magnitude the signal/noise ratio (stellar counts against photomultiplier dark counts) was less than unity. Typical integration times were of the order of 15 minutes but have been as long as 30 minutes for the fainter stars.

We have determined (Imbert et al., 1984) the radial velocity of the center of gravity of four of the observed cepheids; observations are still needed.

CORAVEL RADIAL-VELOCITIES ACCURACY AND ZERO-POINT

The mean standard error for one observation is $<\varepsilon_1> = 2.00 \pm 0.18$ km s$^{-1}$ and $<\varepsilon_1> = 1.27 \pm 0.04$ km s$^{-1}$ for F-K5 and K5-M SMC supergiants respectively. As observations have been repeated only in very few cases, no direct determination of the standard error on the mean stellar velocity is significant and we admit that the above mentioned values of $<\varepsilon_1>$ reflect the reproducibility of CORAVEL measurements. These estimates are in good agreement with those given for velocities of LMC stars (Prévot et al., 1985); they show that the percentage of variables among these late-type (later than F0) stars is probably rather low.

During each CORAVEL observing night radial-velocity standard stars were observed. The velocities given for the present programme stars are on the system defined by the faint ($m_V > 4.3$) IAU standard stars (Pearce, 1957) as discussed by Mayor and Maurice (1985).

RESULTS

The present observations confirm with more precision the findings previously obtained from spectrographic radial velocities of (mainly) early-type stars. In the plane of the sky, stars show a tendency to cluster in groups of apparent mean size of the order of 400 pc and with radial-velocity dispersion of the order of 5 km s$^{-1}$. Groups with very different velocities happen to be very near-by or even to overlap in the plane of the sky; this result is strongly in favour of a great depth of the SMC. The depth of the SMC in the line of sight has been suspected a long time ago (Johnson, 1961; Hindman, 1967; Florsch, 1972b) and confirmed from spectrographic and photometric studies of the brightest supergiants in both Clouds by Ardeberg and Maurice (1977, 1978). Recently Matthewson (1984) even admitted that two parts of the SMC are now separating.

Similar results had been obtained for early-type stars but with velocity dispersion probably significantly larger; this velocity-dispersion difference between early and late type stars has also been found in the LMC (Prévot et al., 1985). A possible interpretation could be that a higher percentage of variables or close binaries is found among early-type supergiants than among late-type supergiants in the SMC.

The surface distribution of radial-velocity groups in the SMC is also quite similar for OBA and for K5-M type stars; thus where low-velocity groups of OBA stars are found, low velocity late-type stars are also found. The same holds for higher velocity groups. This may indicate that these young stars have formed together in kinematically homogeneous regions.

We obtain for the GSR radial-velocities (using rotation velocity of the Sun of 250 km s$^{-1}$) the following mean velocities and dispersions for the same central region of the SMC (OBA stars velocities from Maurice, 1979):

- for   61   OBA-type stars $<V_{GSR}>$ = +4.5 ± 2.4 km s$^{-1}$;   $\sigma$ = 16.5 km s$^{-1}$
- for   18   F5-K type stars $<V_{GSR}>$ = -6.6 ± 4.7 km s$^{-1}$;   $\sigma$ = 20.1 km s$^{-1}$
- for 162   K5-M type stars $<V_{GSR}>$ = -3.8 ± 1.6 km s$^{-1}$;   $\sigma$ = 20.4 km s$^{-1}$

The double peaked velocity distribution previously found for OBA stars is no longer clearly visible but the overall dispersion is the same for the three spectral-type ranges.

The velocity difference between spectrographic early-type star and CORAVEL later-type star velocities is also found in the LMC (Prévot et al., 1985); it indicates that the spectrographic velocity zero-point previously adopted has probably to be corrected by approximately -8 km s$^{-1}$.

Finally we have compared our stellar velocities with those of neutral hydrogen. If we admit that each of the components of the HI profiles originates in (three) different independent complexes, only one of them shows a velocity variation which could be interpreted in terms of rotation. However correlation of this HI complex with stars is rather doubtful.

All these questions will be more thoroughly discussed in a forthcoming paper (Maurice et al., 1985).

REFERENCES

Ardeberg, A., Maurice, E.: 1977, Astron. Astrophys. 77, 277

Ardeberg, A., Maurice, E.: 1978, Compt. Rend. Acad. Sci. Paris, 286, Série B, 375

Azzopardi, M., Vigneau, J.: 1979, Astron. Astrophys. Suppl. Ser. 35, 353

Azzopardi, M., Vigneau, J.: 1982, Astron. Astrophys. Suppl. Ser. 50, 291

Baranne, A., Mayor, M., Poncet, J.L.: 1979, Vistas Astron. 23, 279

Feast, M.W., Tackeray, A.D., Wesselink, A.J.: 1960, Monthly Notices Roy. Astron. Soc. 121, 337

Feast, M.W., Tackeray, A.D., Wesselink, A.J.: 1961, Monthly Notices Roy. Astron. Soc. 122, 433

Florsch, A.: 1972a, Pub. Obs. Astron. Strasbourg, Vol. 2, No. 1

Florsch, A.: 1972b, Compt. Rend. Acad. Sci. Paris 275, Ser. B, 763

Hindman, J.V.: 1967, Australian J. Phys. 20, 147

Imbert et al.: 1984, Observations in progress

Johnson, H.M.: 1961, Pub. Astron. Soc. Pacific 73, 20

Matthewson, D.S., Ford, V.L.: 1984, in IAU Symp. No. 108, Structure and evolution of the Magellanic Clouds, S. van den Bergh and K.S. de Boer, eds. Reidel, p.125

Maurice, E.: 1979, Thesis, Université de Provence

Maurice et al.: 1985, Astron. Astrophys. Suppl. Ser. in preparation

Mayor et al.: 1983, Astron. Astrophys. Suppl. Ser. 54, 495 in IAU Coll. No. 88, Stellar Radial Velocities

Mayor, M., Maurice, E.: 1985, A.G. Davis-Philip, Ed. L. Davis Press, Schenectady,
USA, in press

Pearce, J.A.: 1957, Report of Sub-Commission 30a in Trans. IAU IX, P. Th. Oesterhoff
ed., Cambridge University Press, p. 441

Prévot et al.: 1983, Astron. Astrophys. Suppl. Ser. 53, 255

Prévot et al.: 1985, Astron. Astrophys. Suppl. Ser. (submitted)

Sanduleak, N.: 1968, Astron. J. 73, 246

Sanduleak, N.: 1969, Astron. J. 74, 877

# PHOTOGRAPHIC SURFACE PHOTOMETRY OF THE ANDROMEDA GALAXY

R.A.M. Walterbos
Sterrewacht, P.O. Box 9513, 2300 RA Leiden
The Netherlands


R.C. Kennicutt, Jr.
University of Minnesota

## 1. INTRODUCTION

Recently, the spiral galaxy M31 has been the subject of several detailed investigations. The Westerbork telescope has mapped the distribution of the neutral hydrogen (Brinks and Shane, 1984) and the 21 and 49 cm radio continuum emission (Walterbos et al., 1984; Bystedt et al., 1984) at high resolution. Infrared maps at four wavelengths have been obtained with IRAS (Habing et al., 1984). To complement these radio and infrared data, we have obtained calibrated photographic plates in U, B, V, R, and Hα of M31, using the Burrell Schmidt telescope at Kitt Peak and the Palomar Schmidt. The plates have been scanned at Leiden with the ASTROSCAN reticon densitometer at a resolution of 4".

The combined data set will provide information on the current and recent star formation, the properties of the interstellar dust, and the spiral structure of the galaxy. Also the question of a possible warp in the stellar distribution can be addressed. Here we will discuss some of the specific reduction techniques used to process the large amount of optical data, and a few preliminary results regarding the warp and the interstellar dust distribution will be presented.


## 2. PHOTOGRAPHIC PHOTOMETRY

The large size of the galaxy (4°x2°) introduces several problems not usually encountered in photographic surface photometry. A very large area (10x20 cm, 2000x4000 pixels) had to be scanned to be able to determine the sky level. Over such areas variations in sky background occur due to inherent plate variations, vignetting, scattered light, differential dehypersensitisation (the "Malin" effect), and a possible intrinsic sky brightness variation (M31 is at galactic latitude -20°).

We have used a specific sky subtraction technique to correct for these effects as much as possible. A local background level was determined in small submatrices of

32x32 pixels, using an iterative κ-σ clipping routine (e.g. Newell, 1979), to correct for foreground stars. The background values of all submatrices were combined in one map with 64x128 pixels. On this map a sky fit was determined using only regions outside the image of M31. The fit was blown up to the original map size and then subtracted from the full resolution data, after conversion to intensity. Various fit functions were tested. Comparison with published photoelectric data shows good agreement down to $\mu_B \simeq 25.5-26$ magn. per arcsec squared. A full description of the reduction procedure can be found elsewhere (Walterbos and Kennicutt, in prep.).

## 3. SOME RESULTS

The compact maps obtained as described above are not only useful for background fitting. Due to the data compression, the noise is reduced by a large factor and the maps therefore extremely well show weak extended emission. In contrast to a standard convolution, there is no contamination by foreground stars (except a few strong ones). We have converted the compressed maps to intensity and subtracted sky. Figure 1 shows the sums of four red and four blue plates processed this way. In both colours the deviation of the outer contours from the major axis is apparent. This effect was seen before, though not as convincingly, by Innanen et al. (1982). The deviation seems to be too large to be explained by spiral structure. A real warp in the stellar distribution seems to be present. It is in the same direction as the HI warp (Newton and Emerson, 1977), but sets in somewhat closer to the centre at roughly 18 kpc. Our photometry is not accurate enough to say much about the colour or the absolute light level's of the warp. Comparison with published photometry shows that it is below $\mu_B = 26$.

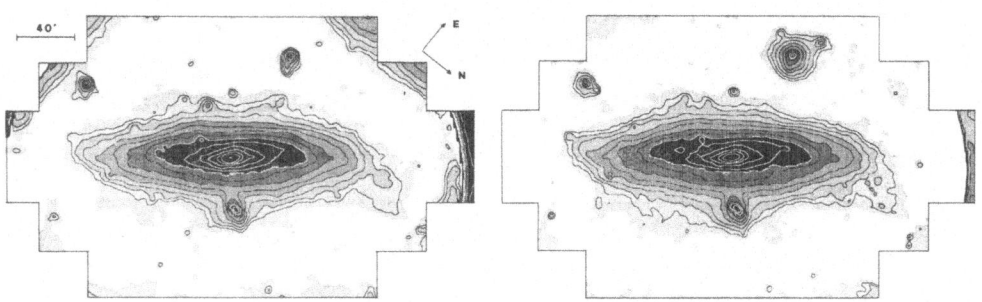

Fig. 1. (Left): Average of four red images of M31, processed as described in the text. The blemishes at the edges are due to the filtre and plate edge effects. Some bright stars with their scattered light halos are also visible. The extension towards the bottom is the elliptical companion NGC205. The increment between contour values is $0.^m5$, the zero point is chosen arbitrarily. The brighter intensities (dark grey values) are not accurate due to the data processing. (Right): Average of four blue images of M31. The bright star towards the top is ν And.

The compact image can also be used in (digital) unsharp masking techniques. We have divided a high resolution image in the B band by a blown up "smoothed" image. Both maps were in intensity units and had sky subtracted. The process is similar to the photographic unsharp masking technique, but having the data in digital format enables a quantitative analysis. Figure 2 shows the image obtained this way. The large scale brightness gradient is effectively removed and the dust and bright spiral arm regions clearly show up. Figure 2 also shows the integrated HI map obtained by Brinks and Shane (1984) at the same scale. Taking into account the fact that the dust on the near side of the galaxy (bottom side in the figure) can be seen better than the dust on the far side, a striking correlation between dust features and HI

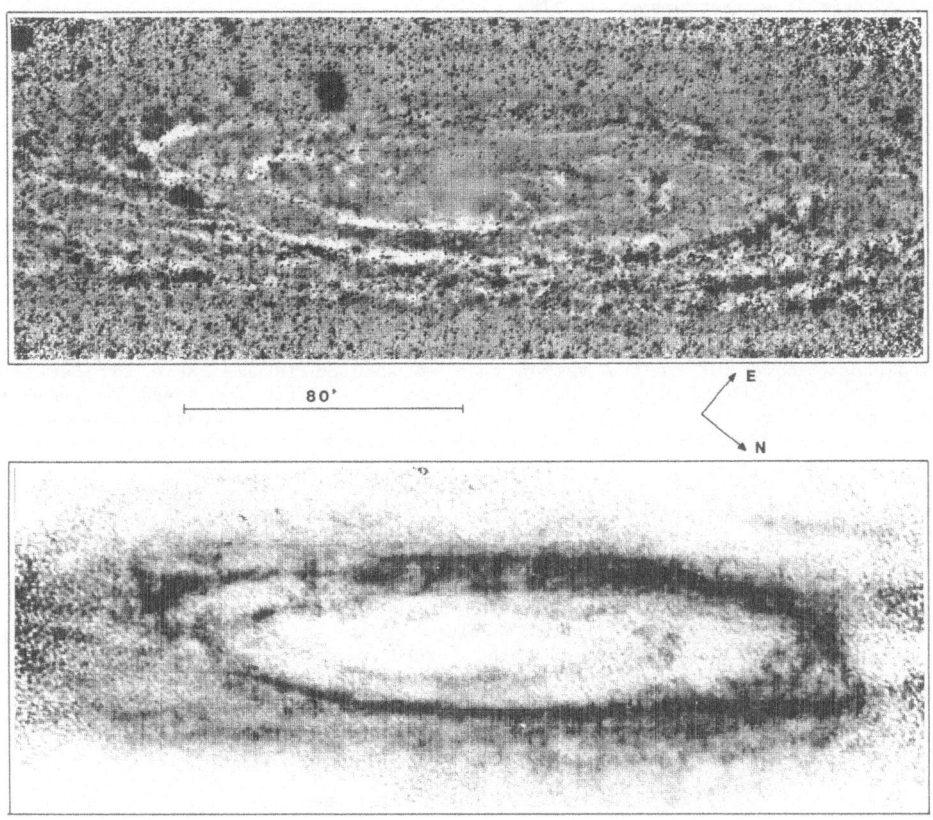

Fig. 2. (Top): Digitally unsharp masked image of M31 in the B band (see text for details). Dust lanes and patches appear as white and light grey. Bright stellar associates and spiral arm segments are dark grey or black. The nuclear region of M31 is not represented accurately due to plate saturation. (Bottom): The integrated HI map from Brinks and Shane (1984) at the same scale and orientation. Notice the striking correlation between HI and dust features.

emission is apparent, down to the smallest scales. Using the multi colour maps, a quantative analysis of the relation between extinction and HI column density will be made in the near future.

ACKNOWLEDGEMENTS: The Burrell Schmidt telescope is operated by the Kitt Peak National Observatory and the Case Western Reserve University. R.A.M.W. was financially supported by the Netherlands Foundation for Astronomical Research (ASTRON), which is supported by the Netherlands Organization for the Advancement of Pure Research (ZWO). R.C.K. acknowledges a visiting grant from ZWO during the initial stages of this project.

REFERENCES

Brinks, E., Shane, W.W.: 1984, Astron. and Astrophys. Suppl. Ser. **55**, 179
Bystedt et al.: 1984, Astron. and Astrophys. Suppl. Ser. **56**, 245
Habing et al.: 1984, Astrophys. J. **278**, L59
Innanen et al.: 1982, Astrophys. J. **254**, 515
Newell, E.B.: 1979, in "Image processing in astronomy", eds. G. Sedmak, M. Capaccioli, R.J. Allen, 100
Newton, K., Emerson, D.T.: 1977, Monthly Notices Roy. Astron. Soc. **181**, 573
Walterbos et al.: 1984, submitted to Astron. and Astrophys. Suppl. Ser.

DISCUSSION

J.-L. Nieto: What is the size of the sky on your plates relative to the size of the galaxy? In other words, what is the accuracy that you estimate for your sky determination?

R. Walterbos: About 50% of the region that we scanned is available for background fitting. From the residuals in the region where the fit is based on, and from comparisons between different plates in the same band and comparisons between our photometry and published photoelectric data, we estimate the error in sky to be 0.01-0.02 in density.

D. Carter: Have you compared the variations of the background you fit underneath the galaxy with what you expect from vignetting, scattering and differential dehypersensitisation?

R. Walterbos: For plates in the V and R bands the gradient due to the galactic background dominates the sky nonuniformity. The background fit for B plates qualitatively looks like what one would expect from vignetting and dehypersensitisation, but a detailed comparison is not possible, because the dehypersensitisation variation is not known and likely varies from plate to plate. Scattered light is important below about $\mu_B$ = 26-26.5 (only along the minor axis).

B. Elmegreen: The correlation between HI and dust cannot be as good as you seem to indicate because the gas on the far side of the spiral arms should not appear as prominent dust features, and the dust clouds containing molecules should not necessarily appear as strong HI features. How much have you studied this correlation in detail, on a cloud by cloud basis?

R. Walterbos: We only report what we see. So far, in M31 CO emission has only been detected in places where there is a lot of HI. The dust on the far side is indeed not as clearly seen as on the near side, and our dust picture is not a direct measure of the amount of dust. We will do a detailed quantative analysis of the correlation HI/dust in the near future.

# HI HOLES IN THE INTERSTELLAR MEDIUM OF MESSIER 31

E. Brinks
ESO, Karl-Schwarzschild-Strasse 2
D-8046 Garching, F.R.G.

INTRODUCTION

Recently a high resolution survey of Messier 31, the Andromeda galaxy, was completed in the 21-cm line of neutral hydrogen (Brinks 1984; Brinks and Shane 1984). The angular resolution of $\Delta\alpha \times \Delta\delta = 24" \times 36"$ corresponds at the distance of M31 to a linear resolution of $80 \times 120$ pc. The velocity resolution measures 8.2 km s$^{-1}$. With this resolution the HI can be studied in detail comparable with that available for studies of the Milky Way galaxy, but from a more advantageous perspective. This has led to the detection of structures in the interstellar medium which can best be referred to as HI holes. These holes are characterized by an absence of HI, they have dimensions varying typically from 100 pc to 1 kpc, and they appear roughly circular on the HI maps. This latter fact indicates that their three-dimensional shape is roughly spherical.

The HI holes in M31 show a clear resemblance to the HI shells and supershells which have been found in our Galaxy by Heiles (1979), and to the superbubbles in the LMC described by Meaburn (1980). More recently HI shells of the same kind as found in M31 are also seen in a WSRT survey of M33, done with a comparable resolution (Deul, private communication). This leads to the conclusion that HI holes seem to be quite a common phenomenon in spiral galaxies.

RESULTS

We have systematically searched the M31 data set for HI holes and determined their characteristics. This search was done much in the same manner as the search described by Heiles for the shells in our Galaxy, i.e. by inspection of the single channel maps. After identifying the location of each hole, more detailed maps were made, such as position-velocity maps through the centre of the hole and various cross-cuts, which were used to measure its properties. We

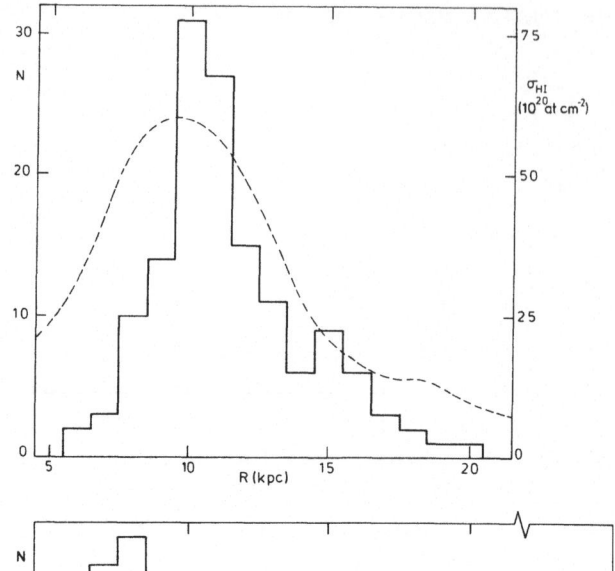

Figure 1: Radial distribution of HI holes. The dashed line shows the average HI surface density, $\sigma_{HI}$, as a function of radius.

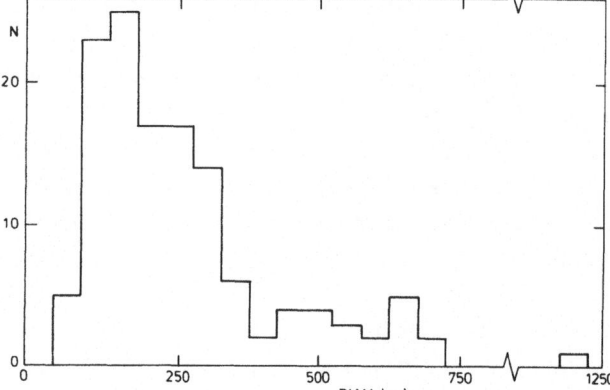

Figure 2: Size distribution of the HI holes.

recorded its position, the velocity of the channel map at which it was seen most clearly, and the average brightness temperature of the surrounding HI. We also assigned a quality factor to each hole. The hole is furthermore fitted by an ellipse to determine its size and orientation. From the position-velocity maps we could, for 80% of the holes, derive an expansion velocity.

The observed properties can be used to derive properties like the indicative age, the energy required to produce a hole, and the mass which is estimated to be absent at the position of a hole. The results can be represented in graphical form. Figure 1 shows the radial distribution of the holes in M31. For comparison the HI column density, $\sigma_{HI}$, as a function of radius is plotted, derived by averaging the total surface brightness maps in rings in the plane of the galaxy. The radial distribution of the holes follows the HI distribution and

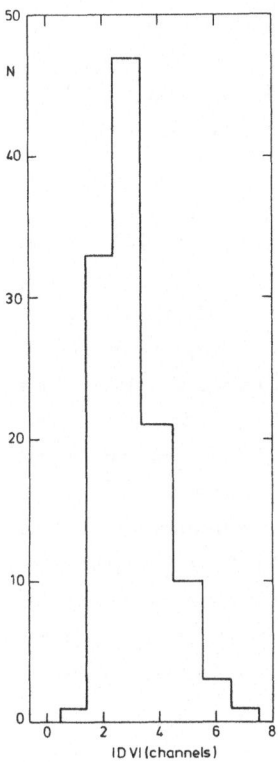

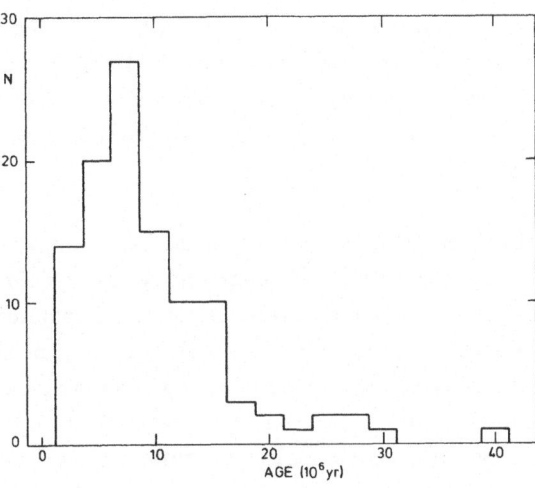

Figure 4: Age distribution in units of $10^6$ yr of the HI holes.

Figure 3: Frequency distribution of the expansion velocities expressed in number of channels. The channel separation is 4.1 km s$^{-1}$.

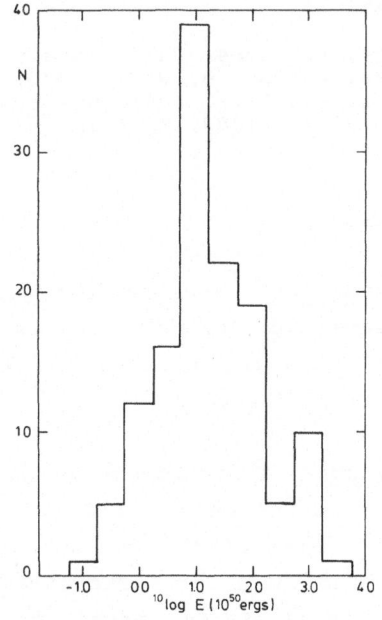

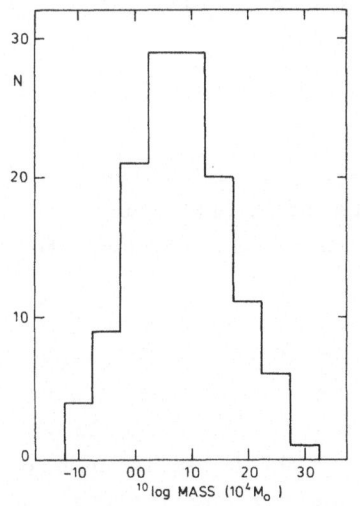

Figure 5: Frequency distribution of the HI mass which is absent from the location of the hole, in units of $10^4$ M$_\odot$.

Figure 6: Frequency distribution of the energies required to produce the holes in units of $10^{50}$ ergs.

peaks around 10 kpc. Figure 2 gives the size distribution of the holes. The steep drop below 100 pc is caused by the resolution limit of our survey. The expansion velocities peak sharply around 12 km s$^{-1}$. This is illustrated in Figure 3 which shows a histogram of the measured values.

Based on the measured size and expansion velocity one can define an indicative age for each hole which is merely the ratio of these two values. The ages derived in this manner should not be considered as an absolute value for the age of an individual hole but they provide an indication, in a statistical sense, for the time scales involved. The results are plotted in Figure 4. The ages run from about 2.5 to 30 10$^6$ years, covering ZAMS lifetimes of stars which range from spectral type O5 to B0. The last two figures, Figure 5 and 6, show the frequency distribution of the amount of HI which is absent from the location of the hole and the amount of energy which is typically needed to produce a hole. These energies range from the equivalent energy of one supernova event for the smaller holes to many super-novae, or perhaps a more copious energy source, for the larger ones.

REFERENCES

Brinks, E.: 1984, Ph.D. Thesis (University of Leiden).
Brinks, E., Shane, W.W.: 1984, Astron. Astrophys. Suppl. 55, 179.
Heiles, C.: 1979, Astrophys. J. 229, 533.
Meaburn, J.: 1980, Monthly Notices Roy. Astron. Soc. 192, 365.

DISCUSSION

B. Elmegreen: Your contour plots for holes appear elongated. Is this corrected for galactic inclination? Are the holes elliptical in the plane? Does the tilt of the ellipse make sense with respect to the direction of shear?

E. Brinks: The contour map shows the outline of the HI holes as they were measured on the individual channel maps without any correction for the beam of the telescope or the inclination of the galaxy. Be-cause the intrinsic three-dimensional shape of the holes is unknown it is not possible to correct the observed shape for the inclination of M31. About the effect of shear on the holes, the only statement which I can make at this moment is that the holes at larger galactocentric distances seem to be rounder, possibly due to less shear.

# HIGH RESOLUTION CO OBSERVATIONS IN M31

F. Boulanger[1,2], J. Bystedt[3], F. Casoli[1], F. Combes[1]

(1) Ecole Normale Supérieure, Paris Ve
    and Observatoire de Meudon 92190 Meudon, France

(2) Kapteyn Laboratory, Postbus 800, 9700 AV, Groningen

(3) Stockholm Observatorium, Saltsjöbaden, Sweden

Abstract : CO observations towards M31 made with the Onsala 20m telescope, reveal a high degree of inhomogeneity of the emission on the 100 pc scale. When comparing the CO observations made towards the same spiral arm but with lower resolution, two kinds of emission appear : with the small beam (33") we detected a narrow and bright component, corresponding to giant molecular complexes ; a fainter and broader emission appears only at low resolution, or when adjacent spectra are averaged. This wider emission is interpreted as coming from a lot of small clouds spread all over the beam and thus sharing all the velocities expected from the HI emission.

## I. Introduction and observations

The large scale distribution of molecular clouds in our galaxy is difficult to study because of distance ambiguities, velocity perturbations and blending of different features. It is therefore of great interest to observe nearby spiral galaxies in the CO line. With the limited resolution of present millimeter wave single dish telescopes, the nearest spiral galaxy Messier 31 is the best candidate for this study. This galaxy was first observed in CO by Combes et al (1977 a, b), who observed the global distribution along the major axis. Boulanger, Stark, Combes (1981, BSC), and Linke (1981) obtained a complete CO map of a large HI complex in the South-Eastern part of the galaxy, with the Bell Labs 7 m telescope (resolution 1.7' at 115 GHz).

Here, we present new observations of this region made with the Onsala Space Observatory 20 m telescope, which has a resolution of 33" at 115 GHz, corresponding to a linear size of 100 pc along the major axis of M31. The observations were performed during March 1984. The single sideband receiver temperature was 380 K. The pointing was accurate within 5" r.m.s. We used a 512 x 1 MHz filter bank ; which gives a resolution in velocity of 2.6 km/s. We observed 24 points, with a spacing of 30". Their positions are shown superimposed on the integrated HI emission observed with the Westerbork interferometer (Brinks and Shane 1984) in figure 1. Each of the points was observed for at least 3 hours, yielding a typical r.m.s. noise level of 100 mK in the 1 Mhz channels.

## II. Results and Discussion

Some of the spectra are displayed in Figure 2. The emission within the spiral arm is very inhomogeneous, on the scale of 100 pc, while the atomic gas distribution is much smoother. The CO and HI lines velocities are the same within the HI velocity resolution (8 km/s). To deduce the CO column densities and the $H_2$ masses, we use a ratio $NH/\int T_A$ (12 CO) $dV = 5\ 10^{20}$ cm$^{-2}$/(K.km/s) which is within the currently adopted values. The $H_2$ masses within one 33" beam are then obtained between 2 and $3\ 10^5$ M⊙, in the center of the spiral arm. Outside the central part of the spiral arm, we can set a limit to the $H_2$ mass of a few $10^4$ M⊙ : giant molecular complexes are not observed in the interarm regions, since their emission would be well above our sensitivity limit.

The main features observed towards the center of the spiral arm have a linewidth of 15–20 km/s (velocity dispersion 6–9 km/s r.m.s.) and are brighter in antenna temperature that the emission observed with lower resolution (Bell Labs spectra of 1.7' beam). In the latter, the diluted bright component is associated with a wider (40 km/s) and fainter emission. The velocity spread of this "diffuse" component is similar to the HI emission and could come from a large number of small clouds spread all over the spiral arm. Since the bright and narrow emission seems extended relative to the 33" beam, we deduced a mean size for the giant complexes between 100 and 200 pc. The fact that these complexes seem equally extended along the major and minor axis, though the projection factor along the minor axis multiplies the distances by a factor 5, means that the height above the plane of the complexes are of the same order (100 pc) as their size. In each spectrum where are observed the complexes, they always appear at the same radial velocity which supports the hypothesis of their forming one entity.

Physical properties of the Giant complexes : Within the 100 pc scale complex, it is possible to select in velocity the HI emission associated with the molecular one. The corresponding mass is M (HI)/M($H_2$) = 0.4. With a total mass of $1.4\ 10^6$ M⊙, the gravitational energy of the complex can be estimated at $4\ 10^{51}$ ergs. Since the kinetic energy associated with the turbulent motions present within the complexes is $E_c = 2$–$5\ 10^{51}$ ergs (for a velocity dispersion of 6–9 km/s), the whole complex is gravitationally bound within the uncertainties on the derivation of the masses and the geometry of the cloud.

The mean density of the complex, estimated from the column density $2\ 10^{21}$ $H_2$ cm$^{-2}$, and a line of sight of ~ 100 pc, is found to be: 7 $H_2$ cm$^{-3}$, lower than the average density of GMC in our own Galaxy : 50–100 $H_2$ cm$^{-3}$. This suggests a larger dilution factor for the molecular clumps in the complex, or different physical properties for the clouds themselves.

Figure 1 : The observed points are represented by crosses (23" in size) and the CO detections by circles on this integrated HI emission map from Brinks and Shane (1984). The display of the HII regions detected in H$_\alpha$ by Pellet et al (1978) shows that no giant HII region is associated with the CO emission.

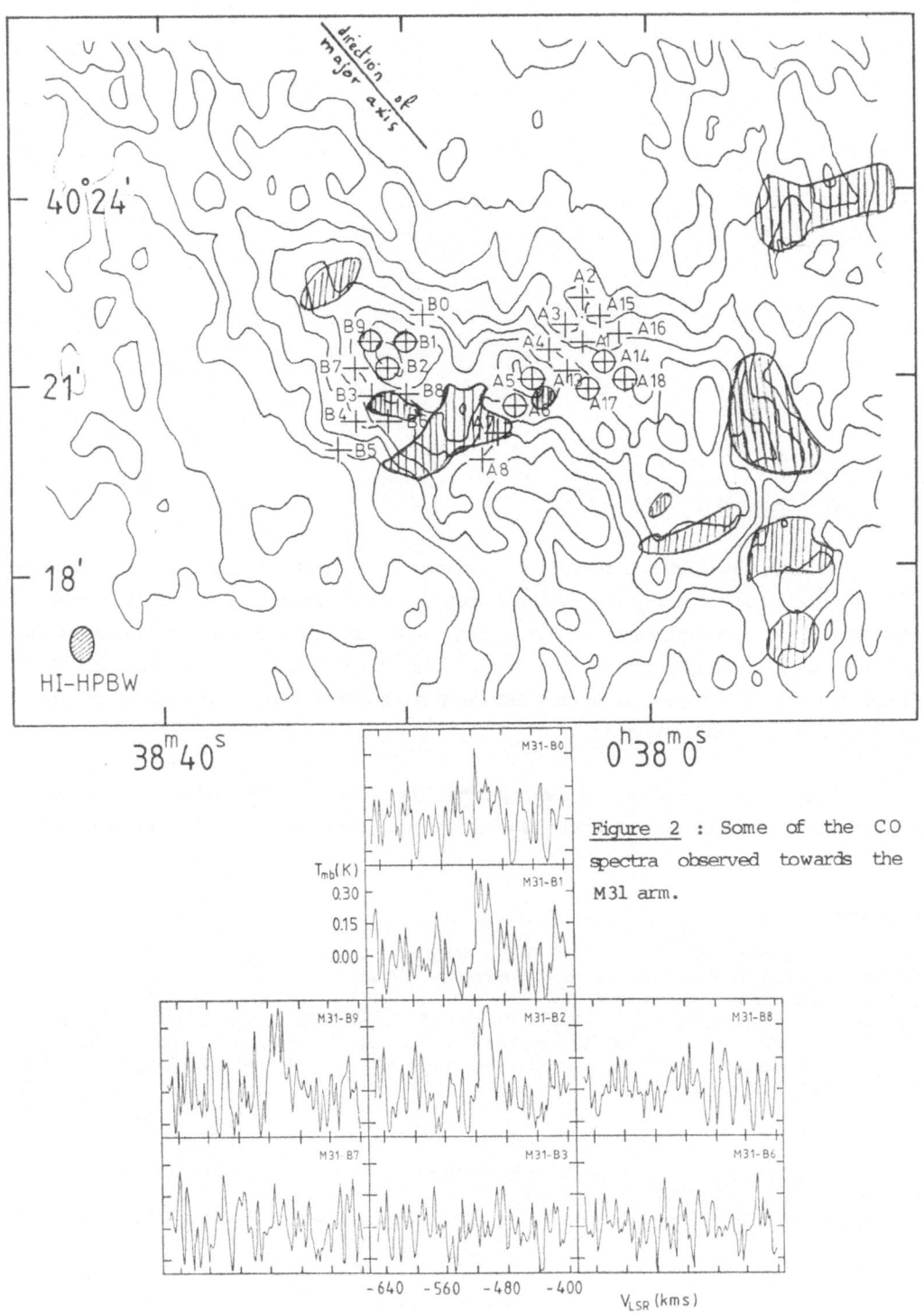

Figure 2 : Some of the CO spectra observed towards the M31 arm.

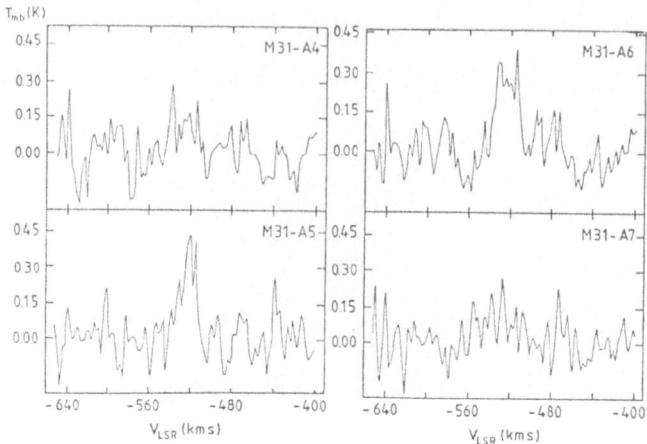

## III – Conclusion

We have detected two kinds of CO emission in the Southern spiral arm of M31 : first a broad velocity and spatially extended component, corresponding to a smooth background of small mass clouds ; second, narrower bright components just resolved by the 33" beam, emitted by giant complexes of size 100 pc and mass $10^6$ M⊙. These complexes are associated with the HI emission peaks and appear to be gravitationally bound. They must be short-lived to account for their absence outside the arms.

Onsala Space Observatory is operated by Chalmers University of Technology, Göteborg, with financial support from the Swedish Natural Science Research Council (NFR).

### References

Boulanger F., Stark A.A., Combes, F. : (1981) Astron. & Astrophys 93, L1.
Brinks, E., Shane, W.W. (1984) Astron. & Astrophys. Suppl.
Combes, F., Encrenaz, P.J., Lucas, R., Weliachew, L. (1977) a Astron. & Astrophys. 55 311
Combes, F., Encrenaz, P.J., Lucas, R., Weliachew, L. (1977) b Astron. & Astrophys. 61, L7
Linke, R.A. (1981) Proceeding of "Extragalactic Molecules" N.R.A.O., Green Bank p. 87
Pellet, A., Astier, N., Viale A., Courtès, G., Maucherat, A., Monnet, G., Simien, F. (1978), Astron. Astrophys. Suppl. 31, 439

DISCUSSION:

B. Elmegreen: Could you detect an Orion type HII region in the giant molecular cloud you have found?
F. Combes: No, it would be embedded in the cloud (for H$_\alpha$ ) and too weak to measure in radio.

# MORPHOLOGICAL INVESTIGATION OF ELLIPTICAL GALAXIES

S. Djorgovski[1], M. Davis[1], and S. Kent[2]

[1]) Astronomy Department, University of California, Berkeley, CA 94720, USA

[2]) Harvard-Smithsonian Center for Astrophysics, Cambridge, MA 02138, USA

ABSTRACT:   We describe the two large surface photometry surveys of elliptical galaxies, completed at Mt. Hopkins and Lick observatories.   These comprise the largest
such surveys  of early-type galaxies to this date,  and will provide us with a large
set of uniform quality data  for a variety of statistical and dynamical studies.  We
discuss briefly some results from the Mt. Hopkins survey.

In the recent years,  there was a marked  increase of interest  in the family of
elliptical galaxies.  Paradoxically enough, the amount of good-quality spectroscopic
information has greatly exceeded  the amount of existing  good-quality morphological
information  about these objects, and what is available  is too heterogeneous  to be
useful for most studies of interest.   In order to bridge this gap, we conducted two
large CCD  surface photometry surveys  of early-type galaxies.  We think that it is
important to have a statistically large sample of data of good <u>and</u> uniform quality.

The first survey  covered some 40  semi-randomly selected ellipticals,  all with
$B_T < 14^m$.   The data were obtained with the Mt. Hopkins 24-inch telescope, and an RCA
CCD, in the Gunn $\underline{r}$ bandpass.  Some of the data, methods of reduction, etc., are presented in Kent (1984).   The early results were reported by Davis, Djorgovski & Kent
(1983), and the complete discussion will appear shortly in a full paper.  This was a
pilot-project  for the  larger survey,  completed  subsequently  at Lick Observatory
(Djorgovski & Davis 1983).  The Lick survey covered some 200 ellipticals and some 50
lenticulars.   All galaxies  were selected from  the CfA redshift survey lists,  and
consequently all have the spectroscopic information:  recession velocities, velocity
dispersions, and line strengths.  All galaxies have $B_T < 14^m$, and $-25° < \delta < 70°$.  We have
a complete sample of some 150 ellipticals with $\delta > 0°$  and $|b_{II}| > 30°$; outside these
limits, the survey is incomplete, but completeness-controlled.  The sample of lenticulars is semi-randomly selected within these limits.   The data were obtained with
the Lick 1-m Nickel telescope  and TI and GEC CCDs,  in a red bandpass similar to $\underline{r}$.
The coverage is up to ~100 arcsec in radius,  and down to $\mu$~24 in the $\underline{r}$ (corresponding to $\mu$~25-26 in the B band).

For each galaxy we obtain the following profiles as functions of the semi-major-axis: surface brightness, ellipticity, position angle, elliptically and circularly summed magnitudes, and deviations from the ellipticity (e.g., "boxedness" of the isophotes). Those profiles can be parametrized, so that one has only a small number of variables describing each galaxy, e.g., fiducial surface brightness, ellipticity, ellipticity gradient, isophotal twist rate, steepness (concentration) of the light profile, etc. The error-bars are carefully evaluated for all quantities.

It is necessary to establish some fiducial radial scale for the measurements of representative quantities. One possibility is to fit a de Vaucouleurs' $r^{1/4}$ profile to the surface brightness data, and use $r_e$ from the fit. However, this assumes that all ellipticals can be fitted well with the $r^{1/4}$ law, which is not the case. [Note that a half-light radius cannot be determined satisfactorily, since we never know the total apparent magnitude of a galaxy.] An alternative, model-independent way to establish a metric radial scale is to use Petrosian's (1976) $\eta$ function. Operationally, this function can be defined as a ratio of the local surface brightness and the mean surface brightness above that isophote:

$$\eta(r) = SB(r)/\langle SB \rangle_r , \qquad \eta(0)=1, \qquad \eta(\infty)=0.$$

We can then use the radii (semiaxes) at which $\eta=0.3$, 0.4, etc. This method is very convenient, since it is given by the profile itself, and it does not depend on the surface brightness intensity scale.

The immediate scientific goal of these surveys is to search for the minimal manifold of elliptical galaxies, that is, to find all the statistically significant variables which determine ellipticals as a family (see, e.g., Tonry & Davis 1981, Terlevich et al. 1981, or Efstathiou & Fall 1984). For this purpose we include both the spectroscopic (velocity dispersion, line strengths) and morphological information. A suitable statistical technique is the principal component analysis (PCA), which gives as a by-product all the correlations (and lacks thereof) present in the data. However, in order to apply this technique correctly, several conditions must be satisfied: (1) the selection effects must be well-known and under control; (2) the data set should be a multivariate Gaussian; (3) the relations between the input variables (measurements) and principal components must be linear; (4) the input variables must be scaled correctly. The last point is important: we think that the best scaling is by the measurement r.m.s. error-bars. The commonly used scaling by the sample variance is not necessarily the appropriate choice, in particular if there are selection effects present. In our version, the eigenvalues are automatically expressed as their statistical significance; if an eigenvalue $\lambda_i=n$, then it is "n-$\sigma$" significant. It is not clear that any of the previous studies which employed the PCA satisfied the above requirements. Our Mt. Hopkins survey suffers strong selection effects, and thus we do not take its PCA results very seriously. All that we

can say at this moment is that the ellipticals are at least a two-parameter family.
We will have to wait for the analysis of our Lick survey to be finished, before making a more quantitative statement.

Another scientific goal is to try to narrow the scatter of luminosity-velocity
dispersion and luminosity-line strength relations. The observed scatter exceed⌄s by
far the error-bars, and is presumably due to the presence of "hidden parameters".
If one can improve on these relations, the ellipticals (and lenticulars?) may become
competitive "standard candles" for the mapping of the Local Supercluster velocity
field, complementary to the use of Fisher-Tully relation for the spirals. So far,
we have had a very limited sucess in this venture -- the cause of the scatter is
still unknown.

Other applications include: investigation of 3-dimensional shapes of ellipticals
by using the ellipticity gradients and isophotal twists information; possible improvement in determination of M/L ratios; structure of galactic cores, through seeing
deconvolutions or modeling of seeing effects (Djorgovski 1983); dynamical modeling
of those galaxies for which there already exists detailed kinematical information;
cluster vs. field morphological differences, and mutual distortions in elliptical
galaxy pairs; a search for subsystems in ellipticals, e.g., incipient disks, dust
lanes, etc.; and possibly some other projects.

We will address here briefly two simple results from the Mt. Hopkins survey.
Figure 1 shows a plot of surface brightness measurements at two different fiducial
radii. If all the ellipticals had the same surface brightness intensity scale and
the same profile shape, all the points should fall together on this graph. If we
allow for a spread in intensity scale zero-points, this set will degenerate into a
45° line. If we also allow for the differences in profile shapes, there will be a
spread around that line. This is exactly what is observed. The spread in the surface brightness is some $4^m$-$5^m$; such surface density variations are of direct importance to the theories of elliptical galaxy formation (see, e.g., Silk 1983, or Wyse
& Jones 1984). Another implication is that the isophotal magnitudes provide a biased estimate of galaxian luminosity. The scatter around the 45° line is due to the
differences in profile shapes, and it amounts to ~$1^m$. This suggests that one cannot
rely on the use of "standard" magnitude curves of growth, a method commonly used to
determine galaxian luminosities and radii from the photoelectric aperture photometry
measurements.

Another simple fact is that the light profiles have different shapes. Therefore,
they cannot all be represented by any formula which does not allow for a shape parameter, e.g., Hubble or de Vaucouleurs laws. King models do have a shape parameter,

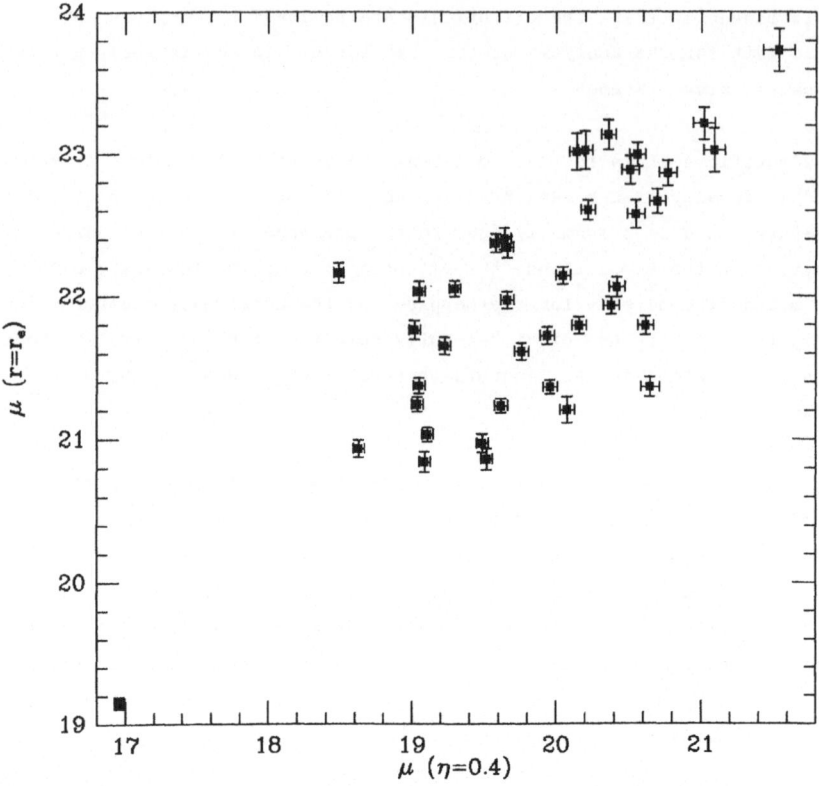

Figure 1. Relation between the isophotal surface brightness measured at two fiducial radii: $r_e$ (from the fits to de Vaucouleurs profile), and r($\eta$=0.4). The errorbars include all sources of error, including the errors of fiducial radii.

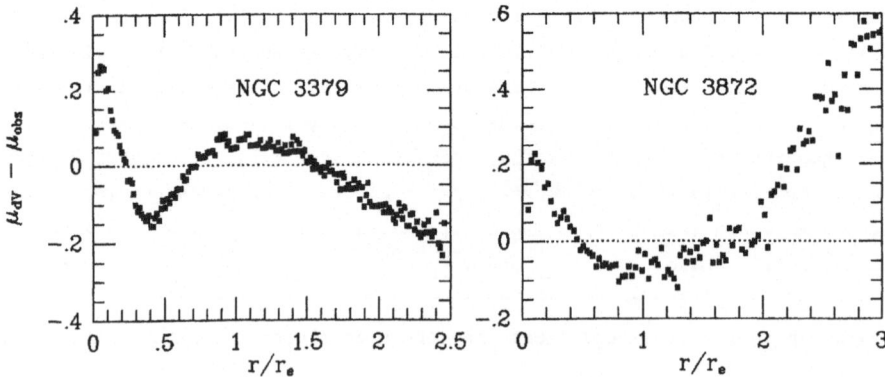

Figure 2. Surface brightness residuals from the best fit de Vaucouleurs profiles for two of the elliptical galaxies from our sample: NGC 3379 (left) and NGC 3872 (right). The residuals for other galaxies show a variety of trends, defying an easy parametrization. A similar situation exists for residuals from the best fit Hubble or King profiles.

but they do not fit all ellipticals either. There is little systematic effect here, although there is a slight tendency for more luminous galaxies to be fitted better by the $r^{1/4}$ law, and for the less luminous ones to be fitted better with King models (this was also indicated by Binggeli et al. 1984). When fitting the profiles, it is important to plot the residuals, as in Figure 2; almost anything looks linear when stretched on a log(SB) vs. $r^{1/4}$ plot! Since none of the common fitting profiles has any substantial physical backing in the terms of galaxy dynamics and structure, it would be a mistake to ascribe any physical meaning to deviations from such empirical and approximate formulas, as it was done ocasionally in the literature. All that can be concluded is that de Vaucouleurs, King, or Hubble profiles do not fit all the elliptical galaxies. Similarly, it may be misleading to state that a galaxy has a luminosity spike in the center, if it shows an excess of light over the seeing-convolved de Vaucouleurs profile. We just do not know how to describe adequately the light profiles for the whole family of elliptical galaxies, but we may learn soon.

This work was supported in part by the NSF grant AST81-18557 to M.D., and the University of California Moore Fellowship and the AAS travel grant to S.D.

REFERENCES:

Binggeli, B., Sandage, A., and Tarenghi, M. 1984, Astron. J. 89, 64.
Davis, M., Djorgovski, S., and Kent, S. 1983, Bull. Am. Astron. Soc. 15, 658.
Djorgovski, S. 1983, J. Astroph. Astron. 4, 271.
Djorgovski, S., and Davis, M. 1983, Bull. Am. Astron. Soc. 15, 932.
Efstathiou, G., and Fall, M. 1984, M.N.R.A.S. 206, 453.
Kent, S. 1984, Astroph. J. Suppl. 56, #1.
Petrosian, V. 1976, Astrophys. J. Lett. 209, L1.
Silk, J. 1983, Nature 301, 574.
Terlevich, R., Davies, R., Faber, S., and Burstein, D. 1981, M.N.R.A.S. 196, 381.
Tonry, J., and Davis, M. 1981, Astrophys. J. 246, 666.
Wyse, R.F.G., and Jones, B.T. 1984, Astrophys. J. 286, 000 (November 1st issue).

DISCUSSION:

D. Carter : Most isophotal magnitudes are quoted to isophotes of 26-27 mag/sq.arcs, much fainter than the characteristic surface brightness which you plot. Errors in the isophotal magnitudes will depend only upon the surface brightness profile below this level, which is impossible to determine from small field CCD images.

S. Djorgovski : It is true that our data do not go as far as $\mu \sim 26-27$, so we cannot make any explicite statements about the magnitudes measured at that level. However, I expect the behaviour to be similar to what we do observe at the lower radii. Even if one has two galaxies with identical profiles, but different intensity zero-points, isophotal magnitudes would provide a biased estimate of their luminosity difference. I think that it is better to use profile—determined metric apertures for photometry.

M. Capaccioli : It is clear that effective parameters in elliptical galaxies may have large errors, because of the peculiar nature of light profiles.

S. Djorgovski : I agree. I think that it is better to use the profiles directly, than to force a fit to an inappropriate formula.

R. Terlevich : You mention measurements of velocity dispersions and line strengths. What is the source of those measurements?

S. Djorgovski : At the moment, we are using the measurements from the CfA survey. Perhaps we will be able to use some better measurements in the future, e.g., those from your collaboration.

J.-L. Nieto : I guess it is clear to everyone now that any analytical formula (de Vaucouleurs' law or King models) cannot fit perfectly every elliptical profile. Two parameters are certainly not sufficient, but as soon as one more parameter is added in an analytical formula, all these parameters are correlated. Elliptical galaxies are more complicated than that: for instance, they may have photometric or dynamical subsystems.

S. Djorgovski : I agree completely. I think that to search for a single empirical formula describing all elliptical galaxies may be a vain venture. Perhaps some multidimensional scheme would work. In any event, it is important to regard the variety of radial light profile shapes globally, for the whole family of elliptical galaxies.

# A COMPARISON OF THE ELLIPTICITY PROFILES OF E AND SO GALAXIES

R. Michard
Observatoire de Nice
B.P. 139, F-06003 NICE CEDEX, FRANCE

Ellipticity profiles, that is the run of the ellipticity $\varepsilon$ of the isophotes against their semi-major axis, have been studied from original observations (Michard 1984) and litterature data (Liller 1960,1966, Leach 1981, Strom et al. 1978, di Tullio 1979). Special attention was given to galaxies of significant apparent flattening, i.e. flatter than SO(3) or E3.

For such SO galaxies the $\varepsilon$-profile is generally found to contain an initial increase, a flat maximum and a final decrease of $\varepsilon$ at low surface brightness (in a few cases the outer decrease is not observed). Such typical $\varepsilon$-profiles are supposedly due to the presence of two components, a disk and a more roundish bulge, with the flatter component dominating the geometry of the isophotes at intermediate radii.

It has been found that the same forms of $\varepsilon$-profiles, that we call "SO-like", are also typical of elliptical galaxies flatter than E3, while for more roundish objects, the $\varepsilon$-profiles show random (?) trends. The proportion of SO-like $\varepsilon$-profiles is 100% for a subsample of 14 well observed nearby flat E objects of the northern hemisphere, 73% for flat E objects in the Coma cluster (Strom et al. 1978), while it falls to 54% for such objects in di Tullio (1979) sample.

This observed analogy between SO and "flat E" galaxies may suggest that elliptical galaxies also are two components systems, although their flatter component is clearly less prominent (less bright ? less flat ? less large ?) than the disk of SO galaxies. Such a possibility has often been mentionned in the litterature.

And since the bulges of SO are clearly oblate structures, with their short axis nearly coincident with the axis of the disk, the present E-SO analogy may also favor oblateness as the normal form of E galaxies.

This work is published in more detail in Astron. Astrophys. (in press).

REFERENCES

Leach R., 1981, Astrophys. J., 248, 485
Liller M.H., 1960, Astrophys. J.,132, 306
Liller M.H., 1966, Astrophys. J.,146,28
Michard R., 1984, Astron. Astrophys. Suppl. Ser., in press
Strom K.M., Strom S.E., 1978, Astron. J. 83, 73
di Tullio G., 1979, Astron. Astrophys. Suppl. Ser. 37, 591

DISCUSSION

M. Capaccioli : How does the result of your fine analysis depend on the assumptions about the true stage of ellipticals ? In other words, what happens if they are triaxial or multisystem objects ?

R. Michard : My results suggest that E galaxies possibly are multi-component objects like SO. On the other hand the triaxiality, or other departures from axial symetry, is needed to explain the twists of isophotes observed in both E and SO galaxies.

S. Djorgovski : To pursue Dr. Capaccioli's point a little further, you can make some empirical tests in your data : for the triaxial systems, there should be a correlation between the ellipticity gradients and isophotal twist rates; on the other hand, for the elliptical (biaxial) systems with a disk component superposed, there should be significant $4\sigma$ residuals around the best fit elliptical isophotes.

R. Michard : I agree.

G. Galetta : In addition to the four ellipticity profile you mentioned, i.e. constant, increasing ellipticity with the radius, decreasing and peaked, there are some elliptical galaxies which show a fifth type of profile, with lower ellipticity at intermediate radius and more flat both in the inner than in the outer parts.

R. Michard : My point was to show that the "peaked" profiles become nearly the rule for galaxies flatter than E3. It is my feeling that this 5th type is scarce, and not found at all among the flat galaxies.

# THE LENTICULAR NGC 3115: A STANDARD FOR GALAXY PHOTOMETRY

Massimo Capaccioli[1], Enrico V. Held[1], and Jean-Luc Nieto[2]

1. Istituto di Astronomia, Università di Padova, Italy
2. Observatoires du Pic du Midi et de Toulouse, France

NGC 3115, a well known bright lenticular galaxy seen edge-on, is an ideal candidate for surface photometry studies due to its proximity and absence of close prominent companions. For these reasons it has been a classical target from the beginning of    extragalactic astronomy (see references in Davoust and Pence, 1982). NGC 3115 was also selected as a photometric standard for its morphological class by the Working Group on Galaxy Photometry and Spectrophotometry of IAU Commission 28. Strange enough, since then few photometric studies were published though the flourishing of the kinematical investigations (Rubin et al. 1980; Illingworth and Schechter 1982). Strom et al. (1977) sampled the main axes giving photoelectric V-magnitudes integrated through 30" apertures. Their results are in good agreement with a similar work by Miller and Prendergast (1968). Tsikoudi (1979) produced an extensive 2-D study based on B and V plates from several telescopes. She gave isophotal parameters and light profiles along the main axes from the very inner region to a distance where $\mu = 28$ B-mss. The reliability of this investigation, however, was questioned by Burstein (1979) who found a strong disagreement with other lenticulars studied by Tsikoudi.

Therefore we thought it worthwhile and timely to produce an independent 2-D study of NGC 3115 taking advantage of the best possible material. Our approach was along the line adopted by de Vaucouleurs and Capaccioli (1979) and Nieto (1983) for another standard galaxy, the elliptical NGC 3379. We report here on the progress of such a work, which so far concerns the B-band only.

In order to map the inner regions we took advantage of the superb resolution of the CFH telescope. The outer regions, instead, were covered with two pairs of very deep plates taken with the ESO and UK Schmidt telescopes respectively. More details are given in Table 1. Our calibrated plate material was scanned with the PDS microdensitometer of ESO at Garching and reduced with the INMP, a software package designed for surface photometry at the Padova Observatory (see Capaccioli et al. 1984). The zero point of the photometric scale was derived by comparing the sky-subtracted intensity maps of the CFH plates (Figure 1) to 39 integrated magnitudes through centered apertures listed in Longo and de Vaucouleurs (1983). The overexposed Schmidt plates were then shifted to match the CFH plates in the overlapping range.

The internal consistency was estimated by the ratios of pairs of intensity frames. It turned out that, in the useful interval ($\mu > 22$ B-mss) and irrespective to the telescope, Schmidt material compares within a maximum standard error of 0.2 mag down to the level of $\mu = 28$ B-mss. CFH plates are more difficult to compare due to seeing; however, in the range from 15.5 to 23 B-mss, they give a standard error of 0.05 mag. In the common range Schmidt and CFH profiles agree within 0.05 mag.

Once the position angle of the major axis was found by interpolating the outer isophotes with ellipses, light profiles along both main axes were produced for each

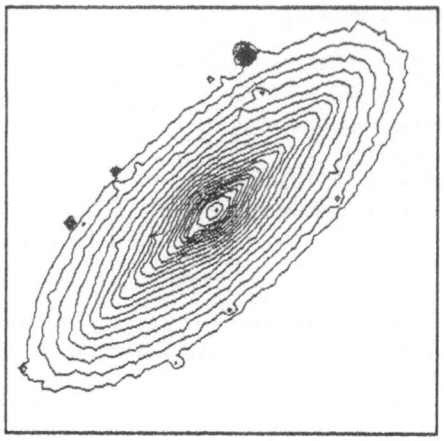

Figure 1. Isophotes of the central part of NGC 3115 from CFH plate No. 1002. The frame cover 7.1 arcmin. The thin structure of the disk is apparent.

TABLE 1. Plate Material for NGC 3115

| Plate No. | Telescope | Scale ("/mm) | Exposure (min) | Emulsion |
|-----------|-----------|--------------|----------------|----------|
| J 7558 | UK Schm. | 67.1 | 32 | IIIa-J |
| J 8323 | UK Schm. | 67.1 | 65 | IIIa-J |
| B 5479 | ESO Schm. | 67.5 | 90 | IIa-O |
| B 5491 | ESO Schm. | 67.5 | 90 | IIa-O |
| 1000 | CFH | 13.9 | 20 | IIa-O |
| 1001 | CFH | 13.9 | 5 | IIa-O |
| 1002 | CFH | 13.9 | 50 | IIa-O |
| 1003 | CFH | 13.9 | 5 | IIa-O |

intensity frame and then combined removing unreliable data. Results, presented in Figure 2 in graphical form, are not seeing corrected; they cover an interval of 12.5 mag corresponding to about 10 arcmin. Figure 3 reports the magnitude rediduals of our major axis light profile to that of Tsikoudi (1979) in the same band; no zero point adjustment was made. It is apparent a disagreement which becomes particularly prominent for all $\mu > 24$ B-mss; the tendency of Tsikoudi's values to be fainter than ours does not confirm quantitatively the trend found by Burstein (1979).

The same Figure 3 reports the comparison with Miller and Prendergast (1968) and Strom et al. (1977); for the latter study the constant value of $(B-V) = 1.0$ was applied. The magnitude residuals refer to integrated values through the right apertures computed on an average intensity map from the four Schmidt plates. There is a clear gradient in the residuals indicating that our data run fainter than the others outwards; we are still trying to interpret this difference.

We wish to thank the Directors of ESO and UKSTU for the generosity in providing one of us (MC) with the Schmidt material.

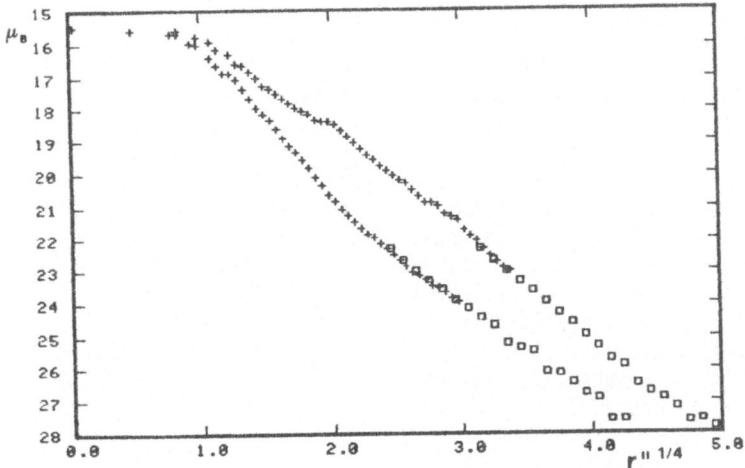

Figure 2. Combined B-band light profiles along the main axes of NGC 3115. Crosses are mean values from CFH plates uncorrected for seeing; squares are from Schmidt material (see Table 1).

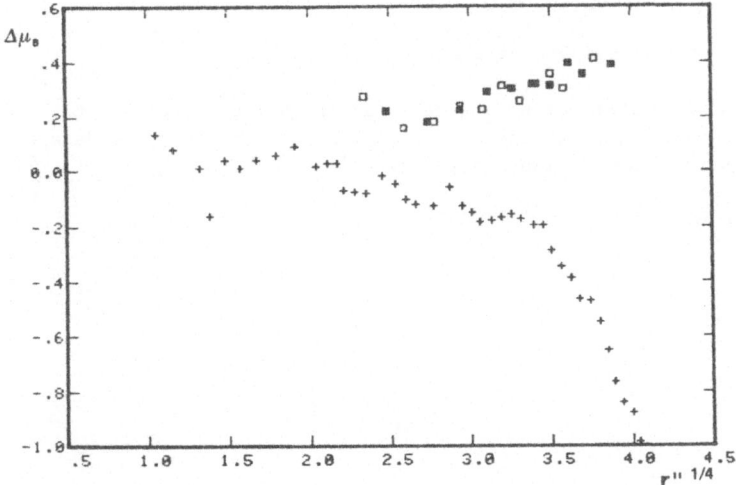

Figure 3. Magnitude differences of our major axis light profile to Tsikoudi (1979: crosses), Miller and Prendergast (1968: full squares) and Strom et al. (1977: open squares), plotted as a function of distance to center in arcsec. No zero point adjustment was made. Strom et al. data were transformed to the B-band by adding a constant (B-V) = 1.0 mag.

REFERENCES


Burstein, D. 1979, Astrophys. J. Suppl. Ser. 41, 435.
Capaccioli, M., Held, E.V. and Rampazzo, R. 1984, Astron. Astrophys. 135, 89.
Davoust, E. and Pence, W.D. 1982, Astron. Astrophys. Suppl. Ser. 49, 631.
de Vaucouleurs, G. and Capaccioli, M. 1979, Astrophys. J. Suppl. Ser. 40, 699.
Illingworth, G. and Schechter, P.L. 1982, Astrophys. J. 256, 481.
Longo, G. and de Vaucouleurs, A. 1983, Univ. Texas Monogr. Astr. No. 3
Miller, R.H. and Prendergast, K.H. 1968, Astrophys. J. 153, 35.
Nieto, J.-L. 1983, Astron. Astrophys. Suppl. Ser. 53, 383.
Rubin, V.C., Peterson, C.J. and Ford, W.K., Jr. 1980, Astrophys. J. 239, 50.
Strom, K.M., Strom, S.E., Jensen, E.B., Moller, J., Thompson, L.A. and
    Thuan, T.X. 1977, Astrophys. J. 212, 335.
Tsikoudi, V. 1979, Asrophys. J. 234, 842.

DISCUSSION

F. Schweizer : A deep IIIa-J plate in the ultraviolet taken of NGC 3115 at the KPNO
4 meters telescope shows faint structures that cross the major axis and look like
spiral arms. Do the two humps on your residual profile from an $r^{1/4}$ law along the
major axis coincide with such arms?

E. Held : We have not yet looked at that.

A. Lauberts : I assume that you propose this galaxy being a standard to be used by
the community. It would be very useful to publish the data in a standardized form,
like circular aperture simulation. Do you agree?

E. Held : Yes, I agree.

# DUST IN ELLIPTICAL GALAXIES - HOW OFTEN, HOW MUCH?

Elaine M. Sadler[1] and Ortwin E. Gerhard[2]

[1] European Southern Observatory, 8046 Garching bei München, F.R.G.
[2] Max-Planck-Institut für Astrophysik,
8046 Garching bei München, F.R.G.

ABSTRACT: We report results showing that dust-lanes are a common feature of nearby elliptical galaxies. About 40% of ellipticals have dust-lanes with diameters of a few kpc and dust masses of $10^4$-$10^6$ $M_\odot$. This result has some possible consequences for galaxy surface photometry, since dust-lanes too small to be seen on photographic plates may produce both "twisting" of inner isophotes and off-centre nuclei.

In this contribution, we are concerned with two specific questions about dust in elliptical galaxies. Firstly, is it common? Secondly, how much dust is likely to be present? As it turns out, the first question can be answered by a simple examination of sky atlas plates; while the second requires the use of calibrated surface photometry for which a CCD is very useful.

Many dust-lane ellipticals are well-known, Centaurus A (NGC 5128) being a famous example, and galaxies of this kind are interesting for a number of reasons. Dust-lanes are a useful probe of the true shape of ellipticals (van Albada, Kotanyi and Schwarzschild 1982; Tohline, Simonson and Caldwell 1982), the appearance of at least some suggests that they may be the result of a merger (Graham 1979, Schweizer 1980) and there is also a suggestive correlation of dust with both the presence and orientation of non-thermal radio sources (Kotanyi and Ekers 1979, Sadler 1982). It is therefore of interest to know whether we are dealing with something rare and exotic, or whether a dust-lane is a rather common feature of early-type galaxies.

An obvious way of estimating the frequency of dust-lanes is to take a complete sample of galaxies, count the number in which dust is visible and divide this by the total. This attempt fails, however, because of a strong selection effect due to resolution - we only recognize a dust-lane on a Schmidt plate if it is either very strong or very nearby. What we have done, therefore, is to select from a magnitude-limited sample of ellipticals (Sadler 1984) a size-limited subsample of bright, nearby galaxies with the criterion a.b > 4.0 arcmin$^2$ (where a and b are the major and minor axis diameters respectively). Of the 44 galaxies in this subsample, 10 (23%) have a dust-lane visible on the ESO(B) sky survey plates. Since a dust-lane is more easily seen when edge-on than face-on, and since we find empirically that we observe only dust-lanes with inclinations between 0° (edge-on) and 35°, i.e. about half the total available solid angle, we multiply the observed fraction of dust-lane galaxies by a factor of two to account for projection effects. This implies that about 40% of nearby ellipticals contain a dust-lane on the scale of a few kpc once the effects of resolution and projection are accounted for. Thus dust is a common, rather than an exotic feature of normal early-type galaxies.

CCD surface photometry in two or more colours provides a powerful method of revealing dust in galaxies. Figure 1 shows a B-R colour map of the central region of IC3370, a nearby elliptical galaxy. It reveals very clearly the presence of a small (10" × 2") nuclear dust-lane which is invisible on photographic plates. Figure 2 shows profiles in B and B-R along a line perpendicular to the dust-lane. The combination of absorption in B and steep reddening in B-R defines the "signature" of a dust feature and confirms that the reddening in figure 1 is due to absorption rather than, for example, an excess of red stars. Given the approximate size of the dust-lane and the mean absorption in B, $A_B$ (both of which can be measured from CCD photometry), and assuming the dust is similar to that in our own galaxy, it is possible to estimate a lower limit for the total mass of dust in nearby dust-lane ellipticals. This ranges from about $2 \times 10^7$ $M_\odot$ for Centaurus A to about $2 \times 10^4$ $M_\odot$ for IC3370. A dust-lane barely visible on a Schmidt plate has a mean absorption of 0.2 to 0.5 mag. in B, depending on its structure, and this corresponds to a dust mass of about $2 \times 10^4$ $M_\odot$ at 25 Mpc for a minimum dust-lane area of 50 arcsec$^2$.

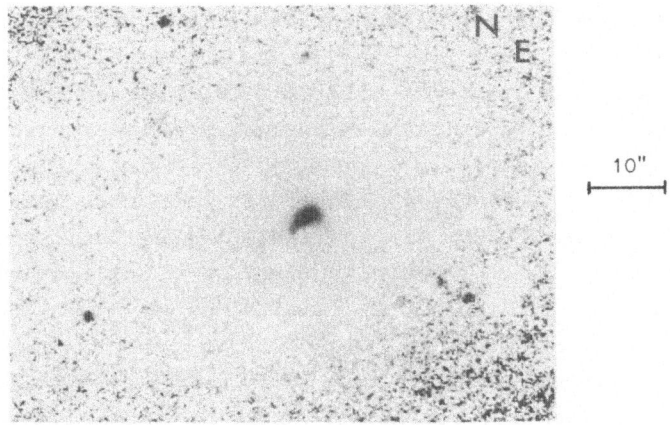

Figure 1:  B-R map of IC 3370, showing the central dust-lane.

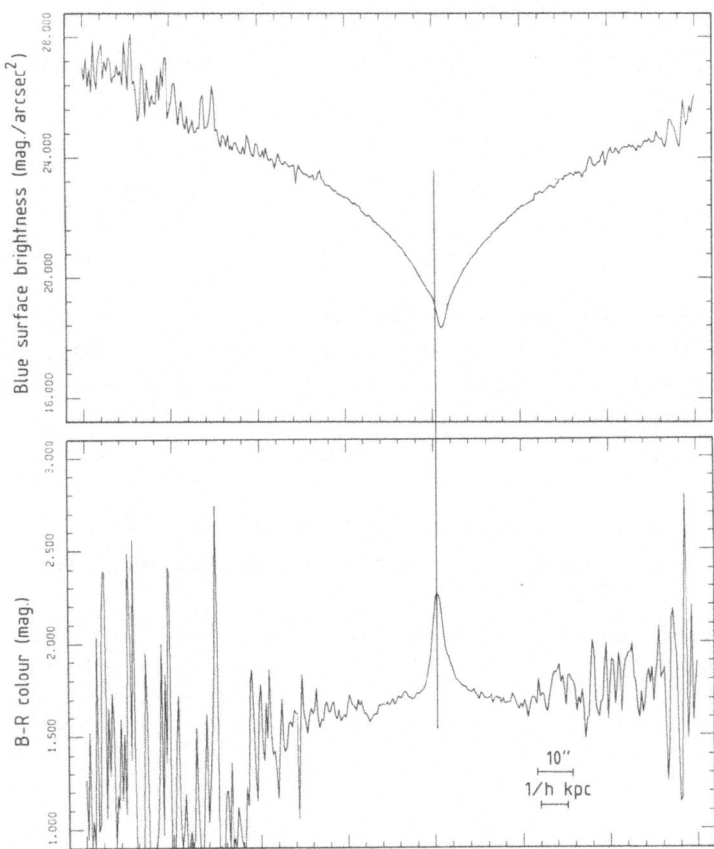

Figure 2:  Profiles in B and B-R through the nucleus of IC 3370. Note the asymmetry
in B at the position of the dust-lane (vertical line).

We therefore expect that many elliptical galaxies contain dust patches or dust-lanes too small to be recognized on a photographic plate, so that the above estimate of 40% could be substantially increased when higher resolution observations of ellipticals become available. In the case of IC3370, the central dust-lane produces both "twisting" of the central isophotes and an off-centre nucleus. True twisting and the effects of dust may be distinguished easily by two-colour photometry. Since most surface photometry of ellipticals has been done in a single band only, it remains to be seen how many "isophotal twists" are really due to dust. Surface photometry of elliptical galaxies has traditionally been done on the assumption that they are free of internal absorption, but in many cases this assumption must now be considered unrealistic.

A more detailed description of some of this work appears in another paper (Sadler and Gerhard 1985), where we also discuss classification problems for dust-lane galaxies.

REFERENCES:

Graham, J.A. 1979, Astrophys. J. 232, 60.

Kotanyi, C.G. and Ekers, R.D. 1979, Astron. Astrophys. 122, 301.

Sadler, E.M. 1982, Ph.D. thesis, Australian National University.

Sadler, E.M. 1984, Astron. J. 89, 23.

Sadler, E.M. and Gerhard, O.E. 1985, Mon. Not. R. astr. Soc., in press.

Schweizer, F. 1980, Astrophys. J. 237, 303.

Tohline, J.E., Simonson, G.F. and Caldwell, N. 1982, Astrophys. J. 252, 92.

Van Albada, T.S., Kotanyi, C.G. and Schwarzschild, M. 1982, Mon. Not. R. astr. Soc. 198, 303.

DISCUSSION:

G. Galletta : Is there always (in your sample) a reddening of the galaxy in correspondence with the dust-lane?

E. Sadler : The presence of a dust-lane seems to have only a small effect on the overall galaxy colours. Certainly there are galaxies such as NGC 5266 which have a strong dust-lane but are nevertheless bluer than many "normal" ellipticals.

R. Terlevich : Two questions: (1) Do you have a suggestion on how to proceed in order to detect extended and diffuse dust distribution? (2) What would be the typical amount of extinction expected in the IUE wavelength range?

E. Sadler : (1) The observations which I described can only detect dust which has a well-defined structure, such as a disk or ring. I think the detection of diffuse dust, if it exists, is difficult because unless the amount of dust is very large there will not be much reddening of the galaxy colours. (2) The mean absorption in the IUE wavelength range is about twice $A_B$, so the expected UV extinction would be 0.5-2 magnitudes in a typical dust-lane.

J.-L. Nieto : A detection rate of 40% is certainly a lower limit, since an improvement in the resolution of the data yields new dust-lane objects: for instance NGC 6251 (Nieto et al., 1982), NGC 6702 (Capaccioli et al., 1984, IAU Colloquium 78) and NGC 1199 (Nieto et al., this colloquium), observed for completely different purposes. Some galaxies, though, resist very much, like M87 for instance.

E. Sadler : I agree with you that better resolution will enable us to find more and more dust-lane galaxies.

M. Capaccioli : In our sample of UV spectra of E galaxies, there is only one object with the clear signature of dust, namely NGC 4374 (M84) which has a well-known and prominent dust-lane crossing the nucleus. The attempt to remove the 2300 Å feature with the standard extinction curves gives an extremely high UV excess for the object, even larger than that in M87 or NGC 4649.

E. Sadler : The differential absorption across the 2300 Å feature is about ½ $A_B$, i.e. 0.1-0.5 magnitudes for most of the dust-lanes we observe, and this is too small to produce a clear feature in the UV spectrum. We have made the assumption that the dust in elliptical galaxies is similar to that in our own (spiral) galaxy but this may not be correct since the physical conditions are quite different and this may affect, for example, the distribution of grain sizes. At least one dusty elliptical (IC4320; Warren-Smith and Berry, 1983) is known to have an extinction curve different from that of our galaxy.

S. Djorgovski : There have recently been two surveys for emission lines in the central parts of elliptical galaxies, with detection rates similar to yours (40%). Did you try to correlate the presence of dust with emission line activity?

E. Sadler : Yes, I was involved in one of those surveys. Our observations included twenty-one galaxies with dust-lanes, and all but one of these showed emission lines of Hα and/or [NII]. Therefore there seems to be a strong association of dust and ionized gas in these galaxies.

# LILLER'S PHOTOMETRY REVISITED

M. Capaccioli and R. Rampazzo
Institute of Astronomy, University of Padova
35100, Padova, Italy

## 1. Introduction

Correlations between photometric and geometric parameters are currently used to inves tigate the intrinsic shape of elliptical galaxies. Data come either from large com pilations (e.g. RC2; see applications in Marchant and Olson, 1979, and Olson and de Vaucouleurs, 1981), or from individual studies (e.g. Richstone, 1979). A classical photographic source is the B-band photometry published by M.H. Liller in 1960 and 1966 (hereafter LI). The two papers report luminosity profiles and isophotal parame ters for a total of 22 ellipticals in the Virgo cluster. Successive investigations overlapping part of Liller's sample (van Houten, 1961; Fraser, 1977; King, 1978 = KI; Burstein, 1979; Benacchio et al., 1984 = BCDSS) have ascertained the good quality of the geometrical data (see also Leach, 1981), but have put into light the presence of a systematic error in the photometric scale. This error increases steadily with the surface intensity and is probably due to an incorrect calibration. In order to recov er such a large body of information (wich refers to several galaxies at the same distance) we attempted a correction of the photometric scale on the ground of KI's study, which is of high photometric quality and has the largest overlap with Liller's sample.

We have established the systematic departure of Liller's photometric scale from KI's directly from the mean trend of the residuals $<\Delta\mu>$(LI-KI) for nine galaxies in com mon, evaluated as a function of the surface brightness $\mu$(LI) (see Fig. 1). Residuals from individual galaxies were arbitrarily shifted in order to minimize the global dis persion in the range 23÷24 B-mss. The mean deviations were then applied to transform $\mu$(LI) into the corrected surface brightness $\mu$(LC) = $\mu$(LI) - $<\Delta\mu>$.

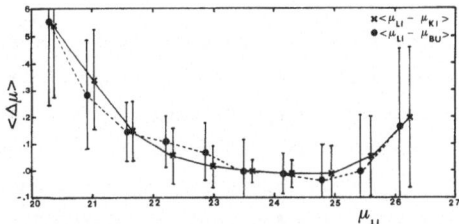

Fig. 1. Mean residuals $<\Delta\mu>$ of Liller's photometric scale from King (1978 = KI) and Burstein (1979 = BU) plotted versus $\mu$(LI).

New zero point calibrations of the corrected data were also computed using the photoe lectric integrated magnitudes listed by Longo and de Vaucouleurs (1983). The photoelec tric sources with largest discrepancies were discarded; the rejection criterion was not always straightforward due to the paucity of data and to their large dispersion (greater than $\pm$ 0.1 mag in the mean). In particular, four galaxies had no photoelec

tric data and were removed from the sample, which thus reduced to 18 objects. The ze
ro point correction has a mean value of -0.02 ± 0.14 mag and never exceeded 0.3 mag.
The actual values of the corrected luminosity profiles (hereafter indicated by LC)
were then used to compute the axis ratios log R(25) = log a/b at $\mu$ = 25 B-mss. The
effective radii $R_e$ and the total magnitudes $B_T$ were derived instead from $r^{1/4}$ interpo
lation of the corrected "equivalent" profiles. Our corrections do not modify the sur
face brightness levels of the outer luminosity profile ( $\mu$ > 23.5 B-mss). Here the
only variation is introduced by the change of the zero point if any. In this latter
case the values of log R(25) are modified only if the ellipticity is not constant.
The effective radii and the total magnitudes are clearly affected by the correction
applied to the inner luminosity profiles and to the zero point.

## 2. Comparison with other sources

To evaluate the reliability of our correction we compared the LC values of log R(25),
of $R_e$ and of total magnitude of each galaxy (see Table 1) to those found in the
literature.

TABLE 1. LI corrected parameters

| NGC | log R(25) | $R_e$ | $B_T$ | NGC | log R(25) | $R_e$ | $B_T$ |
|-----|-----------|-------|-------|-----|-----------|-------|-------|
| 4168 | 0.11 | 39.7 | 12.15 | 4473 | 0.20 | 34.5 | 11.21 |
| 4261 | 0.08 | 56.6 | 11.17 | 4478 | 0.07 | 10.1 | 11.59 |
| 4339 | 0.03 | 27.3 | 12.05 | 4486 | 0.12 | 77.0 | 9.67 |
| 4365 | 0.13 | 73.0 | 10.55 | 4551 | 0.10 | 13.8 | 12.66 |
| 4374 | 0.00 | 78.8 | 10.10 | 4552 | 0.06 | 53.1 | 10.81 |
| 4387 | 0.16 | 7.9 | 12.66 | 4564 | 0.26 | 18.3 | 11.61 |
| 4406 | 0.22 | 148.7 | 9.62 | 4621 | 0.09 | 54.2 | 10.68 |
| 4458 | 0.03 | 17.2 | 12.73 | 4649 | 0.11 | 90.3 | 9.72 |
| 4472 | 0.10 | 123.2 | 9.21 | 4660 | 0.11 | 11.8 | 11.97 |

RC2 represents the most complete source for what concerns the axis ratio. The compar
ison with the corrected values is shown in Fig. 2; any possible trend of the residu
als from the regression line is masked by the large dispersion. Values of log R(25)
interpolated from KI tables compare well to LC except the case of NGC 4261, where LC
is however in agreement with RC2. BCDSS values tend to be slightly larger than LC's.
The largest discrepancy is for NGC 4486; for this galaxy BCDSS give 0.17 ( $\varepsilon$ =1-b/a=
0.3) against 0.12 of LC, KI gives 0.10 and RC2 the much smaller value 0.03 ( $\varepsilon$ = 0.1).
LC values of the effective radius were also compared to data found in the literature
(Fig. 3), namely KI (values extracted from the actual light profiles), BCDSS, RC2,
and Fraser (1977). In the mean $R_e$(LC) values are 1% larger than KI (comparison over
nine galaxies) and 4.5% smaller than BCDSS (comparison over six galaxies), with a
dispersion less than 25%. Fraser's (1977) study, whose quality has already been debat
ed (BCDSS), has 8 galaxies in common with LI; $R_e$ values (not plotted in the figure)

are 35% smaller than in LC, with a dispersion of 26%.

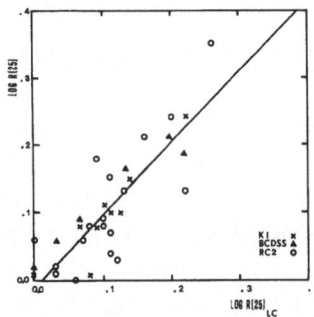

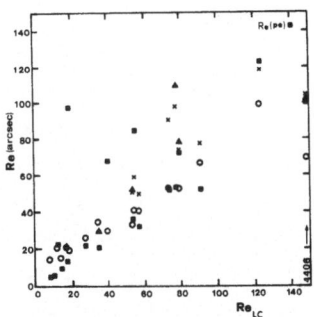

Fig. 2. Corrected values of Log R(25)=
log a/b compared to those from RC2, KI,
BCDSS. The solid line: log R(25)$_{RC2}$ =
1.085 log R(25)$_{LC}$ - 0.012 is the best
fit of LC to RC2.

Fig. 3. Values of the effective radii
from LC plotted versus the same quanti
ties from KI, BCDSS, RC2 and from aper
ture photometry. The symbols are the same
as in Fig. 2. Note the large dispersion.

The effective radius of NGC 4406 plays a key role in fixing  the slope of the regres
sion lines ( R$_e$(LC) versus R$_e$(KI) and versus R$_e$(BCDSS)). If this object is removed
from our sample (Capaccioli et al. 1984 have already remarked the very poor consisten
cy among photographic measurements of this bright galaxy), the agreement of LC with
KI and BCDSS improves significantly (see Fig. 3). The effective radii were derived
from RC2 with the formula log 2*R$_e$ = log A$_e$ - 0.27 (log R(25))$^2$; according to Olson
and de Vaucouleurs (1979), it improves that given in the catalogue. RC2 values are 4%
smaller than R$_e$(LC) in the mean and show a clear trend.

For a further test we also attempted to estimate the effective radii from the photoelec
tric  aperture photometry using the following technique. First we assumed that each
galaxy profile could be approximated by an r$^{1/4}$ law, then, accounting for elliptici
ty  variations, we computed the best value of the slope parameter (from which R$_e$(pe)
is soon obtained) by comparing galaxy model to the actual measurements. The uncertain
ty on the results was estimated empirically to be about $\pm$5%. The photoelectric data
tend to be 24% smaller than R$_e$(LC) in the mean. Altough embedded in the large disper
sion there is a gradient among the smaller values; it becomes even clearer if we cal
culate the relative resuduals with respect to KI, BCDSS and RC2. Among the other NGC
4261 deserves some attention. This galaxy is possibly a misclassified lenticular (Bar
bon et al. 1984); whatever the case, the estimate of its effective radius varies quite
significantly from author to author.

Total magnitudes were derived from the integration of the r$^{1/4}$ law (de Vaucouleurs,
1961) best fitting the LC data; no correction for internal and galactic absorption
has been applied. B$_T$(LC) total magnitudes compare well with the B$_T$(RC2) in the entire
set although systematically brighter (<B$_T$(LC)-B$_T$(RC2)> = -0.15 $\pm$ 0.19).

3. Summary and conclusions

The main goal for a correction to Liller's photometric scale was to gain useful param eters for statistical analysis such as log R(25), $R_e$ and $B_T$ in a sample of galaxies at the same distance.

In the mean the values obtained for log R(25) seem in fair agreement with those of RC2 although the dispersion is very large.

The values of the effective equivalent radii were compared to four sources: KI, BCDSS, RC2 and data from aperture photoelectric photometry. The first two sources agree in side 5%, with a dispersion less than 25%. RC2 values shown a similar agreement in the mean but with a larger dispersion. The comparison with aperture photometry is charac terized by a large gradient and significant deviations for small values of $R_e$(pe). This is likely due to the fact that the attempt to extract $R_e$ from integrated data is hopeless.

Total magnitudes are systematically brighter than those listed in RC2 with a mean difference $<B_T(LC)-B_T(RC2)> = -0.15 \pm 0.19$.

## References

Barbon, R., Benacchio, L., Capaccioli, M., Rampazzo, R. 1984, Astron. Astrophys, 137, 166.

Benacchio, L., Capaccioli, M., De Biase, G., Santin, P., Sedmak, G. 1984, in prepara tion (BCDSS).

Burstein, D. 1979, Ap. J. Supp. Ser., 41, 435.

Capaccioli, M., Davoust, E., Lelièvre, G., Nieto, J.-L. 1984, in "Astronomy with Schmidt-Type Telescopes", ed. M. Capaccioli, Reidel Publ. Co., p. 379.

de Vaucouleurs, G. 1961, in "Problems in Extragalactic Research" , ed. McVittie, MacMillan, p. 3.

de Vaucouleurs, G., de Vaucouleurs, A., Corwin, H. Jr. 1976, "Second Reference Catalo gue of Bright Galaxies", Texas Univ. Press, Austin (RC2).

Fraser, C. W. 1977, Astron. Astrophys. Supp., 29, 161.

King, I. R. 1978, Ap. J., 222, 1 (KI).

Leach, R. W. 1979, Ap. J., 248, 485.

Liller, M. H. 1960, Ap. J., 132, 306 (LI).

Liller, M. H. 1966, Ap. J., 146, 28 (LI).

Longo, G., de Vaucouleurs, A. 1983, " A General Catalogue of Photometric Magnitude and Colors in the U, B, V System", Texas Univ. Press, Austin.

Marchant, A. B., Olson, D. W. 1979, Ap. J., 230, L157.

Olson, D. W., de Vaucouleurs, G. 1981, Ap. J., 249, 68.

Richstone, D. O. 1979, Ap. J., 234, 825.

van Houten, C. J. 1961, B.A.N., 16, 1.

# CONSTRUCTING DISTRIBUTION FUNCTIONS FOR GALAXIES WITH APPLICATION TO M87

Andrew Newton

Department of Theoretical Physics, 1 Keble Road, Oxford OX1 3NP, England

1. **Introduction.** Even a complete set of photometric and kinematic measurements of a galaxy do not uniquely determine the mass distribution and dynamics of the observed tracer particles. Presented with this indeterminate problem, the dynamicist has two choices: He can either (i) make some assumption about the nature of the orbits of the material in the galaxy, (for example that the velocity dispersion tensor is isotropic) and infer the mass distribution which is implied; or (ii) assume some form for the mass distribution (for example that the mass-to-light ratio is independent of radius) and determine the distribution of orbits required by the observations. In this paper I consider the latter option.

The distribution of orbits in a galaxy is described by the phase-space distribution function $f(\underline{x},\underline{v},t)$. Our task is to construct a distribution function which yields surface brightness and kinematics compatible with the observations. On physical grounds the distribution function must satisfy $f \geqslant 0$ over the whole of the phase space. It may prove to be impossible to construct a distribution function satisfying this constraint: the assumed mass distribution may then be ruled out. For a general stellar system, the construction of a distribution function is a formidable task. In certain special cases, in which analytical expressions exist for the integrals of motion, the problem is greatly simplified. In section 2, I discuss a technique that has been developed (Newton & Binney 1984; hereafter NB) for producing distribution functions for spherical systems, and its application to the nearly-spherical elliptical M87. In section 3, I discuss the extension of this technique to a wider class of stellar systems.

2. **Spherical Galaxies.** Complete observations of a spherical galaxy yield at every projected radius R, the surface brightness $\Sigma(R)$, and line-of-sight velocity dispersion $\sigma_v(R)$. Binney & Mamon (1982) (hereafter BM) showed how to obtain from the observations the unique three-dimensional profiles of stellar density $\nu(r)$, the radial and tangential components of velocity dispersion, $\sigma_r(r)$ and $\sigma_t(r)$ and the value of M/L that are required by the hypothesis of a constant mass-to-light ratio. These profiles are obtained by solving the stellar hydrodynamic equation. BM pointed out that to be physically acceptable, they must satisfy $\sigma_r^2 \geqslant 0$, $\sigma_t^2 \geqslant 0$. In addition, they must derive from a distribution function which satisfies $f(\underline{x},\underline{v}) \geqslant 0$ for all $(\underline{x},\underline{v})$. By Jeans' theorem, the distribution function of a steady-state spherical galaxy is a function of the stellar energy E, and angular momentum $L = |\underline{L}|$. We wish to obtain an $f(E,L)$ satisfying this condition which yields moments

$$\nu(r) = 4\pi r^{-2} \int dE \int LdL \ f(E,L) \ v_r^{-1}(E,L) \qquad (1)$$

$$\nu\sigma_r^{2}(r) = 4\pi r^{-2} \int dE \int LdL \ f(E,L) \ v_r^{+1}(E,L) \qquad (2)$$

close to the data. Our method for obtaining approximate solutions for f(E,L) from
equations (1) and (2) is based on an iterative deconvolution scheme developed by Lucy
(1974). Lucy's algorithm obtains approximate solutions to a one-dimensional problem.
Recognising that the moments $\nu$ and $\nu\sigma_r^{2}$ are projections onto real space of the distribution function in phase space, a generalisation of Lucy's method is possible to obtain
an approximate f(E,L) from equations (1) and (2). The outline of the method may be
summarised as follows:

(1) Derive $\nu$ and $\sigma_r$ profiles from the observations using the method of BM.

(2) Choose an initial guess distribution function $f_o(E,L)$ by, for example, least
squares fitting some analytical model to the $\nu$ and $\sigma_r$ profiles.

(3) Generate improved estimates of the distribution function $f_n(E,L)$ by a generalisation of Lucy's iteration scheme.

As shown by Lucy, the non-negativity of $\nu$, $\sigma_r^{2}$ and f is preserved at each iteration.
Tests of the algorithm are described in detail in NB. These tests demonstrate the
ability of the algorithm to reproduce a distribution function f(E,L) from a set of
pseudo-data and illustrate its behaviour in a case where no non-negative f is compatible
with the pseudo-data.

Application to M87. Young et al. (1978) and Sargent et al. (1978) (hereafter YS) made
photometric and spectroscopic measurements of the core of M87. The surface brightness
shows a sharp peak in the central arcsecond, while the line-of-sight velocity dispersion
rises from 250 km/s at 1 arcminute to 350 km/s at 1 arcsecond. Starting from the
assumption of isotropic velocity dispersion, YS derived a mass distribution for M87 and
inferred the presence of a supermassive black hole in the nucleus with a mass of about
$5\times10^{9}$ $M_\odot$. In the YS model, the luminosity spike arises from a dense star cluster drawn
in by the gravitational field of the central mass concentration.

By contrast BM examined the consequences of assuming a constant mass-to-light ratio in
M87. They obtained $\nu\sigma_r$ and $\sigma_t$ profiles which satisfy the conditions $\sigma_r^{2} \geqslant 0$, $\sigma_t^{2} \geqslant 0$.
In the BM picture, the luminosity cusp arises from a central kernel of high velocity
stars on predominantly radial orbits. This structure dominates the light out to about
one arcsecond and is surrounded by the main body of the galaxy: a relatively diffuse
envelope of lower velocity stars on weakly radial orbits. The work of BM left open the
question of whether it was possible to populate the stellar orbits in such a way as to
produce this two-component dynamical structure: in other words, whether it was possible
to construct an f(E,L) which yields the BM $\nu$ and $\sigma$ profiles.

To answer this question, we applied our algorithm to the BM profiles of M87. To approximate the two-component structure of M87, we chose as our initial guess a linear combination of Michie (1963) and Kent-Gunn (1982) distribution functions:

$f_o(E,L) = \alpha\, f_M(E,L) + \beta\, f_{KG}(E,L)$ with parameters obtained by least squares fitting to the BM profiles. The distribution function generated at the twelfth iteration, $f_{12}(E,L)$ yielded moments $\nu_{12}(r)$ and $(\nu\sigma_r^{\,2})_{12}(r)$. These moments were projected onto the sky, a point luminosity source (V = 16.75 mag) was added and a seeing convolution was performed along the lines of de Vaucouleurs & Nieto (1979). The resulting brightness

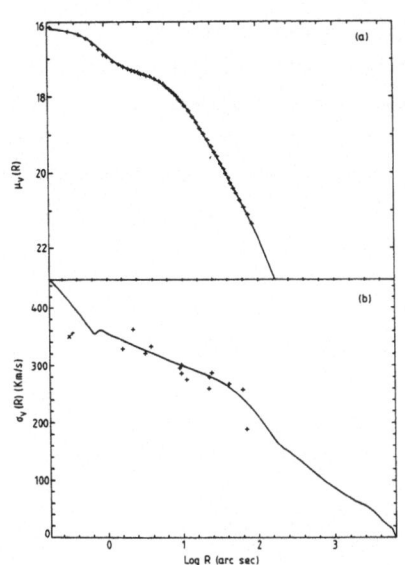

profile (fig.A) fits the YS observations to within 0.04 magnitude, while the line-of-sight velocity dispersion (fig.B) lies within the 10% scatter of the YS data. We conclude that the assumption of a constant mass-to-light ratio in M87 is compatible with the YS observations. Higher resolution photometry will be necessary to determine the mass distribution present in the nucleus of M87.

3. Extension to non-spherical systems.
In reality, of course, very few galaxies exhibit a high degree of spherical symmetry. It is natural to consider whether the algorithm described in the previous section can be extended to other stellar systems. In principle, in any system in which we can calculate the observable moments of the distribution function, it should be possible to set up a Lucy-type iteration scheme for constructing a distribution function from the observed moments.

(a) Photometry and (b) kinematics of M87. The plus signs are measurements made by YS, the full lines are the fit produced by the NB algorithm. In the lower panel, the cross is a measurement made by Dressler (1980).

These calculations are particularly simple for flat, axisymmetric systems where the distribution function depends on E and $L_z$. Up to second order, there are three independent moments: the surface density $\Sigma(R)$, the radial velocity dispersion $\sigma_r(R)$ and the rotation velocity $\overline{v}_\phi(R)$. They derive from $f(E,L_z)$ via equations very similar to (1) and (2) and so an iteration scheme similar to that implemented in the spherical case could be used to produce an $f(E,L_z)$.

In fact, any potential which admits a set of isolating integrals with analytical expressions $I(\underline{x},\underline{v})$ allows us to calculate the Jacobian for the transformation from $(\underline{x},\underline{v})$ space to integral space analytically, which greatly simplifies the calculation of the moments. A wide class of such potentials, are those in which the Hamilton-Jacobi equation separates - the "Stäckel" potentials - which includes spherical and axial symmetry as degeneracies. These potentials have a long history in mechanics, and have been discussed most recently in stellar dynamics by Lynden-Bell (1962) and de

Zeeuw (1984).  It should be straightforward to develop an iteration scheme for constructing distribution functions for triaxial galaxies with separable potentials.

Finally, in a general triaxial potential the Hamilton-Jacobi equation does not separate.
Although numerical orbit calculations demonstrate the existence of isolating integrals
in addition to the Hamiltonian (e.g. Schwarzschild 1979), they are not known in closed
form as $I(x,v)$.  However, these orbit calculations yield the contribution to the density (and other moments) of each individual orbit.  Schwarzschild (1979) used the techniques of linear programming to produce distribution functions for triaxial galaxies as
a set of occupation numbers of individual orbits.  Models for axisymmetric (Richstone,
1980) and spherical (Richstone & Tremaine, 1984) galaxies have subsequently been constructed using Schwarzschild's method.  Lucy's iterative deconvolution method can be
used to construct distribution functions in all these cases, with the advantage over
linear programming that smooth, more realistic distribution functions are produced, as
discussed by Lucy (1974) and NB.

References

Binney, J.J. & Mamon, G.A., 1982. M.N.R.A.S., 200, 361.
Dressler, A., 1980. Astrophys. J., 240, L11.
Kent, S.M. & Gunn, J.E., 1982. Astron. J., 87, 945.
Lucy, L.B., 1974. Astron. J., 79, 745.
Lynden-Bell, D., 1962. M.N.R.A.S., 124, 95.
Michie, R.W., 1963. M.N.R.A.S., 125, 127.
Newton, A.J. & Binney, J.J., 1984. M.N.R.A.S., 210, 711.
Richstone, D.O., 1980. Astrophys. J., 238, 103.
Richstone, D.O. & Tremaine S.D., 1984. Preprint.
Sargent, W.L.W., Young, P.J., Boksenberg, A., Shortridge, K., Lynds, C.R. & Hartwick,
    F.D.A., 1978, Astrophys. J., 221, 731.
Schwarzschild, M., 1979. Astrophys. J., 232, 236.
de Vaucouleurs, G. & Nieto, J.-L., 1979. Astrophys. J., 230, 697.
Young, P.J., Westphal, J.A., Kristian, J., Wilson, C.P. & Landauer, F.P., 1978.
    Astrophys. J., 221, 721.
de Zeeuw, P.T., 1984. Ph.D. thesis, Leiden.

# RADIO STUDIES OF PECULIAR GALAXIES AT 10.7 GHZ

M. Urbanik[1], R. Gräve[2], U. Klein[3]

1) Astronomical Observatory, Jagiellonian University
Fort Skala, ul. Orla 171, 30-244 Krakow, Poland

2) Max-Planck-Institut für Radioastronomie
Auf dem Hügel 69, D-5300 Bonn 1, FRG

3) Radioastronomisches Institut der Universität Bonn
Auf dem Hügel 71, D-5300 Bonn 1, FRG

Four objects from Arp's (1966) "Atlas of Peculiar Galaxies" have been mapped at 10.7 GHz using the Effelsberg 100-m radio telescope of the MPIfR. The galaxies observed represent two distinct classes of interaction with their companions. NGC 520 and NGC 3448 represent highly disturbed objects with long optical tails. The first of them most likely forms a close pair of objects (Stockton and Bertola, 1980) while the second one is interacting with the nearby galaxy UGC 6016.

The two other objects in the present study, NGC 2276 and NGC 3627, are probably interacting galaxies within looser groups. Tidal perturbations of their spiral patterns are accompanied by highly peculiar HI distributions and kinematics (Peterson and Shostak, 1974; Haynes et al., 1979).

The integrated 10.7 GHz flux densities of all four galaxies were used to compute the average disk brightness temperature and the radio index R defined as the ratio of the radio to optical luminosities (e.g. Klein, 1982).

Estimates of the contributions from central sources have been made using extrapolations of available high resolution data at other frequencies. The 10.7 GHz flux densities have also been combined with published total intensity data at lower frequencies to obtain the integrated radio spectra of all four objects. Details of the observing procedure and the above mentioned computations will be given elsewhere (Urbanik et al. (in preparation)).

Contour maps of NGC 3448, NGC 2276, and NGC 3627 are presented in Figures 1a,b, and c respectively. The integrated radio spectra of all objects observed are shown in Figure 1d. Table 1 summarizes the most important integral parameters of the galaxies studied.

The two morphologically similar galaxies NGC 520 and NGC 3448 were found to have

284

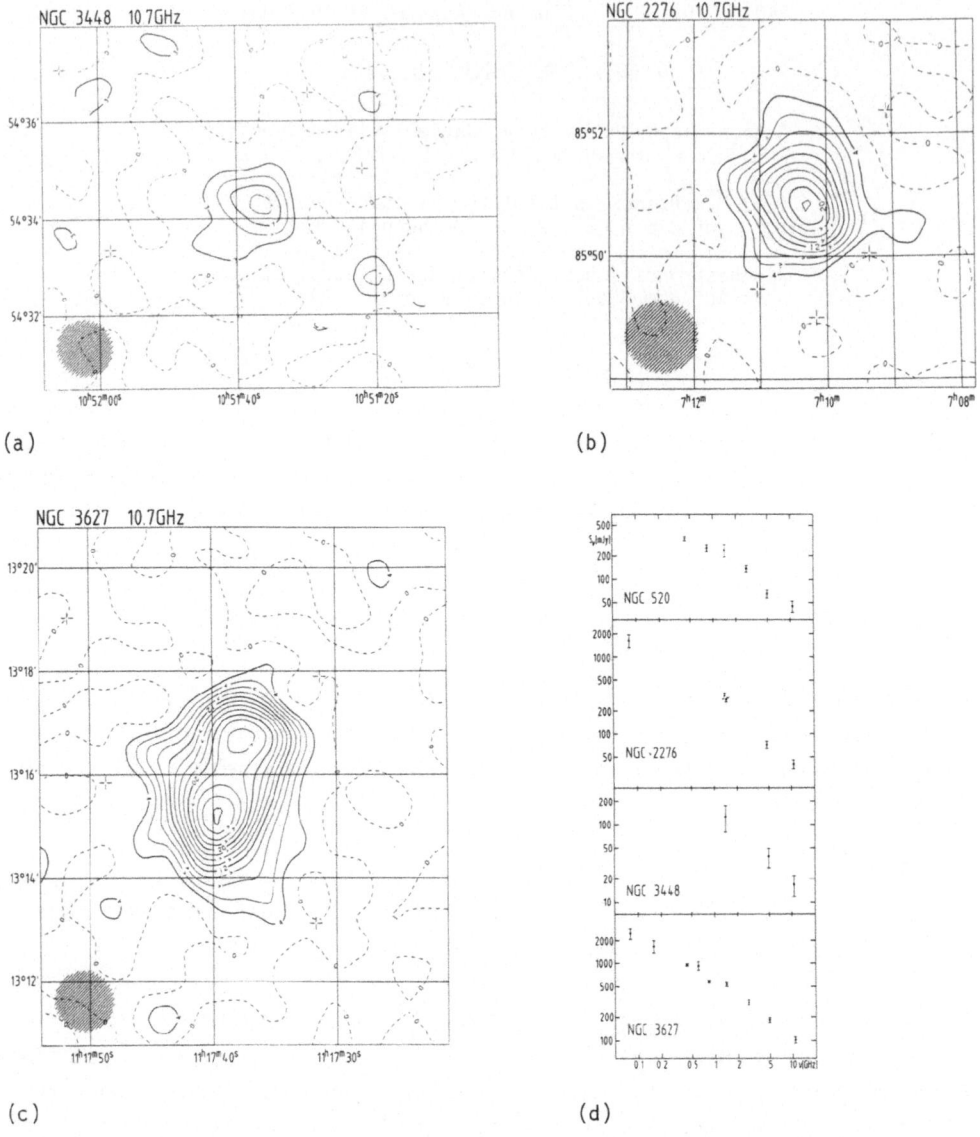

(a)

(b)

(c)

(d)

Fig. 1a,b,c:  Contour maps of NGC 3448, NGC 2276 and NGC 3627 at 10.7 GHz.  Zero contours are dashed.  The first solid contour corresponds to 3 mJy/beam area for NGC 3448 and to 4 mJy/beam area for the other galaxies.  The contour interval is 2 mJy/beam area in all cases.

Fig. 1d:  Integrated radio spectra of all galaxies in this study.

Table 1: Integral parameters of the observed galaxies as derived from the 10.7 GHz observations

| Name | Distance[+] [Mpc] | $S_{10.7}$ [mJy] | Radio index | $\dfrac{S_{nucleus}}{S_{tot}}$ | $T_b disk$[++] [mK] | Spectral index $\alpha$[+++] |
|---|---|---|---|---|---|---|
| NGC 520 | 23 | 45 ± 7 | 18.7 | ∿ 100% | — | -0.62 ± 0.06 |
| NGC 3448 | 11 | 17 ± 5 | 7.1 | < 40% | — | -1.0 ± 0.5 |
| NGC 2276 | 23 | 40 ± 5 | 15.0 | ∿ 2% | 24.4 | -0.68 ± 0.05 |
| NGC 3627 | 7 | 100 ± 10 | 5.2 | — | 6.1 | -0.65 ± 0.03 |

[+] based on $H_0 = 100$ km s$^{-1}$ Mpc$^{-1}$

[++] corrected to face-on

[+++] $S \propto \nu^{\alpha}$

very different radio properties. The radio emission from the first of them is completely dominated by a powerful unresolved central source. No extended disk emission has unambiguously been detected in that object. In contrast to that, NGC 3448 is a rather weak radio emitter. Its emission at 10.7 GHz is clearly resolved and extends slightly in the direction of the optical disk. The contribution from the central source as extrapolated from Reakes' (1979) data does not exceed 40%. Although the optical and UV data (Bertola et al., 1984) imply a rather high star formation rate in NGC 3448, its overall level of emission as measured by the radio index R does not exceed that of late-type galaxies.

The Sc galaxy NGC 2276 was found to have a very bright radio disk with only a small contribution from the central source. The average disk brightness temperature and the radio index are more than three times higher than those of the radio-brightest nearby spirals. This result agrees well with that of Condon (1983).

The disk brightness of the Leo Triplet spiral NGC 3627 is comparable to that typical for Sb galaxies. Its radio structure (Fig. 1c) is dominated by a bright bar-like feature with two prominent emission peaks close to the ends of the optical bar. The maxima of the radio emission coincide roughly in position with large groups of HII regions found there by Hodge (1974).

The radio spectra of all four objects can be reasonably well fitted by a simple power law ($\alpha \cong -0.7$), with the flux densities available for NGC 3448 being rather poor.

The 10.7 GHz observations of the four objects presented above have shown two features possibly related to their interactions: The powerful central source in NGC 520 and a bright radio disk in NGC 2276. Excess of radio emission from central regions is known to be common in interacting galaxies (Hummel, 1981; Stocke, 1978). NGC 2276 also possesses a bright optical disk. The UBV colours of the galaxy and its Hα emission (Kennicutt and Kent, 1983) suggest enhanced star formation activity.

The unusual radio structure of NGC 3627 does not require interactions in order to be understood. Bright bar-like radio features are known to exist in some barred spirals (e.g. Sancisi and Ekers, 1978; Klein and Emerson, 1981; Klein et al., 1983). A high degree of gas compression accompanied by enhanced star formation close to the bar ends is also predicted by hydrodynamical calculations (e.g. Huntley, 1980; Tubbs, 1982). Strong projection effects, particularly important in the regions discussed here, can also produce similar effects (Urbanik and Otmianowska (in preparation)). Thus NGC 3627 does not show any radio peculiarities obviously related to its inter-action. A similar conclusion can be drawn for NGC 3448.

## References

Arp, H.C.: 1966, "Atlas of Peculiar Galaxies", California Institute of Technology, Pasadena
Bertola, F., Casini, C., Bettoni, D., Galletta, G., Noreau, L., Kronberg, P.P.: 1984, Astron. J. 89, 350
Condon, J.J.: 1983, Astrophys. J. Suppl. 53, 459
Haynes, M.P., Giovanelli, R., Roberts, M.S.: 1979, Astrophys. J. 229, 83
Hodge, P.W.: 1974, Astrophys. J. Suppl. 27, 113
Hummel, E.: 1981, Astron. Astrophys. 96, 111
Huntley, J.M.: 1980, Astrophys. J. 238, 524
Kennicutt, R.C., Kent, S.M.: 1983, Astron. J. 88, 1094
Klein, U.: 1982, Astron. Astrophys. 116, 175
Klein, U., Emerson, D.T.: 1981, Astron. Astrophys. 94, 29
Klein, U., Urbanik, M., Beck, R., Wielebinski, R.: 1983, Astron. Astrophys. 127, 177
Peterson, S.D., Shostak, G.S.: 1974, Astron. J. 79, 767
Reakes, M.: 1979, Monthly Notices Roy. Astron. Soc. 187, 509
Sancisi, R., Ekers, R.D.: 1978, Astron. Astrophys. 67, L21
Stocke, J.T.: 1978, Astron. J. 83, 348
Stockton, A., Bertola, F.: 1980, Astrophys. J. 235, 37
Tubbs, A.D.: 1982, Astrophys. J. 255, 458

# THE VELOCITY FIELD OF NGC 1365

Steven Jörsäter

European Southern Observatory

Karl-Schwarzschild-Str. 2,

D-8046 Garching bei München

The velocity field of the barred spiral NGC 1365 has been mapped using a large number of optical spectra. The data acquisition and reduction is described elsewhere (Jörsäter et al. 1984b). This galaxy also shows an abnormal velocity field in high excitation lines such as [OIII] and [NeIII] (Jörsäter et al. 1984a) but only the normal velocity field as derived from the Balmer lines and [OII]λ3727 will be discussed in the following.

Fig. 1 shows a contour plot of this galaxy. The inclination is 55° and the PA of the line of nodes is 228° (both found from faint outer isophotes). Isovelocity contours are shown in Figs. 2 and 3, and the basic rotation curve in Fig. 4. The innermost 7" has a solid body rotation (straight parallel contours in Fig. 3) but outside this region quite a lot of non-circular motion is present. The rotation curve in Fig. 4 consists of three straight segments, a steep slope in the solid body region, $V = 270$ km s$^{-1}$ outside of 80" (deprojected to the plane of the galaxy) and a straight line joining these two regions. It is obviously not possible to define the rotation curve very precisely, but fortunately its exact drawing has only a small effect on the conclusions.

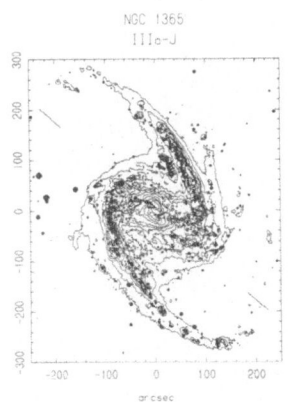

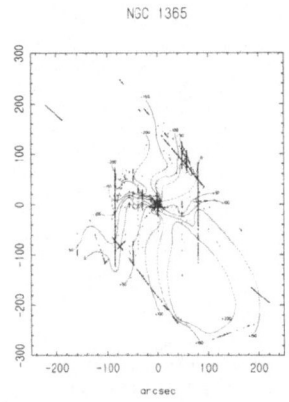

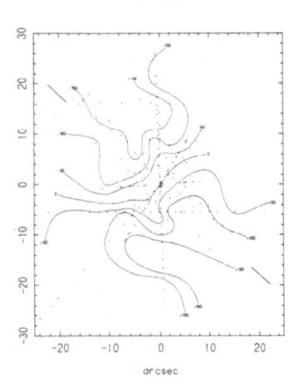

Fig. 1: Isophotes.

Fig. 2: Isovelocities. Dots denote measured points.

Fig. 3: An enlargement of the central portion in Fig. 2.

288

Figs. 5 and 6 show the velocity residuals obtained when subtracting the above velocity curve. Several interesting features are seen: The spiral arms appear to be receding from the nucleus of the galaxy indicating that the arms are located outside of corotation. This is in agreement with theoretical results that bars should end at corotation. Large non-circular motions are seen in the bar in the region of the dust lanes. Velocity residuals along the line of sight change by over 200 km s$^{-1}$ over as little as 15" implying a very strong shock. The observed velocities seem to agree well with those predicted by Roberts et al. (1979).

The bright gas in the nuclear region forms a tiny nuclear spiral inside the bar (Fig. 6). The arms of this spiral are expanding with velocities of the order of 120 km s$^{-1}$ in the plane of the galaxy.

REFERENCES

Jörsäter, S., Lindblad, P.O., Boksenberg, A. 1984a. Astron. Astrophys. in press.

Jörsäter, S., Peterson, C.J., Lindblad, P.O., Boksenberg, A. 1984b. Astron. Astrophys. Suppl. in press.

Roberts, W.W., Huntley, J.M., van Albada, G.P. 1979, Astrophys. J. 233, 67.

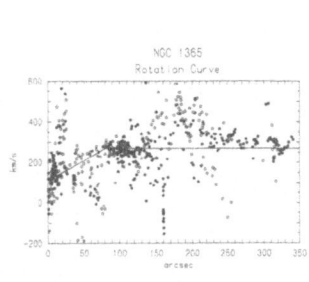

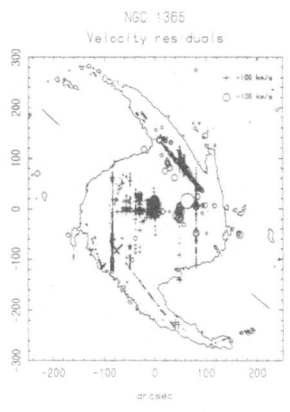

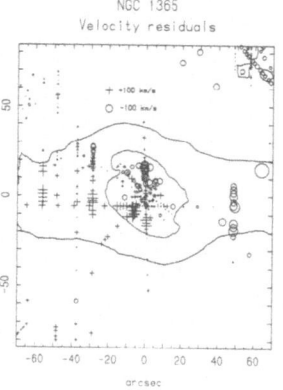

Fig. 4: The rotation curve. Filled circles are velocities from SW, open are from NE.

Fig. 5: Velocity residuals. The size of the symbols is proportional to the magnitude.

Fig. 6: An enlargement of the central portion in Fig. 5.

MORPHOLOGY AND RADIAL VELOCITY OF GALAXIES

USING COLOUR-COLOUR DIAGRAMS

G.I. Thompson, K. Nandy

Royal Observatory Edinburgh

Edinburgh, EH9 3HJ  U.K.

It is known that the position of a galaxy in a colour-colour diagram contains information on its morphology and radial velocity. We are starting a program to exploit this fact, to measure colours of faint galaxies on Schmidt and AAT plates, preferably to the plate limit, to add another dimension to work that has already been done on the distribution of faint galaxies. This note addresses one of the problems involved, that of measuring the colour of faint galaxies on photographic plates. Although small, the image has structure. It is ill-defined and its extent depends on exposure level so a magnitude measured in an area defined by an isophote or threshold above sky may measure different areas of the galaxy on different plates and so falsify the colour obtained. If it were certain that all the plates had the same point spread function it would be meaningful to measure intensity within a specified area to produce a true colour. The suggestion under test is to try to make this possible by finding convolutions, which, when applied to the better resolved plates gives their images the same shape as those on the poorest resolved.

To this end it is necessary to use stars to find the convolution and use them to show that this method of measuring colour works. A set of three U.K. Schmidt plates, approximately B, V and R, were selected and digitized by COSMOS. Figure 1a shows the B,V,R diagram of the stars in which magnitudes were obtained in the usual way to a low threshold above sky. The sloping line represents the photoelectric B, V, R relation for stars; it is not intended as a fit, but is used as a guide to the eye and reproduced in all other diagrams.

The images on the V plates had the widest point spread function and convolutions were found to spread the B and R images to match. No details of this process are given here. Essentially the stars are used to measure the P.S.F. for each plate, and the convolved images of stars used to check the correctness of the convolution and that it does not change total magnitude. In all diagrams in Figure 1, except 1a, the V magnitude is measured in the unconvolved image while the B and R magnitudes are measured in the convolved images. In Figure 1b the V magnitude used is identical to that used in Figure 1a. The perimeter of the measured area on the

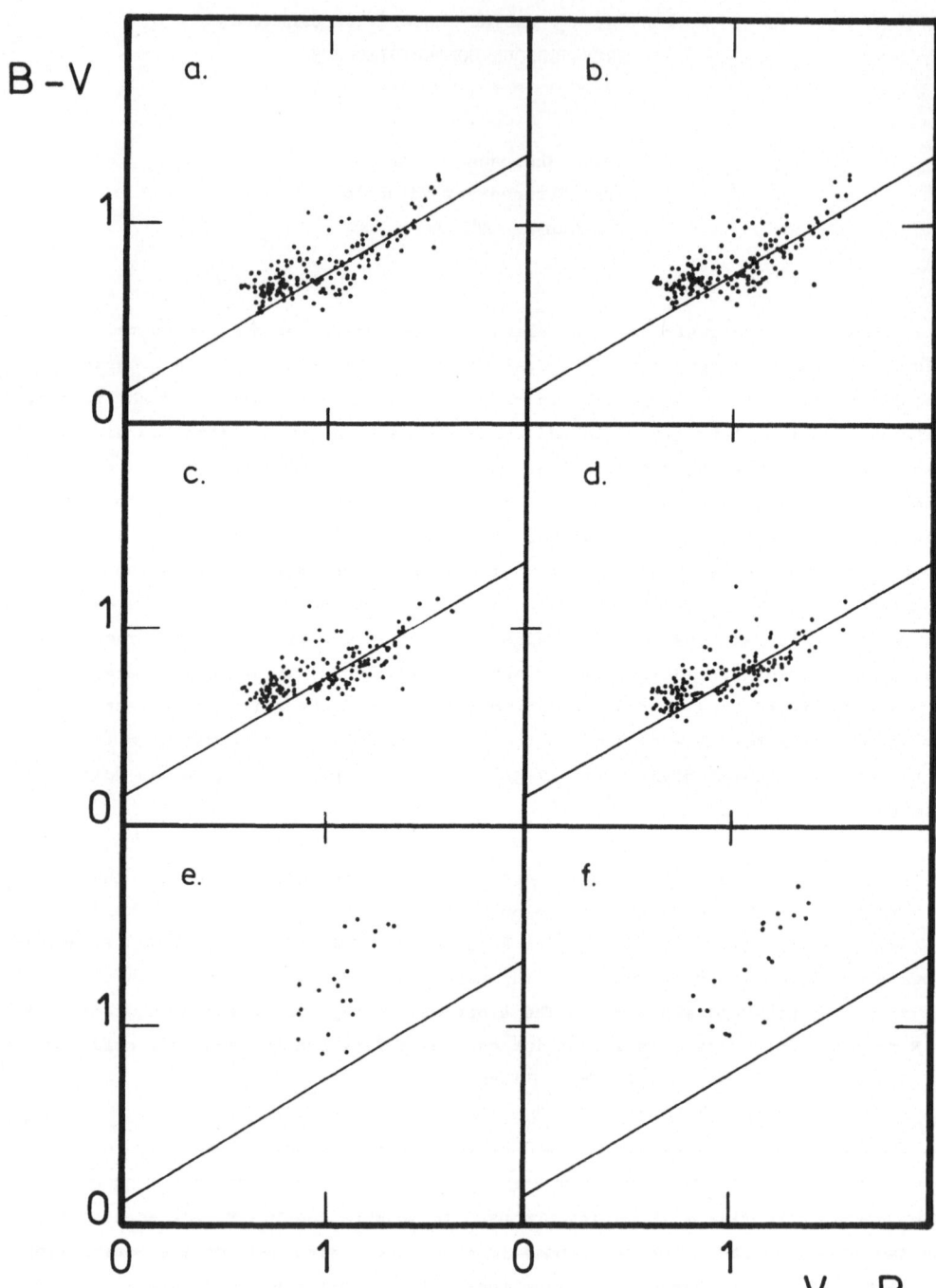

Figure 1. (B-V). v. (V-R) for stars and galaxies

V plate was transformed to the B and R plates and intensity within that area measured in the convolved images. This diagram shows that the process makes little difference to the colours. It is possible to pick out patterns of points which reproduce from Figure 1a to Figure 1b showing that these stars have hardly moved at all. This, however, merely proves that the convolution is correct. In Figures 1c and 1d the perimeter of the measured area was defined at a signal-to-noise ratio of 1.5 and 2.5 respectively in the unconvolved V image. That perimeter was tranformed to the B and R convolved images. The enclosed areas were smaller than the original by factors of two to four. Of course the magnitudes so measured are meaningless, but evidently the colours are essentially correct.

It is this property, that it is possible to measure colour using just bits of images, that is important. There is no point in using it for single stars, but for non-point sources such as galaxies it will be of value. We have tried applying it to some probable E0 galaxies on the test plates used here. Figure 2 shows the expected distribution of colour according to the computations of Coleman et al (1980). Although the colours of the test plates are not exactly equivalent to B,V,R, this diagram serves for orientation. In Figure 1e the signal-to-noise ratio on the perimeter of the measured area on the V image is two, in Figure 1f it is three. There is a factor two difference in the measured area but the colour distribution is essentially the same.

Comparison with Figure 2 suggests that this set of galaxies has radial velocites from z = 0.0 to 0.2. They lie in the magnitude range 18 to 19.5. It was not possible to go deeper with this set of plates due to the limited penetration of the V plate. However, it is evident that a knowledge of colour can improve studies of the clustering properties of galaxies.

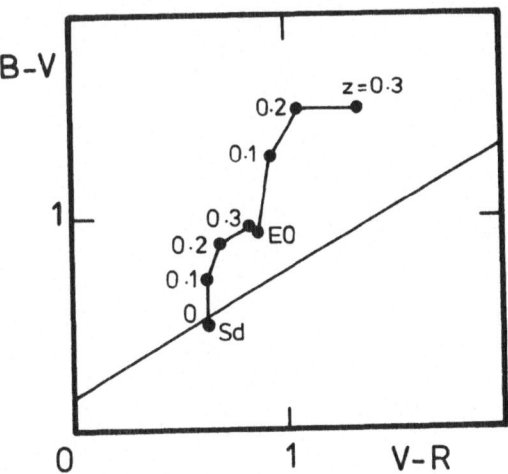

Figure 2. (B-V). v. (V-R) for galaxies from Coleman et al. (1980).

The main point is that this method, artificially matching the point spread functions of the plates and measuring intensity within the same metric area is the best for obtaining colours of faint galaxies. It will be extensively used in our program to measure large numbers of galaxies on Schmidt and AAT plates.

References

Coleman, Wu and Weedman, 1980, Astrophy. J. Suppl., <u>43</u>, 393.

Discussion

<u>G. Bruzual</u> : How faint can you get using this method on Schmidt plates? Have you thought about applying the same technique to 4m plates? In that case you should be able to reach 23rd - 24th magnitude that would be much more relevant to cosmology. Are you aware of Koo and Kroh's work?

<u>G. Thompson</u> : I agree the cosmology is at the limit of large telescopes rather than of Schmidts and I am working with AAT plates at present. The interpretration of the position of a galaxy in a colour-colour diagram is ambiguous, so I see the usefulness of this work is in forming colour-colour diagrams of clusters where all the galaxies are assumed to have the same z and the diagram is interpreted as a distribution in morphology. There is need for a large number of colour-colour diagram at low redshift for comparison with those at high redshift.

# THE Z-DISTRIBUTION OF RADIO CONTINUUM EMISSION IN NGC 891

R.J. Allen and F.X. Hu
Kapteyn Astronomical Institute, Postbus 800,
9700 AV Groningen, The Netherlands.

Determination of the three-dimensional distribution of Galactic radio emission is severely hampered by our unique observing position inside the Galaxy. Despite the impressive detail and high accuracy of the recent radio surveys of the Galactic background (e.g. Haslam et al., 1982), the models are in certain important aspects only weakly constrained by the data. In particular the distribution of emission at heights of 1 kpc and greater above the plane of the Galaxy is very uncertain; it is confused by the contribution from the thin disk material which is closer, stronger, and unfortunately along the same line of sight. However, accurate knowledge of the distribution of intensity and spectral index at great heights above the Galactic plane is essential for an understanding of the origin and propagation of cosmic rays. For these reasons, observations of external galaxies which are seen nearly edge-on play a key role in advancing our knowledge about these problems. NGC 891 is an especially interesting choice in this regard, because it is thought to resemble our Galaxy in many respects. Since the first detection of z-extension in the radio continuum of NGC 891 (Allen, Baldwin and Sancisi, 1978; hereafter ABS), this galaxy has been repeatedly observed with the Westerbork Synthesis Radio Telescope (WSRT) at various frequencies and with increasing sensitivity and resolution.

We have now combined the most recent observations at 6-cm made by Strong and Allen (1982) with the highest resolution 21-cm data obtained from new HI synthesis observations by Sancisi (private communication), in order to derive the best possible results on the z-distribution of the radio continuum emission in NGC 891. The resulting cleaned radio maps have been smoothed to a resolution of 20" × 20" (gaussian) and averaged in strips 200" long, parallel to the major axis and symmetric about the minor axis of the galaxy. The observed strip distributions (averaged for both sides of the galaxy) are shown in Figure 1 for the two wavelengths. The immediate impression is that these plots show evidence for two components; a narrow component confined to within z less than about $0.5$, and underneath this a broader component which extends to the limit of the observations at about 2'. Such a decomposition was earlier proposed by ABS based on the older 6-cm WSRT observations which reached to about 1' above the plane. The present results con-

firm and extend this suggestion, and also indicate that the two components have
different spectral indices; the thin component is relatively stronger at 6-cm.
Another new piece of information is that the spectral index is constant above
0.5; Figure 1 shows that in this range the ratio of surface brightness at 6 and
21-cm is constant to within the noise.

These two features of the observations have been put into a simple two-component
model for the z-distribution of the radio emission from NGC 891 as shown in Figure
2. The thin disk component is a gaussian with FWHM = 10" (700 pc for an assumed
distance of 14 Mpc) and a spectral index of -0.47. The thick disk component is
constant in brightness from z = 0 to about z = 10", and decreases exponentially
thereafter with a scale height of 26" = 1.8 kpc; its spectral index is -0.95 and
constant with increasing z. This simple model gives a surprisingly good fit to the
observations; after smoothing with a 20" gaussian beam and subtracting form the
original data, the residuals are everywhere within a few percent of the signal or
a few times the rms noise.

It is interesting to compare these results with the best models for the
Galaxy which are currently available in the literature. The z-distribution for the
Galaxy as it would be seen edge-on by a distant observer has been derived from
model fits to the 408-Mhz survey data by Phillips et al. (1981). Their best-fit
model is sketched in Figure 3. The correspondence with our model for NGC 891 in
Figure 2 is quite striking. However, Phillips et al. caution that the precise
shape above 1 kpc is very uncertain; they have also obtained reasonable fits for
models which would predict simply a constant surface brightness from 1 to 9 kpc.
The difference in scale height between what one may estimate from Figure 3 as
about 5 kpc for the Galaxy, and the much more accurate value of 1.8 kpc from
Figure 2 for NGC 891 may therefore not be significant.

The results for NGC 891 in Figure 2 do not suffer from model-dependent
assumptions, and furthermore provide valuable new information on the constancy of
the spectral index with z. What does this tell us about cosmic rays? One thing is
that current models are all based on the assumption that cosmic rays are produced
in the thin disk near z = 0 and diffuse to higher z, perhaps with some reaccelera-
tion, but ultimately losing energy along the way. The spectral index in all of
these models must steepen with increasing z; the only way to avoid this is to have
very short propagation times (cf. for example Strong and Allen, 1982). A more
detailed confrontation between theory and observations is now under way.

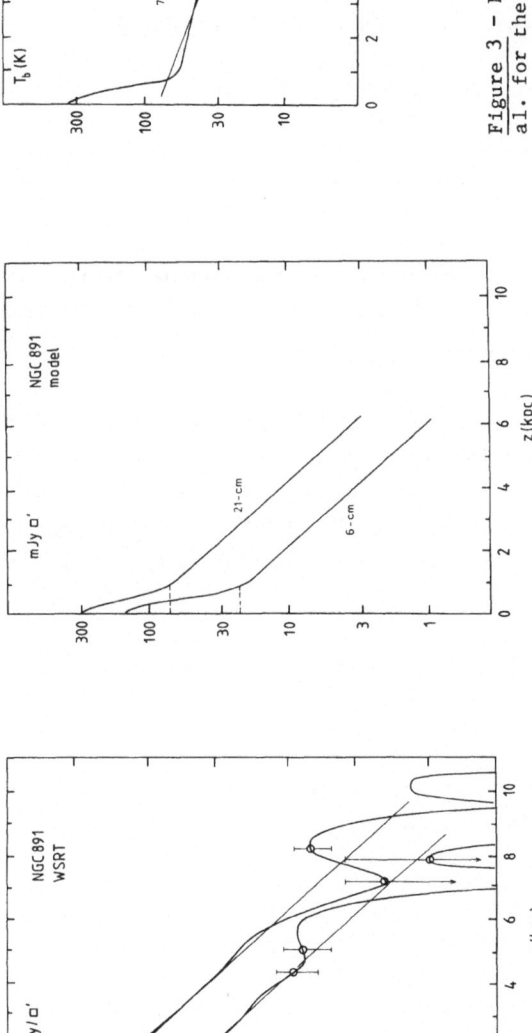

Figure 1 - z-distributions of radio surface brightness in NGC 891 as observed with the WSRT at 21-cm and 6-cm wavelength, with a 20" beam. Error bars with a total length of twice the rms noise are shown at a few selected positions.

Figure 2 - Two-component model for the intrinsic z-distribution of radio surface brightness in NGC 891, calculated for 21-cm and 6-cm wavelengths.

Figure 3 - Model by Phillipps et al. for the z-distribution of radio surface brightness for the Galaxy at 75-cm wavelength.

References

Allen, R.J., Baldwin, J.E., Sancisi, R., 1978, Astron. Astrophys., 62, 397.

Strong, A.W., Allen, R.J. 1982, in "Proceedings of the 17th International   Cosmic
    Ray Conference", Vol. 2 (C.E.N. Saclay; Gif-sur-Yvette, France), p. 248

Haslam, C.G.T., Salter, C.J., Stoffel, H., Wilson, W.E., 1982, Astron. Astrophys.
    Supp. 47, 1

Phillips, S., Kearsey, S., Osborne, J.L., Haslam, C.G.T., Stoffel, H., 1981,
    Astron. Astrophys. 103, 405.

Discussion

A. Bosma: This work nicely shows that improvement in observational data can change
the questions one addresses in the interpretation. Previous data had it that the
spectral index increases with Z for several galaxies.

F.X. Hu: From the observations at 2.8 and 49 cm of NGC 891 and NGC 4631 Klein et
al. (1984, Astron. Astrophys. 133, 19) also found: "The steepening of spectral
index in Z is less than previously believed." We also have found a good correlation between U light and radio 21 and 6 cm emission of NGC 891 in the range of Z =
30" to 70". All of these results will provide more constraints on the models than
before.

# SOME METHODS FOR PHOTOMETRY OF GALAXIES IN CLUSTERS

A. Bijaoui*, G. Mars*, E. Slezak* et O. Lefèvre**

* Nice Observatory, B.P. 139 - 06003 NICE CEDEX, FRANCE

** Pic du Midi and Toulouse Observatories - 14, av. E. Belin

31400 TOULOUSE, FRANCE

We have developped some digital methods to process photometric observations on clusters of galaxies for photographic and electronographic emulsions. These methods allow to obtain global magnitudes and radial profiles of all detectable objects.

## REAL TIME ANALYSIS

Using a PDS microdensitometer linked to a PDP 11/40-computer, we have adapted a COSMOS-like technique of plate analysis.

The machine reads line by line the plate. We choose a spot size and sampling steps of 20μm x 20 μm for a Schmidt Palomar plate. We compute the skybackground with an iterative 3-σ rejection for a sample of lines. We achieve a real-time fields labelling (1) allowing to extract some global characteristics of the objects formed by the connected pixels for which the density is greater than the threshold above the background. A smoothing can be achieved to detect the faintest objects.

In fact, taking into account the computing time, we have extracted only the position and density of the maximum of the field, the area and the integral of the density.

Using the classical diagram log (Flux)/log(Area) (2) we obtain a Bayesian classifier allowing to extract the non-stellar objects.

On a Schmidt Palomar plate, in the Coma field, we have detected about 10 000 galaxies on 12 000 objects. After a visual inspection, the rejection rate is small (7%).

## OBJECT BY OBJECT ANALYSIS

Taking into account the large amount of time needed to process a plate, and the faint number of studied objects, we have also developped a method using a scan object/object, followed by a full automated process to compute the background, reduce into intensity, smooth, determine the ellipticity parameters of a given isophote level, integrate radially, eliminate superposed stars and obtain global magnitudes.

An application is running now on a sample of 600 galaxies in A194 cluster.

## GLOBAL OFF-LINE ANALYSIS

Using a Systime 8/750, we have developped an analysis method using a global scan of a large field (typically 4096 x 4096 pixels). This method mixes the possibilities of the real time analysis (detections segmentation, recognition, exhaustivity) with the advantage of the object/object analysis (smoothing, photometry, structural parameters, radial profile).

We are processing the Coma field Schmidt Palomar plate with this method, building classifiers related to radial profiles (3) or moments (4).

This method has been also applied on electronographic plates taken with the CFH telescope for faint clusters of galaxies (0.2 > z > 0.05). Here we have introduced a preprocessing software (flat field and distorsion reduction). This allows to process 4 clusters in 2 colors, with about 300 objects/plate.

With the use of mini-computers with virtual memory and large size disks, the problem of large fraction of Schmidt plate processing is solved by off-line analysis.

A laterly, off-line analysis allows to extract other structural features (separation of components, gaussian fitting, model parameters,...).

## REFERENCES

(1) A. Rosenfeld, 1969 in "Picture Processing by Computer", p. 82, Academic Press, New York

(2) MacGillivray, H.T. and Dodd, R.J. : 1979, M.N.R.A.S. **186**, p.69

(3) Butchins, S.A. : 1982, Astron. Astrophys. **109**, p. 360

(4) Jarvis, J.F. and Tyson, J.A. : 1979, Proc. SPIE v. 172, p. 422

## DISCUSSION

<u>H. MacGillivray</u> : What pixel size do you use to scan your Schmidt plates and how much CPU time do you require on your Vax 11/750 to process the data from a whole Schmidt plate ?

<u>A. Bijaoui</u> : We have scanned the plate with a 20 x 20 μm spot and steps. We needed about 30 mn of CPU to process a 4096 x 4096 frame (smoothing, skybackground mapping and fields detection and measurements).

# HI OBSERVATIONS OF THE VIRGO CLUSTER GALAXIES

W.K. Huchtmeier

Max-Planck-Institut für Radioastronomie
Auf dem Hügel 69, 5300 Bonn 1

Global properties of a complete sample of Virgo cluster galaxies are derived containing all spiral and irregular galaxies brighter than $m_B=14.2$ within $10^o$ of M87 and a large number of galaxies brighter than $m_B=14.95$ in that area. Data were taken from the HI-catalogue of Huchtmeier et al., 1983 and completed with new observations with the 100 m radiotelescope at Effelsberg. 177 of the 456 galaxies are detected (39%) to a limit of the order of $10^8$ $M_\odot$ assuming a distance of 21.8 Mpc for the Virgo cluster (Sandage and Tammann, 1976).

Noncluster comparison samples (i.e. nearby galaxies) in the blue magnitude system (Shostak, 1978, Li and Liu, 1981, and Bottinelli et al., 1982) were compared with the nearby sample ($v_o$ <500 km/s, named hereafter KKT, Kraan-Korteweg and Tammann, 1979). The agreement for the distance - independent HI-mass to luminosity ratio ($M_H/L_B$) is very good. The large proportion of late-type dwarf galaxies results in slightly higher $M_H/L_B$ values for the KKT sample compared to the others. A more detailed comparison between the Virgo cluster sample and the KKT sample is made for the $M_H/L_B$ ratio and the quasi-surface density $\sigma_H$. The first of these quantities is known to be luminosity dependent, the second is not. In view of the type and luminosity segregation in the Virgo cluster this is very important in connection with the discussion of the HI deficiency.

In Fig. 1 the HI-mass $M_H$ is given as a function of absolute magnitude $M_B$. Mean values over magnitude intervals one magnitude wide ($\bullet$) mark a tight correlation. Correspondent values for the Virgo sample (o) agree with these data as does a sample of Sa-galaxies (Huchtmeier, 1982) ($\square$). Part of the low upper limits are due to SO galaxies but some are due to spiral galaxies, too, indicating "missing" HI. For a discussion of the HI deficiency the sample was divided into concentric rings half a degree wide. Values of relative HI content ($M_H/L_B$ and $\sigma_H$) are averaged in these rings. The amount of galaxies per morphological type and ring were taken to construct an "expected" value, which was used for normalizing the relative HI content (Fig. 2). There is a pronounced HI-deficiency in the clusters' central part by a factor 5 within $1^o$ and a factor of 1.5 between $2^o$ and $3.5^o$ from the clusters' centre as demonstrated by both the $M_H/L_B$ ratio and $\sigma_H$ in Fig. 2. For a smaller number of galaxies Chamaraux et al. (1980) derived a deficiency of a factor of 2.2 within $2^o$ of the Virgo centre.

The observed HI-line widths were used to derive the Tully-Fisher (1977) relation (TF). In Fig. 3a the calibrating sample of nearby galaxies (Richter and Huchtmeier, 1984) yields $M_B^{o,i}=-7.17$ x $(\log \Delta v_i)$ - 2.12. This relation fits well to the present

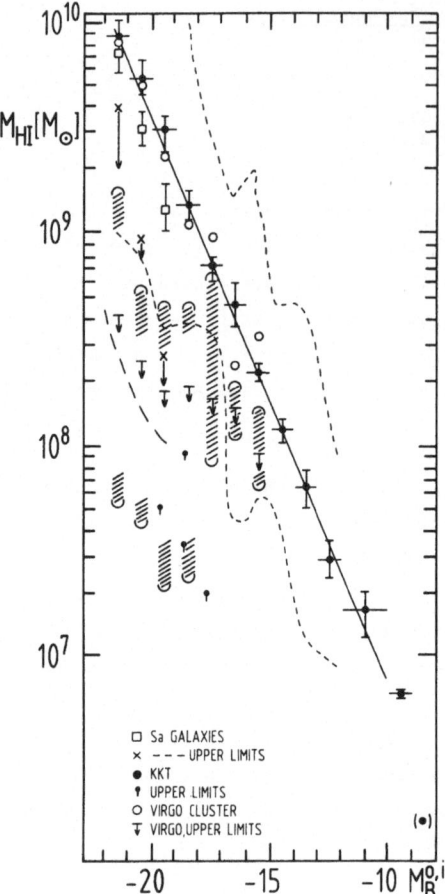

Fig. 1: The correlation of total HI mass with absolute magnitude for different samples. In case of the KKT sample the range of detected galaxies is given by broken lines. There are also a few upper limits for S0 galaxies. Hatched areas indicate the upper limits from the Virgo sample.

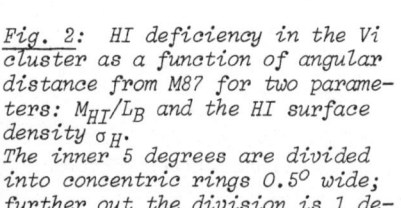

Fig. 2: HI deficiency in the Virgo cluster as a function of angular distance from M87 for two parameters: $M_{HI}/L_B$ and the HI surface density $\sigma_H$.
The inner 5 degrees are divided into concentric rings 0.5° wide; further out the division is 1 degree.

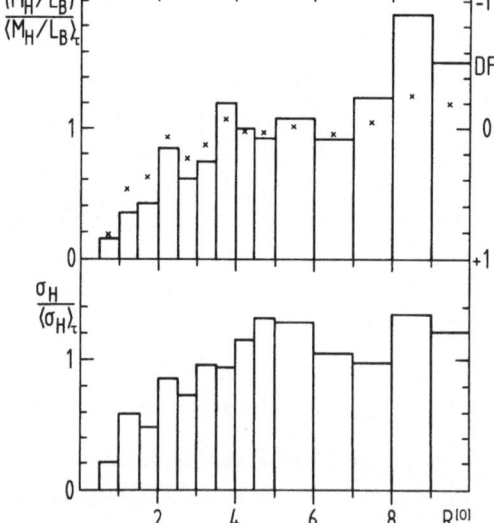

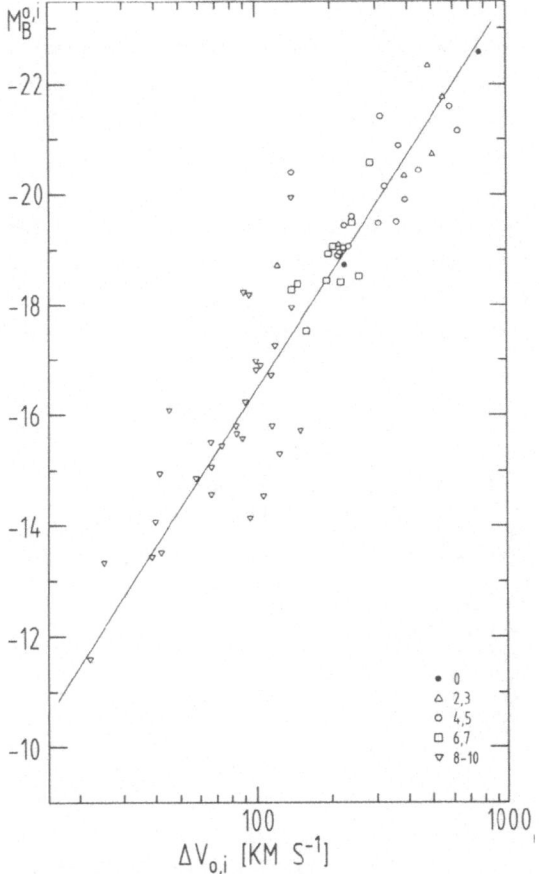

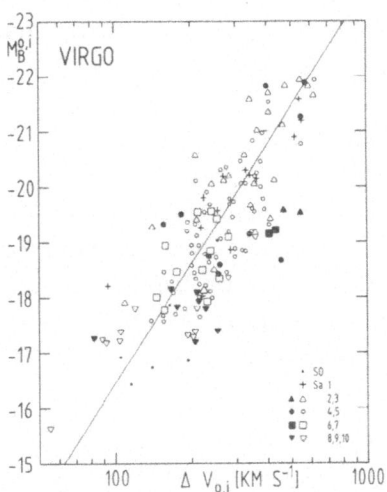

Fig. 3: Absolute magnitude as a function of corrected 21 cm line width
a) the calibration sample of nearby galaxies (Richter and Huchtmeier, 1984),
b) the present Virgo sample. Galaxies with low inclination (i ≤30°) are given by black symbols.

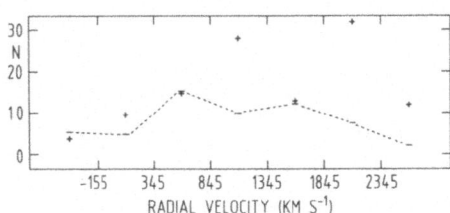

Fig. 4: Distribution of Virgo galaxies in radial velocities for near (-) and far (+) distances as derived from the TF relation.

Virgo sample (Fig. 3b) where galaxies of low inclination (i <30°) are
marked by filled symbols. These galaxies of low inclination contribute only a small
amount to the considerable scatter of the sample. A distance of 23±4 Mpc is derived
for the Virgo cluster. This implies a Hubble constant of the order of 50 km/s/Mpc.
When applying the TF to the whole sample to derive individual distances we will consider only the information nearer (-) or further away (+) for a first check. In Fig.4
the number of galaxies per radial velocity interval is given for nearer (-) and far
(+) galaxies. The distribution of the nearer galaxies is regular and symmetric. But
the far (+) galaxies dominate at high radial velocities showing part of the southern
extension. Applying the TF to galaxies from the W-group we derive a distance about
twice that of the cluster.

## References

Bottinelli, L., Gouguenheim, L., Paturel, G.: 1982, Astron. Astrophys. 112, 61
Chamaraux, P., Balkowski, C., Gérard, E.: 1980, Astron. Astrophys. 83, 38
Huchtmeier, W.K.: 1982, Astron. Astrophys. 110, 121
Huchtmeier, W.K., Richter, O.-G., Bohnenstengel, H.-D., Hausschildt, M.: 1983, ESO
    preprint No. 250
Kraan-Korteweg, R.C., Tammann, G.A.: 1979, Astron. Nachr. 300, 181
Li Zong-Yun, Liu Ru-Liang: 1981, Chin. Astron. Astrophys. 5, 205
Richter, O.-G., Huchtmeier, W.K.: 1984, Astron. Astrophys. 132, 252
Sandage, A., Tammann, G.A.: 1976, Astrophys. J. 210, 7
Shostak, G.S.: 1978, Astron. Astrophys. 68, 321
Tully, R.B., Fisher, J.R.: 1977, Astron. Astrophys. 54, 661

## DISCUSSION

J.-L. Nieto: There is missing hydrogen within 3° of the Virgo cluster. Isn't it an
effect related to a segregation of morphological types among spiral galaxies? What
happened to this effect if you add the contribution of radio ellipticals such as M84
and M87?

W.K. Huchtmeier: The average relative HI-content ($M_H/L_B$ and $\sigma_H$) for different morphological types was determined from the nearby galaxies (KKT sample). The mean properties were used to construct a "comparison sample" with the same number of galaxies
per morphological type as for the Virgo sub-samples. As a consequence the normalized
relative HI content of Fig. 2 is free from the effect of type segregation. Elliptical
galaxies have not been considered for this comparison at all as there are generally
only upper limits available for this galaxy type.

# AN HI AND OPTICAL STUDY OF THE GAS POOR VIRGO CLUSTER SPIRAL NGC 4571

J. M. van der Hulst
Netherlands Foundation for Radio Astronomy
Postbus 2, 7990 AA Dwingeloo, The Netherlands.


R. C. Kennicutt
Department of Astronomy, University of Minnesota
116 Church St. SE, Minneapolis, MN 55455, USA.

## I. Introduction

Several studies have indicated that the galaxies in the Virgo cluster core region have low HI contents and low star formation rates compared to galaxies of similar morphological type in the field. (Giovanardi et al. 1983, Giovanelli and Haynes 1983, Kennicutt 1983, Stauffer 1983, Van Gorkom et al. 1984, Warmels and van Woerden 1984). The most favored interpretation for this effect seems to be stripping by ram pressure when galaxies move through the dense intergalactic medium in the Virgo cluster core. Yet very few galaxies have shown signs of current stripping, two possible exceptions being NGC 4438 (Kotanyi et al. 1983) and NGC 4654 (Warmels, private communication). Kennicutt (1983) identified a few galaxies with extreme properties: galaxies with morphological type Sc, but colors, HI contents and star formation rates which are more typical for field Sb galaxies. We chose one such anemic (van den Bergh 1976) galaxy, NGC 4571, for a detailed study using the National Radio Astronomy Observatory Very Large Array in the 1 km configuration to study the HI and the Mt. Lemmon, Mt. Palomar and Kitt Peak National Observatory  facilities to obtain optical continuum and Hα data.


## II. Observations

The VLA observations made in June 1983 provided 16 channel maps with a velocity resolution of 25 km/sec and an angular resolution of 41". From these we determined the distribution of HI column density and the radial (intensity weighted mean) velocity field. These are shown in Figure 1 and 2, superposed on a print of a high resolution plate obtained with the 2.5 m DuPont telescope at Las Campanas. The derived rotation curve (adopting $V_{systemic}$= 340 km/sec, inclination = 34° and position angle of the line of nodes = 40°) is shown in Figure 3. In the

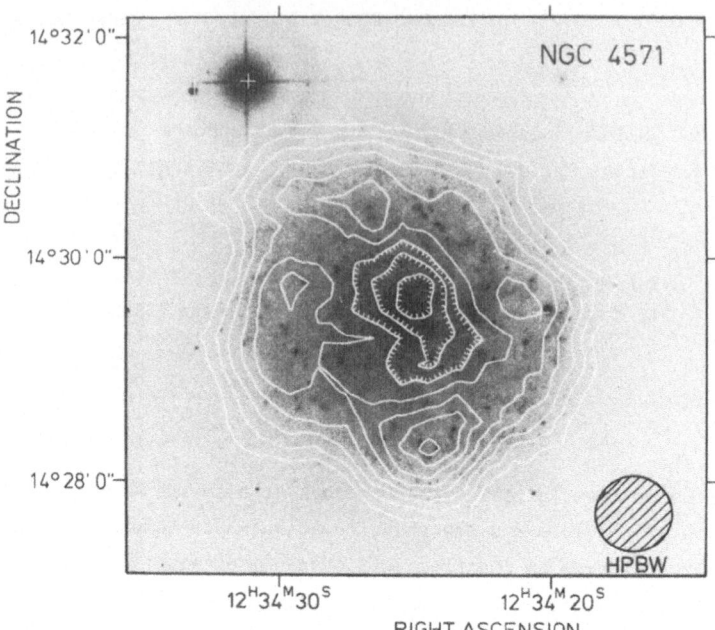

Figure 1. HI column density distribution. The HI contours are 0.6, 1.2, 1.7, 2.3, 2.9, 3.5 and $4.1 \times 10^{20}$ atoms/cm². Hatched contours in the central regions indicate decreasing column density.

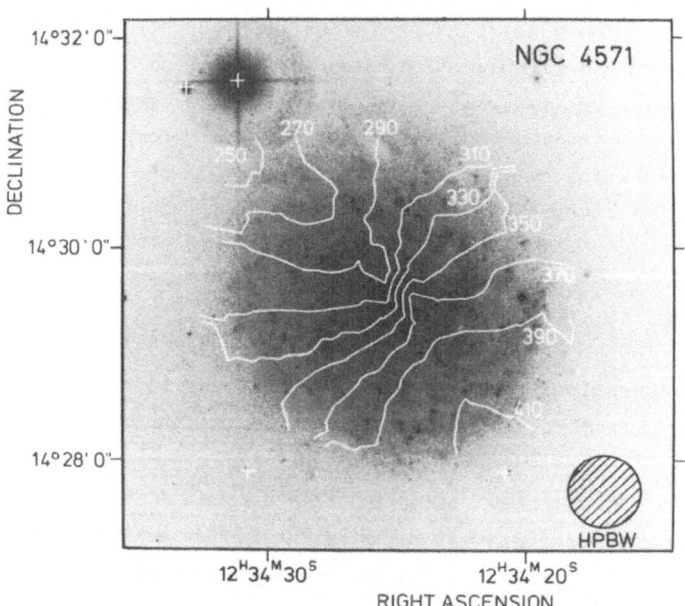

Figure 2. Radial velocity field. The velocity contours are labelled in km/sec and represent heliocentric velocities.

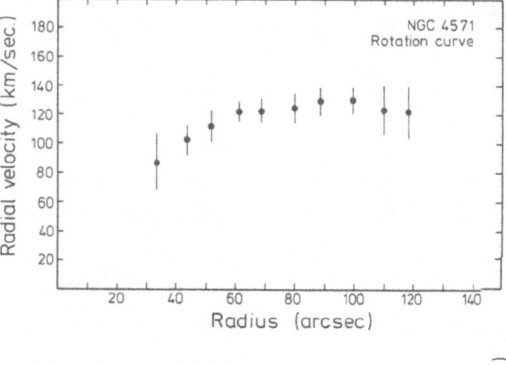

Figure 3. Rotation curve derived from
the radial velocity field of Fig. 2
assuming the gas layer exhibits
planar, circular motions.

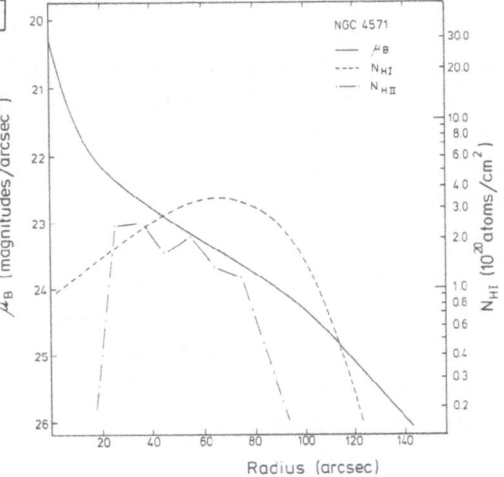

Figure 4. Radial distribution of
blue optical surface brightness
($\mu_B$), HI column density ($N_{HI}$),
and surface density of HII
regions ($N_{HII}$).

inner part the rotation curve is not well defined because of our poor resolution
and because of lack of measurable HI.

Figure 4 summarizes some of the optical data: the blue luminosity profile
obtained from surface photometry from a sky limited IIIa-J plate taken in 1979 on
the 1.3 m Palomar Schmidt telescope, and the radial distribution of HII regions
obtained from an Hα plate taken with the Carnegie image tube direct camera on the
2.1 m telescope at KPNO through a 20 A wide interference filter. The zero point
of the surface photometry was determined from photoelectric BVR photometry obtained in May 1982 using the Schmidt Two-holer photometer on the Mt. Lemmon 1.5 m
telescope.

## III. Results

NGC 4571 is classified as an Sc galaxy. This is supported by its luminosity
profile which is disk dominated and has only a small nuclear or bulge component
contributing ~4% to the integrated luminosity. Also the rotation curve, e.g. the
implied mass surface density distribution, supports the Sc classification. The

slow rise and low maximum velocity of the rotation curve are quite typical for late type galaxies like M33 (Newton 1980) and NGC 2403 (Wevers 1984).

In all other respects, however, NGC 4571 is not a typical Sc galaxy at all as already pointed out by Kennicutt (1983). The current starformation rate in the disk is quite depressed. Both the total Hα luminosity and the characteristic luminosities of the brightest HII regions are more than a factor 3 lower than normal for an Sc galaxy. The broadband colors determined from our BVR photometry are unusually red (B-V ≈ 0.70), consistent with this low starformation rate and suggesting that the stellar disk is evolved and probably has not been subject to active starformation in the past 1-2 billion years. Also the HI properties are quite unusual for an Sc galaxy. There is an HI deficiency in the inner parts (more typical for Sa-Sb galaxies, cf. Bosma 1981, Wevers 1984) and more striking, the column densities are very low everywhere in the disk. Peak values reach only $4 \times 10^{20}$ atoms/cm$^2$, a factor 5 below values normal for Sc galaxies and more typical for the outer parts of spiral galaxies and the disks of S0 galaxies (Van Woerden et al. 1983). The total amount of neutral gas is $1.5 \times 10^9$ M$_\odot$ for an assumed distance of 21.9 Mpc. The HI mass to total mass ratio is 0.03. The overall low gas density is, however, consistent with the low starformation rate and red disk colors, and confirms the findings of Giovanelli and Haynes (1983) and Giovanardi et al (1983) that anemic galaxies in the Virgo cluster have low gas surface densities. However, this relation between gas surface density and anemia is not always seen in more distant clusters (Kennicutt et al. 1984).

A very striking property of the HI disk, not yet mentioned but very clear in Fig. 4 is its small size compared to the optical diameter. The HI size at $1.83 \times 10^{20}$ atoms/cm$^2$ (Bosma 1981, Warmels and van Woerden 1984) is 3.2' while the De Vaucouleurs diameter measures 3.7'. This is very unusual for galaxies in the field but not uncommon for galaxies in the central region of the Virgo cluster (Warmels and van Woerden 1984, Van Gorkom et al. 1984) and is what one expects from simple models of stripping by ram pressure. Stripping may be occurring at some level though the regularity of the HI distribution and the velocity field, as well as the red color of the disk suggest that very recent stripping is not the dominant process for gas removal.

The HI and optical properties are, however, consistent with a slow gas removal process or one which took place at least a few billion years ago (~one crossing time) so that the disk had time to readjust. It is of interest to note that if a large fraction of the gas was removed a few billion years ago, stellar mass loss might have been able to replenish some fraction of the observed gas distribution over most of the disk, except maybe the very central (bulge) part.

An alternative explanation for the peculiar properties of NGC 4571 is just an inhibition or suppression of the disk formation process. Whether such would result from being in a cluster environment is not absolutely clear, since anemic galaxies have also been found outside clusters of galaxies (Bothun and Sullivan 1980).

We gratefully acknowledge travel support from NATO grant No. 0592/82. RCK acknowledges in addition NSF grant AST 81-11711-01. Mt. Lemmon is supported in part by NSF grant AST 82-14422. NRAO is operated by Associated Universities Inc., under contract with the  National Science Foundation. KPNO is operated by Associated Universities for Research in Astronomy Inc.,  under contract with the National Science Foundation.

## References

Bosma, A. 1981 A.J. 86, 1825.
Bothun, G. C., Sullivan, W. T. 1980 Ap.J. 242, 903.
Giovanardi, C., Helou, G., Salpeter, E. E., Krumm, N. 1983 Ap.J. 267, 35.
Giovanelli, R., Haynes, M. P. 1983, A.J. 88, 881.
Kennicutt, R. C. 1983, A.J. 88, 483.
Kennicutt, R. C., Bothun, G. D., Schommer, R. A. 1984 A. J. 89, 1279.
Kotanyi, C. G., van Gorkom, J. H., Ekers, R. D. 1983 Ap.J. 273, L7.
Newton, K. 1980 M.N.R.A.S. 190, 689.
Stauffer, J. R. 1983, Ap.J. 264, 14.
van den Bergh, S. 1976 Ap.J. 206, 883.
van Gorkom, J. H., Balkowski, C., Kotanyi, C. G. 1984 in "Clusters and Groups of Galaxies", ed. F. Mardirossian et al., Reidel Publ. Co., p.293.
van Woerden, H., van Driel W., Schwartz, U. J. 1983 in "Internal Kinematics and Dynamics of Galaxies", IAU Symp. 100, ed. E. Athanassoula, Reidel Publ. Co.
Warmels, R. H., van Woerden H. 1984 in "Clusters and Groups of Galaxies" ed. F. Mardirossian et al., Reidel Publ. Co., p.251.
Wevers, B. M. H. R. 1984 Ph.D. Dissertation, University of Groningen.

Discussion.

B. Elmegreen: What are the masses of the 4 large HI clouds?

Van der Hulst: I estimate the mass of each of the large HI complexes to be about $10^7 M_\odot$.

R. Terlevich: If HI is systematically removed in cluster Sc's (compared with "field" Sc's) you may expect some changes in the chemical evolution of the disk. For example, the HII regions may be systematically underabundant (compared with those in "field" Sc's) or disk gradients may be somehow different. Do you have information about metal content in the HII regions of stripped galaxies that can clarify this point?

Van der Hulst: No, we do not yet have information on the abundances in the HII regions in NGC 4571 or any other HI poor galaxy. There are some indications from other work (W. Romanishin et al. 1982, Ap.J. 258, 77 and G. D. Bothun et al. 1984 A. J. 89, 1300) that low luminosity galaxies and especially low surface brightness galaxies possibly have low metallicity disks.

S. Jorsater: How well defined is really the concept of a "field" galaxy?

Van der Hulst: The concept of "field" galaxies is in my opinion a difficult one. One usually refers to galaxies outside clusters as "field" galaxies but probably should use some measure of the local galaxy density as for example was introduced by A. Dressler (Ap. J. 236, 351, 1980). Also true "field" galaxies or truly isolated galaxies are quite rare.

T. Jaakkola: Is NGC 4571 at the near side or at the far side of the cluster? The question has significance for such questions as an inflow into the cluster, the mean redshift of Virgo (and H), and the velocity difference between the spirals and ellipticals, of course only when observations are made for numerous low redshift spirals.

Van der Hulst: If one applies the Tully-Fisher relation to all galaxies in the Virgo cluster for which HI line widths are available one finds that NGC 4571 is farther away than the mean distance to the cluster. This argument, of course, depends heavily on whether one can blindly apply the Tully-Fisher relation to obtain cluster depth information.

# INFLUENCE OF THE K-EFFECT ON THE COLOURS AND THE $\lambda_o$ 4000 BREAK AMPLITUDES OF GALAXIES IN DISTANT CLUSTERS

E. Laurikainen and T. Jaakkola

Observatory and Astrophysics Laboratory

University of Helsinki

ABSTRACT

The colour distributions of the Coma cluster and DC 0329-52 shifted to z = 0.39 and z = 1.0 are simulated, in regard to the influence of the K-term on the morphological and colour distributions of 50 brightest galaxies. A good fit with the colours of actual distant clusters is obtained, except that the observed colours are slightly redder than the predicted ones. Also Spinrad's data on the z-dependence of the $\lambda_o$ 4000 break amplitude can be understood via the K-effect. The results do not support the hypothesis of cosmological evolution of galaxies.

## 1. COLOUR DISTRIBUTION

Butcher and Oemler (1978) and Kron (1977) have paid attention to blue tails in the colour histograms of clusters Cl 0024+1654 (z = 0.39), 3C 295 (z = 0.46) and Cl 1305-29 (z = 0.95). They argue that this indicates an excess of blue galaxies in distant clusters and is evidence of cosmological evolution. We examined this interpretation by calculating how the K-efect affects the colour distributions of the nearby clusters Coma (z = 0.023) and DC 0329-52 (z = 0.057) seen at redshifts z = 0.39 and z = 1.0. The simulated colours were then compared with the observed ones.

Starting will all galaxies of Coma and DC 0329-52 present in Dressler's (1980) catalogue, we picked up 50 brightest K-dimmed galaxies of these clusters seen as redshifted to z = 0.39 and z = 1.0. The K-corrections are from Coleman et al. (1980). Galactic reddening is determined on the basis of the 0.25 cosec b law. The results are shown in Fig. 1, giving also observations for clusters at z ≈ 0.4; for comparison of simulation at z = 1.0 with data of Cl 1305-29, see Fig. 10 of Spinrad (1977).

As expected, the K-effect weakens preferentially the early type galaxies, which strengthens the blue tails. Contrary to the evolutionary interpretation, the actual distant clusters appear even slightly <u>redder</u> than expected. The data are inconsistent with cosmic colour evolution whether the comparison is made with a moderately spiral-rich cluster DC 0329-52 or even with such an extremely spiral-poor cluster as Coma.

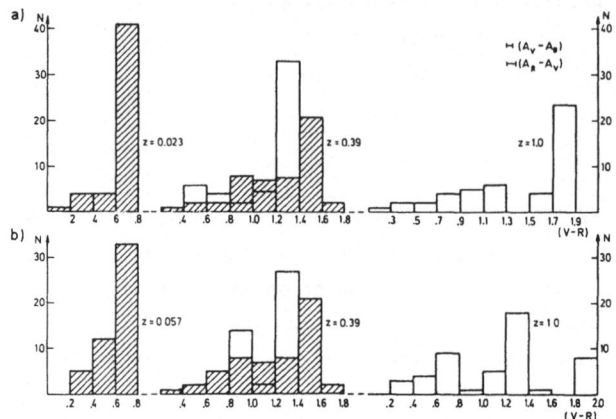

Figure 1. V-R colour histograms for 50 brightest galaxies in clusters. (a) The real distribution for Coma (left) and simulated ones at z = 0.39 and z = 1.00. The hatched columns at z = 0.39 give the observed distribution for 50 brightest galaxies of Cl 0024+1654.
(b) The same for DC 0329-52 and data for 3C 295.

## 2. OTHER SPECTRAL EVIDENCE

Spinrad (1980) has pointed out that the $\lambda_0$ 4000 discontinuity is a useful parameter for studies of cosmic evolution of galaxies. He gives data for E galaxies with the amplitude of the discontinuity decreasing with increasing z and interprets this as due to an increasing proportion of the light arising in the turnoff stars of types F8-G2V in the younger (distant) galaxies.

However, the K-effect may be present also in this result. It makes the progressively bluer galaxies with at the same time lower break amplitudes and even perhaps with later morphology more pronounced and thereby selected for observation. Galaxy morphology in distant clusters is not well known, and determinations of the morphological contents are usually based on colour distribution. Therefore the influence of the K-effect through the intrinsic spreads of the colour and the $\lambda_0$ 4000 break amplitudes within the E-class probably affects Spinrads data and at large z even non-ellipticals may enter the data. We examined the problem in the following manner.

The mean values of the Hubble type indices ($<T>$) were first calculated for n brightest galaxies, with n = 3, 5, 9 and 20, at different redshifts using the galaxies of Coma and DC 0329-52 affected by K. The ($<T>$-z)-relations can be transformed into ($F^+/F^-$-z)-relations if the amplitudes $F^+/F^-$ of the $\lambda_0$ 4000 discontinuity are known for each type. $F^+/F^-$ for type E (T = -5) is from Whitford (1975). The other data are from Pence (1976). For E-S0 (T = -4), $F^+$ = F(4060 Å) and $F^-$ = F(3840 Å). For the later types, $F^+$ = F(4040 Å) and $F^-$ = F(3620 Å), the difference in $\lambda_0$ for $F^-$ following from the steeper slope of the discontinuity in E-S0. We made a smooth curve delineating the observed points, from which the $F^+/F^-$ - values can be read for each type $<T>$. Using this data the "theoretical" ($F^+/F^-$-z)-relations are obtained, as shown in Fig. 2.

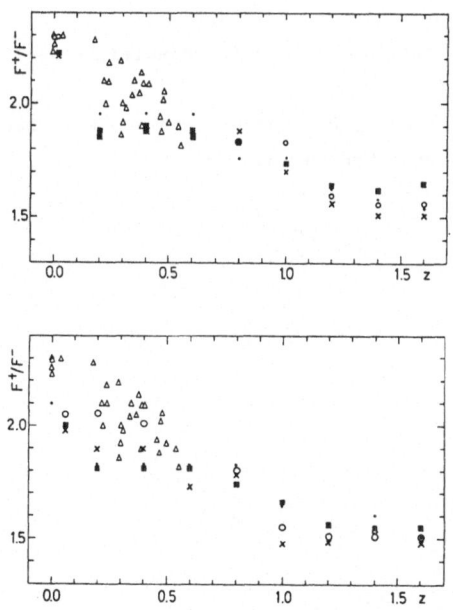

Figure 2. Relation between the theoretical mean $\lambda_0$ 4000 break amplitude and z for the n brightest galaxies, with n = 3 (dots), 5 (crosses), 9 (circles), and 20 (filled squares). The oserved data (triangles) are from Spinrad (1980). The upper figure is for DC 0329-52 and the lower one for Coma.

The observed values of $F^+/F^-$ are slightly higher and the slope steeper than those predicted. However, this follows rather from the procedure (inaccuracy of the types and of the magnitudes) than from evolution : in principle the theoretical and the observed values should be identical at the smallest redshifts. The fit at z = 0.5 and the fact that for higher redshifts the theoretical slope is the same as the one observed at smaller redshifts is perhaps more significant. We conclude that evolution is not necessarily the reason to Spinrad's relations.

REFERENCES

Butcher,H. and Oemler,A.: 1978, Astrophys. J. <u>219</u>, 18.

Coleman,G.C., Wu,C.-C., and Weedman,D.W.: 1980, Astrophys. J. Suppl. <u>43</u>, 393.

Dressler,A.: 1980, Astrophys. J. Suppl. <u>42</u>, 565.

Kron,R.G., Spinrad,H., and King,I.: 1977, Astrophys. J. <u>217</u>, 951.

Pence,W.: 1976, Astrophy. J. <u>203</u>, 391.

Spinrad,H.: 1977, in B. Tinsley and R.B. Larson (eds.), The Evolution of Galaxies and Stellar Populations, Yale, p. 301.

Spinrad,H.: 1980, in G.O. Abell and P.J.E. Peebles (eds.), Objects of High Redshifts, IAU Symp. 92, p.39.

Whitford,A.E.: 1975, in A. Sandage, M. Sandage, and J. Kristian (eds.), Galaxies and the Universe, Chicago, London, p. 159.

POSTER SESSION

# MASS-TO-LIGHT RATIO OF ELLIPTICAL GALAXIES

R. BACON

Observatoire de Lyon (France)

## 1. GEOMETRIC AND PHOTOMETRIC MODELS

- The M/L ratio (f) is constant throught the inner part of the galaxy.
- Isodensities are similar ellipsoids of axial ratio q, (apparent axial ratio $q_a$).
- The distribution of light follows the $r^{1/4}$ law with parameters $l_e, r_e$.

## 2. DYNAMICAL MODELS

2.1 Isotropic model (method 1) : In the simplest case of a spherical galaxy with isotropic velocity dispersions, the knowledge of the photometric parameter ($l_e r_e$) and the central velocity dispersion $\sigma_o$ sets the value of f (Poveda 1960). For a flattened isotropic galaxy the tensor virial theorem gives $f l_e r_e / \overline{\sigma^2}$, as a function of q and $q_a$. A detailed isotropic model (Bacon,1984) shows that $\overline{\sigma^2}$, the weighted mean value of $\sigma$ in all space, can be replaced by $\sigma_o^2$ with a correction for the flattening. As q is generally unknown, the most probable value of $f l_e r_e / \sigma_o^2 (q_a)$ has been calculated (Fig.1) using the statistics of Binney and de Vaucouleurs (1981).

2.2 Anisotropic and isotropic models (method 2) : To take into account the anisotropy of some very bright ellipticals, we need both stellar rotation (V) and dispersion ($\sigma$). The tensor virial then gives the

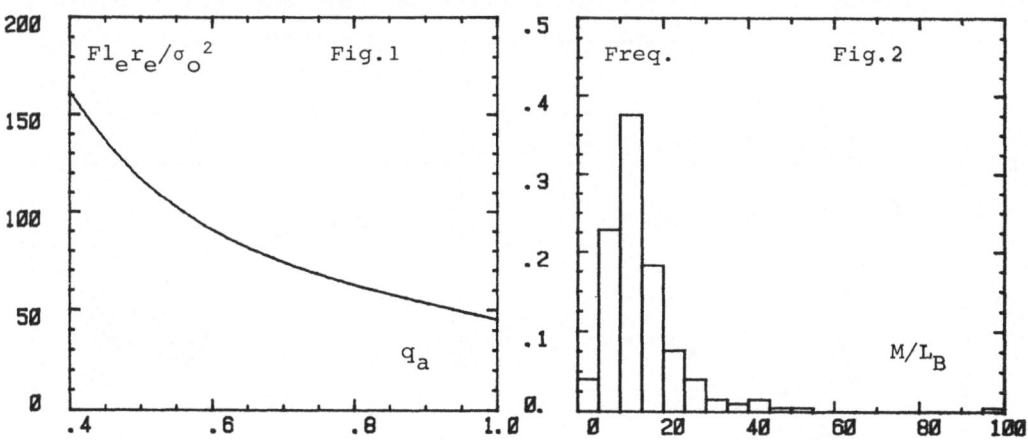

Fig.1 : Most probable value of $f l_e r_e / \sigma_o^2$ versus the apparent axial ratio $q_a$. f is the mass-to-light ratio in solar units, if $l_e$ is in $L_\odot pc^{-2}$, $r_e$ in pc and $\sigma_o$ in $km.s^{-1}$ .
Fig.2 : $M/L_B$ histogram of frequencies from sample 1 .

value of $fl_e r_e / \overline{\mu^2}$, where $\overline{\mu^2} = \overline{\sigma^2} + \overline{V^2}$ . Anisotropic and isotropic models show that $\overline{\mu^2}$ is fairly well approximated by $\mu_o^2 = \sigma_o^2 + Vmax^2$, which is nearly independant of q and $q_a$.

$$fl_e r_e / \mu_o^2 = 53$$

The scatter is between 16 to 25 % , depending on the degree of aniso-tropy.

## 3. SAMPLE

The main sample is composed of 197 ellipticals with $\sigma_o$ values from the catalogue of Davoust et al (1984), with photometric parameters from RC2 or UGC catalogues, and distance moduli from de Vaucouleurs and Olson (1982) or redshift. A subsample of 30 ellipticals have Vmax va-lues in Davies et al (1983).

## 4. APPLICATION

The $M/L_B$ ratio f has been calculated following the two methods, with estimation of errors.

4.1 Main sample : Method 1 gives, for 195 ellipticals :

$$f = 12.2 +/- 0.25 \quad \sigma(\log f) = 0.22$$

See histogram in Fig.2 .

4.2 Subsample : Method 2 gives f = 9.6 +/- 0.4 $\sigma(\log f) = 0.19$ . A si-milar value of f = 10.8 is derived from method 1. Comparison of the $M/L_B$ ratios from the two methods is shown in Fig.3. Note that anisotro-py (which is taken into account in method 2) lowers the $M/L_B$ values.

## 5. IS THE M/L RATIO CORRELATED WITH LUMINOSITY ?

Using spherical and isotropic approximations Faber and Jackson (1976), Michard (1980,1983) found positive correlation between luminosity and $M/L_B$, while Schechter and Gunn (1979), Schechter (1980), Tonry and Davies (1981) found no correlation. We find no correlation in the main sample (Fig.4) and in the subsample.

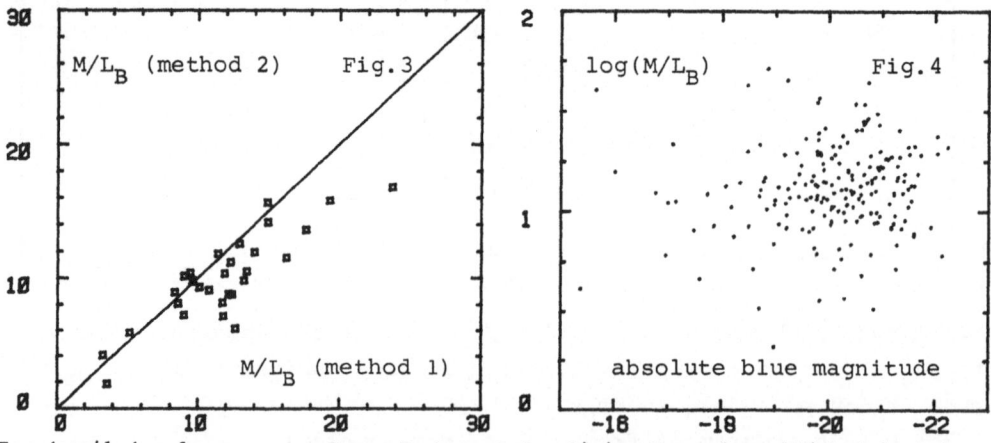

For detailed references see Bacon R.,Monnet G., Simien F.,to be published in A.A.

# DECONVOLUTION OF THE ELLIPTICAL NUCLEUS OF M 31

O. Bendinelli, G. Parmeggiani and F. Zavatti
Dipartimento di Astronomia, Università di Bologna

In previous works a method for the deconvolution of circular images from the PSF has been applied to some galaxies and quasars (see, e.g. Bendinelli et al., 1981, 1984; Zavatti et al., 1984). We extend here the method to elliptical images assuming constant both the axial ratio and the position angle of the semi-major axis. In this case the relation between the observed intensity profile $f(r,\omega)$ along a diameter with azimuthal angle $\omega$ and the semi-major axis intensity distribution $\psi(\rho)$ can be expressed by

$$f(r,\omega)=(\pi\sigma^2)^{-1} \exp(-r^2/2\sigma^2) \int_0^\infty \rho F(r,\rho,\omega) \psi(\rho) \, d\rho \qquad (1)$$

where $\rho = r/\overline{k^{-2} \cos^2\omega \sin^2\omega \, 7}^{1/2}/k$, $k=1-\varepsilon$, $\sigma$ is the dispersion parameter of the Gaussian describing the PSF, and

$$F(r,\rho,\omega)=\int_0^{2\pi} \exp\Big\{ -(2\sigma^2)^{-1} \ \underline{/\rho^{-2}(\cos^2\theta +k^2\sin^2\theta)}$$
$$-2r\rho \ (\cos\theta\cos\omega +k\sin\theta \sin\omega)\underline{/}\Big\} \, d\theta \qquad (2)$$

The discretization of Eq. 1 leads to an ill-posed linear algebraic system, which, as in the spherical case, is solved by a regularization method (see Tikhonov and Arsenin, 1977). As application we derive the deconvolved brightness distribution in the nucleus of M 31 using Kent's (1983) profile. This was obtained up to 5 arcsec with a resolution of 0.314 arcsec/pixel by a CCD attached at one of the MMT mirrors. Fig. 1 shows Kent's profile ($\varepsilon = 0.20$; $\sigma = 0.55$ arcsec) and its deconvolution from the PSF. The comparison (see Fig. 2) with the composite profile of Light et al. (1974), which is based in the very inner part on the extra-atmospheric data collected by Stratoscope II, shows that we derive a nucleus about 1 mag sec$^{-2}$ fainter at the central peak and more smoothly connected to the bulge. Nevertheless the qualitative agreement is good, and the quantitative differences could be due to the data processing.

In effect the Stratoscope data were calibrated by means of Johnson's (1961) profile observed with an estimated $1''.3 \ \sigma$ seeing. If the seeing was really $0''.9$, as inferred by Tremaine and Ostriker (1982), a central lowering of about 0.7 mag sec$^{-2}$ results from convolution effects. Furthermore Kent's (1983) profile has been restored from one Gaussian approximation of the seeing, and hence its central value may raise of some tenth of mag sec$^{-2}$, if the wings are considered. In conclusion the pointed differences found between the deconvolution of ground-based CCD photometry and the extra-atmospheric data, may be due only to the scanty knowledge of observing conditions. A more extensive discussion on the nuclear structure of M 31 is reported elsewhere (Parmeggiani et al., 1984).

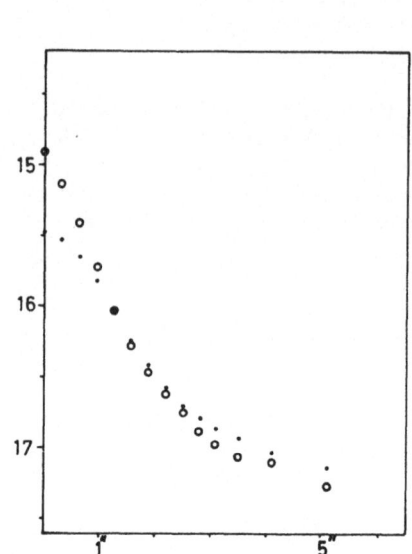

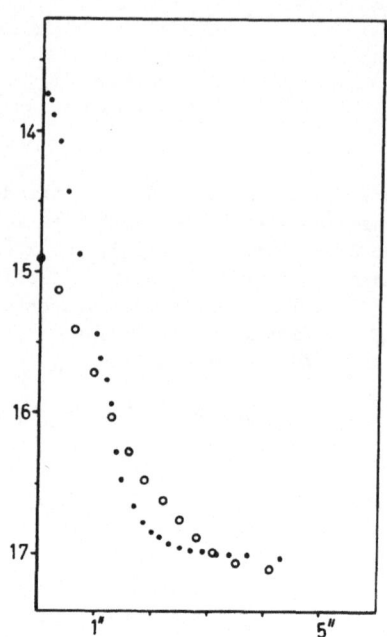

Fig. 1. Observed inner profile of M 31 (•), and its deconvolution (o).

|Fig. 2. Comparison between Light et al. profile (•) and the deconvolution (o) of Kent's data.

REFERENCES

Bendinelli, O., Parmeggiani, G., and Zavatti, F.: 1981, in Proc. ESO Conf. "Scientific Importance of High Resolution at Infrared and Optical Wavelenghts", ed. M.H. Ulrich, ESO, Garching, p. 115.

Bendinelli, O., Lorenzutta, S., Parmeggiani, G., and Zavatti F.: 1984, Astron. Astrophys, 138, 337.

Johnson, H.M.: 1961, Astrophys, J. 133, 309.

Kent, S.M.: 1983, Astrophys. J. 266, 562.

Light, E.S., Danielson, R.E., and Schwarzschild, M.: 1974, Astrophys. J. 194, 257.

Parmeggiani, G., Bendinelli, O., and Zavatti, F.: 1984 submitted to Astrophys. Space Sci.

Tikhonov, A.N., and Arsenin, V.Y.: 1977, Solution of Ill-Posed Problems, Winston, Washington D.C.

Tremaine, S.D., and Ostriker, J.P.: 1982, Astrophys. J. 256, 435

Zavatti, F., Bendinelli, O., and Parmeggiani, G.: 1984, Astron. Astrophys. 134, 99.

# DISTANCE OF GALAXIES FROM THE METHOD OF SOSIES

L. Bottinelli (1), L. Gouguenheim (1), G. Paturel (2),
P. Teerikorpi (3) and I. Vauglin (2)
(1) Observatoire de Meudon, DERADN, 92195 Meudon Principal
(2) Observatoire de Lyon, 69230 St Genis Laval
(3) Observatoire de Turku, Tuorla, SF 21500 Pukkio, Finland

The Faber-Jackson relation (FJ, 1976) connects the absolute
magnitude to the velocity dispersion. In order to use it for distance
determination, we have prepared a compilation of velocity dispersions.
So we collected all the raw measurements available (880 measurements
for 546 galaxies) and reduced them to an homogeneous system (Davoust,
Paturel, Vauglin, 1984).

To bypass the main difficulties (value of the slope, correction for inclination effect, morphological type dependence...) we
have selected 'sosie' galaxies (Bottinelli, Gouguenheim, Paturel, 1983)
having the same apparent axis ratio, the same morphological type and
the same velocity dispersion. From the concept subjacent to the FJ
relation, these galaxies have the same absolute magnitude. If one of
these galaxies is a calibrator whose distance is known, the distance
of its sosies can be derived directly from the difference of their
apparent magnitudes or from the ratio of their apparent diameters.
A first application is made from 31 calibrating galaxies in the Virgo
cluster leading to the distance modulus of 97 galaxies.

Recently, Teerikorpi (1984) suggested that a Malmquist bias
can affect all distance determinations based on a linear relationship,
including the method of sosies. So it is interesting to compare our
distance moduli $\mu(S)$ to unbiased reference distance moduli. These
reference moduli $\mu(P)$ are derived through the Peebles'model (Peebles,
1976) from radial velocities corrected for the local anisotropy.
Both distance moduli, $\mu(S)$ and $\mu(P)$, are obviously based on the same
zero point (i.e. $H_o=100 kms^{-1} Mpc^{-1}$). They are compared in Figure 1 ;
the slope of the mean regression line is significantly different
from 1. On the other hand in the range of the calibrating distance
moduli (about 30 to 31) both scales are in good agreement. Out of
this range the departure is in the sense expected from a Malmquist
bias (for distance smaller as well as greater than those of
calibrators).

    This result can be understood in two alternative ways :
(i) The parameters of the model are not appropriate to describe the
local velocity field. But preliminary results showed that it is
difficult to find a reasonable infall velocity to force both scales
to be identical.
(ii) The method suffers from a Malmquist bias in agreement with
Teerikorpi predictions.

REFERENCES

Bottinelli, L., Gouguenheim, L., Paturel, G. : 1983, C.R. Acad. Sc.
     Paris, 296, 1725
Davoust, E., Paturel, G., Vauglin, I. : 1984, Astron. Astrophys.
     submitted
Faber, S.M., Jackson, R.E. : 1976, Astrophys. J. 204, 668
Peebles, P.J.E. : 1976, Astrophys. J. 205, 318
Teerikorpi, P., 1984, Astron. Astrophysics. (in press)

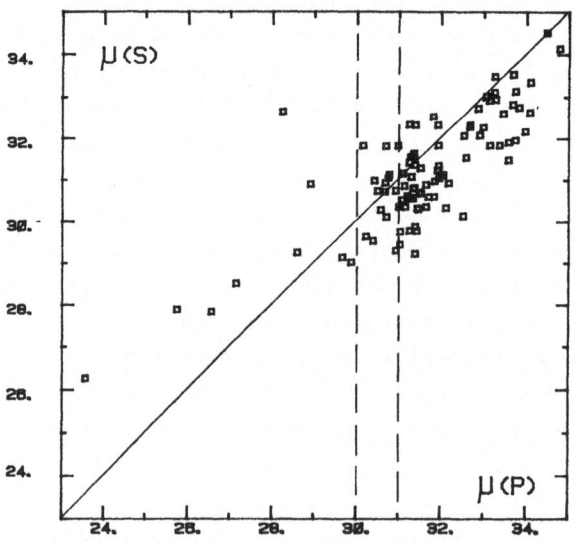

Figure 1 : Distance moduli μ(S) from sosie galaxies versus distance
           moduli μ(P) from Peebles' model.

# MASS TO LUMINOSITY RATIO OF BARS IN SPIRAL GALAXIES

M.F. DUVAL[1], G. MONNET[2]

[1]Observatoire de Marseille, 2 Place Le Verrier, 13248 Marseille Cedex 04, France

[2]Observatoire de Lyon, 69230 Saint-Genis-Laval, France

Comparison of the M/L ratio for bars and for the surrounding material (disk) is fundamental to assess the origin of bars. From B surface photometry, we have derived the luminosity of the total components: bulge, bar, centre disk or lens superposed on an extended disk. The mass densities of the axisymmetrical components come from the axisymmetrical "rotation curve" of the ionized gas after removal of the perturbation effect of the bar.

Mass of the bar has been determined in five galaxies ranging from SBb to SBm by various methods: an hydrodynamical model of the gas flow in NGC 5383 (Duval and Athanassoula, 1983), a model fitting of the deviations from pure circular motions at the end of the bar for NGC 3359, NGC 7479, NGC 7741 (Duval, 1984), use of the position of the Lagrangian neutral points in the LMC coming from its decentred bar potential (Duval, 1984). Table 1 gives the following data: Column 1) name of the galaxy, 2) the absolute galaxy magnitude, 3) the radius of bar using the 25 mag/$\square$" radius as an unit, 4, 5 and 6) the M/L ratios of the bar, of the central disk and the extended disk corrected from galactic and internal absorption, 7 and 8) the $M_{Bar}/M_{central\ disk}$ and $M_{Bar}/M_{25}$ ratios. The M/L ratios ($\sim$ 4) of these bars are intermediate between the (M/L) for the central disks (< 1) and for the extended disks (> 10). The low values for the central disks are probably explained by a vigorous star formation induced by the bars. The high values for the extended disks are usually interpreted as an indication for massive haloes. Bars would thus likely appear as density waves in the true disk population, with no strong preference for either young or old stars.

Table 1 also shows that the bar lengths and masses are highly variable and uncorrelated with the galaxy absolute magnitude.

Table 1

| Object | $M_T^{\circ}$ | $\rho_{Bar}/\rho_{25}$ | $(M/L)_{Bar}$ | $(M/L)_{cd}$ | $(M/L)_{ed}$ | $M_{Bar}/M_{cd}$ | $M_{Bar}/M_{25}$ |
|--------|------|------|------|------|------|------|------|
| NGC 3359 | -19.60 | 0.18 | 4.4 | 0.5 | 14 | 0.41 | 0.02 |
| NGC 5383 | -20.14 | 0.95 | 6.5 | 9.3 | 27 | 0.94 | 0.27 |
| NGC 7479 | -20.95 | 0.45 | 2.1 | 0.8 | 20 | 0.86 | 0.03 |
| NGC 7741 | -18.59 | 0.37 | 3.4 | 0.4 | 11 | 0.70 | 0.05 |
| LMC | -18.40 | 0.41 | 4.0 | - | - | - | 0.28 |

"S" shape distorsion in the isoradial velocities has been clearly seen in all of these objects (Duval, 1984). This effect will be presented in next papers (Duval, Monnet, submitted and in preparation).

### References

Duval, M.F., Athanassoula, E.: 1983, Astron. Astrophys. 121, 297
Duval, M.F.: 1984, Thèse d'Etat, University Aix-Marseille I

# SYSTEMATIC ERRORS IN ZWICKY'S MAGNITUDES

G. Fasano
Osservatorio Astronomico
5, Vicolo dell'Osservatorio
35100  PADOVA

Zwicky's magnitudes $m_{zw}$ (1961-68; ZGC) are compared with $B_T$ magnitudes given by Sandage and Tammann in the 'Revised Shapley-Ames Catalog of Bright Galaxies' (Sandage and Tammann, 1981; RSA). All galaxies in our sample are in the region of sky defined by: $|b| > 30°$; $\delta > -2°$. Moreover only objects of RSA with magnitudes $B_T$ less than 13.2 have been included in the analysis. For the total sample of 663 objects it results:

$$\langle B_T - m_{zw} \rangle = \langle \Delta B \rangle = -.243 \pm .020 \tag{1}$$
$$\sigma_{\Delta B} = .496 \pm .014$$

In the following we check for the possible dependence of $\Delta B$ on all available parameters:

a) <u>mean surface brightness</u>

Figure 1 shows that $\Delta B$ well correlates with mean Zwicky's surface brightness, defined by:

$$m'_{zw} = 5 \, \text{Log} \, D_{25} - 2.5 \, \text{Log} \, R_{25} + m_{zw} + \text{const.} = \mu_{zw} + \text{const.}$$

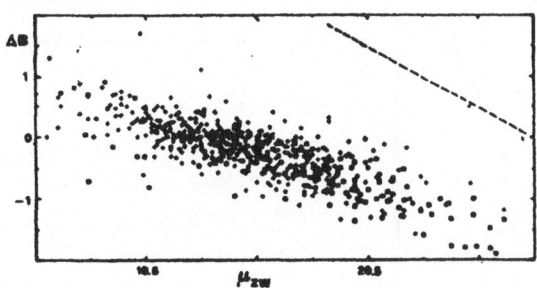

Figure 1 - The correction $\Delta B = (B_T - m_{zw})$ vs. the Zwicky surface brightness $\mu_{zw}$.

The linear regression:

$$\Delta B = a + b \mu_{zw} \tag{2}$$

yields:    $a = 10.94 \pm .29$ ;  $b = -.574 \pm .015$.

The standard deviation of the residuals is:

$$\sigma_{\Delta(\Delta B)} = .288 \pm .008 \qquad (3)$$

much smaller than the value given by relations (1).

b) <u>coordinates</u>

Even if not physically understood, the statistical dependence of $\Delta B$ on $\delta$ is unquestionable (Paturel,1977; Peterson,1979; deVaucouleurs and Pence,1979; Auman et al., 1982). $\langle\Delta B\rangle$ increases by about $0^m.5$ when $\delta$ varies from $-2°$ to $90°$ and by about $3^m.0$ as $\mu_{zw}$ decreases from $22^m.0$ to $17^m.0$. Therefore the correlation between $\Delta B$ and $\delta$ is of the second order with respect to that between $\Delta B$ and $\mu_{zw}$. It might be worth testing whether relation (2) shows a systemetic variation of the zero point a'=a+b$\mu_{zw}$

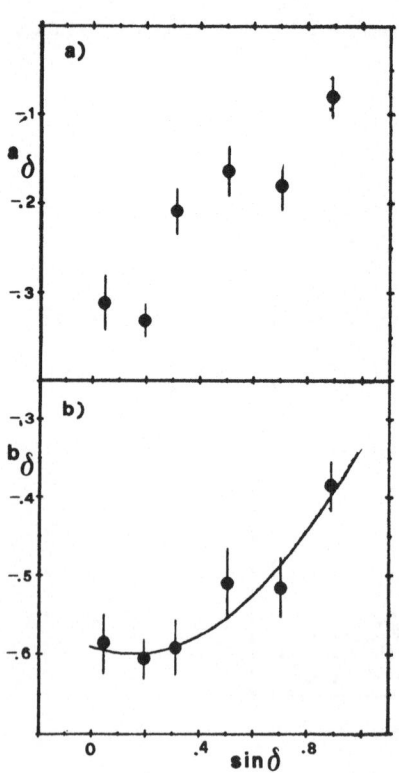

Figure 2 - Analysis of the relation (2) at varying sin $\delta$.

when different intervals of sin $\delta$ are considered. Figure 2 shows, for each given sin$\delta$ interval, the computed values of the coefficients a' and b, and their errors. It results:

$$b_\delta = -.592 - .111\sin\delta + .369\sin^2\delta$$
$$a_\delta = 11.15 + 2.53\sin\delta - 7.26\sin^2\delta \qquad (4)$$

c) <u>ZGC volume number</u>

The claimed dependence of $\langle\Delta B\rangle$ on the particular ZGC volume number is discussed. By comparing Fig. 2a to Fig. 3, we think it's plausible that the observed variation of $\langle\Delta B\rangle$ with the ZGC volume number reflects the above mentioned dependence on $\delta$. In fact, no residual dependence on the ZGC volume is found when the correction of $m_{zw}$ is made taking into account only the dependence of coefficients a and b on $\delta$.

d) <u>morphological type</u>

The important point arising from analysis on T is the strong correlation existing between morphological type and slope of equation (2) (see Figure 4). This in fact is considerably greater in spirals of early type than in late type spirals and irregulars. No much difference can be seen among S0 and elliptical galaxies. They behave in a way intermediate to that of the other types, so that the same coefficients a and b can be assigned to these objects. They are:

$$a_T = 9.32 \ ; \ b_T = -.495 \qquad (T\leqslant0) \qquad (5a)$$

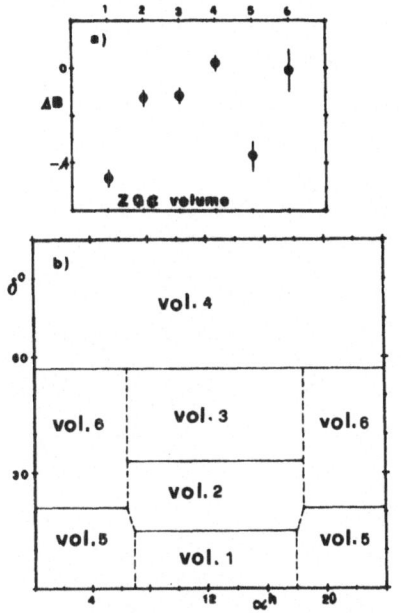

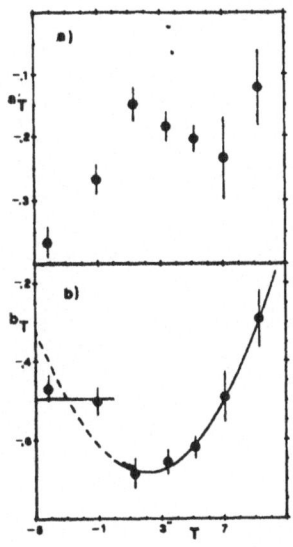

Figure 3 - ⟨ΔB⟩ for each ZGC volume(a)
Area of the sky covered by each ZGC
volume (b).

Figure 4 - Analysis of the relation (2)
at varying T.

Therefore the quadratic best fit of coefficients a and b for spirals and irregular galaxies is:

$$a_T = 12.60 + .542T - .140T^2$$
$$b_T = -.653 - .030T + .007T^2 \qquad (T>0) \qquad\qquad (5b)$$

In the light of the above results, we suggest the following relation for the correction ΔB:

$$\Delta B = a(T, \sin \delta) + b(T, \sin \delta)\mu_{zw} \qquad\qquad (6)$$

in which the main dependence is on the surface brightness, even if a residual dependence on both T and sin δ has also to be taken into account. We adopt analytical relations for a and b in which T and sin δ are two separate parameters. The correction
ΔB then becomes:

$$\Delta B = k_a a_T a_\delta + k_b b_T b_\delta \mu_{zw} + k \qquad\qquad (7)$$

where $k_a$, $k_b$ and k are normalization parameters. We find: $k_a$ = .894 ± .009 ;
$k_b$ = -1.59 ± .06 ; k = -.048 ± .160 and correspondingly: ⟨Δ(ΔB)⟩ = -.006 ± .01 ;
$\sigma_{\Delta(\Delta B)}$ = .244 ± .007. The standard deviation of the residuals (now even lower than the
value given by (3)) turns to be almost independent on both total magnitudes and type.

REFERENCES

Auman,J.R., Hickson,P., Fahlman,G.G. : 1982, P.A.S.P., 94, 19
Paturel,G. : 1977, Astron. Astrophys., 56, 259
Peterson,S. : 1979, Astrophys. J. Supplts, 40, 527
Sandage,A., Tamman,G.A. : 1981 ; "A revised Shapley-Ames Catalog of Bright Galaxies",
     Carnegie Institution of Washington
de Vaucouleurs,G., Pence,W.D. : Astrophys. J. Suppl. 39, 49
Zwicky,F. et al. : 1961-1968, Catalogue of Galaxies and Clusters of Galaxies,
     California Institute of Technology

# STOCHASTIC STELLAR ORBITS IN GALAXIES

Ortwin E. Gerhard

Max-Planck-Institut Astrophysik

Karl-Schwarzschild-Str.1, Garching , FRG

We have studied the appearance of stochastic orbits caused by perturbations of two-dimensional integrable galaxy potentials, using a
canonically invariant form (Gerhard 1984b) of a method due to Melnikov
(1963) and Holmes & Marsden (1982). The unperturbed integrable potentials are separable in elliptic confocal coordinates and have the
characteristic box/loop orbits found for planar motion in non-spherical galaxies (de Zeeuw 1984). Under perturbations of these potentials, a stochastic layer forms near orbits that separate loop and
box orbits. Such an orbit is shown
in Figure 1. Its $t_{\pm\infty}$ limit is the
unstable y-axial orbit and it may be
thought of as a product of an
oscillation in some radial-like
coordinate, and a pendulum motion in
an azimuthal-like coordinate which
has just enough energy (action) to
reach its top on the y-axis. There
is a oneparameter ($\theta_0$) family of
these socalled <u>homoclinic</u> orbits
depending on the relation of
oscillation and pendulum phase at
$t = 0$.

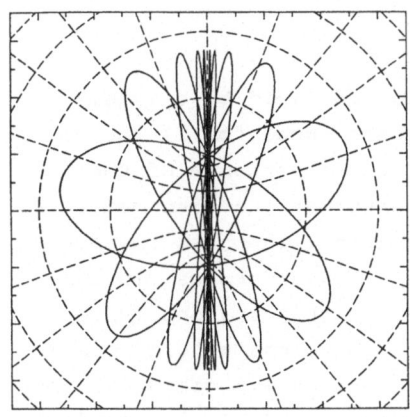

Fig. 1: The y-axial and one
of its asymptoic orbits in $\Psi_c$.

The Melnikov integral as a function of $\theta_0$ is given by $M(\theta_0) =$
$-(\partial J_1/\partial I) \int_{-\infty}^{\infty} dt\{I,H_1\}$ where $I$ is the unperturbed second integral, $J_1$
the oscillation action, $\epsilon H_1$ is the small perturbation Hamiltonian, and
the integral of the Poisson bracket which represents the change of $I$
under the perturbation is taken along an unperturbed homoclinic orbit
parametrized by $\theta_0$. If $M(\theta_0)$ has simple zeroes then no second global
integral of motion exists in the perturbed case. Then $\epsilon$ max $M(\theta_0)$
gives a qualitative estimate for the size of the stochastic layer near
the homoclinic orbits, or in other words for the rapidity with which
the regular orbital structure of the integrable potential is destroyed
by the perturbation.

Specific results were obtained for one integrable potential $\psi_C$ with density $\propto r^{-2}$ at large radii, and a comparison with numerical work was made. A typical result for a computed normalized $\overline{M}(\Theta_0) \propto \max M(\Theta_0)$ curve (Gerhard 1984a) is shown in Fig. 2. From comparing $\overline{M}$ for various perturbations of $\psi_C$ the main result is that those perturbations that are consistent with the photometry of early-type galaxies approximately

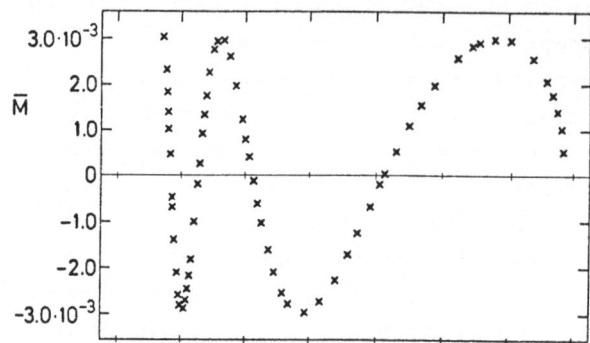

Fig. 2: Typical curve $\overline{M}(\Theta_0)$ with simple zeroes implying stochasticity.

preserve the regular orbital structure of $\psi_C$ and lead to only small stochastic layers. Specifically, this has been found for i) cos m$\phi$ perturbations with m = 0,2,1,4 in order of the importance of the resulting stochastic layer ii) potential perturbations with moderate ellipticity gradients iii) small figure rotation. In these cases, galaxy models may be constructed using only regular orbits. In contrast, other perturbations lead to large stochastic regions and a rapid breakdown of the regular structure of phase space; e.g. cos m$\phi$ perturbations m = 3 or m $\geqslant$ 5. The fact that m=3 or m$\geqslant$5 perturbations are inconsistent with galaxy observations might be due to a time-evolution of the distribution function caused by non-ergodic stochastic orbits which would cause a system to evolve until it is close to an integrable potential.

Comparison of the above perturbation-theoretical results with numerical work suggests that one may understand the regular orbit structure of several galaxy potentials in terms of their proximity to integrable Stäckel potentials.

de Zeeuw, T., 1984, Mon. Not. R. astr. Soc., submitted.
Gerhard, O.E., 1984a, Astron. Astrophys., submitted.
Gerhard, O.E., 1984b, Astron. Astrophys., submitted.
Holmes, P.J., Marsden, J.E., 1982, Comm. Math. Phys. __82__, 523.
Melnikov, V.K., 1963, Trans. Moscow Math. Soc., __12__, 1.

# RELATIONS AMONG OPTICAL, RADIO AND X-RAY PROPERTIES OF ABELL CLUSTERS OF GALAXIES

M. Kalinkov, K. Stavrev, and I. Kuneva
Department of Astronomy and National Astronomical Observatory
Bulgarian Academy of Sciences, 72 Lenin blvd.
Sofia 1184, Bulgaria

A data bank, containing optical, radio and X-ray data for all Abell clusters of galaxies has been compiled. The relations between various parameters, presenting global or specific characteristics of the clusters have been determined. Distance-limited and luminosity-limited samples, containing radio sources or/and X-ray sources, together with non-limited samples have been processed for various richness groups, Bautz-Morgan types, etc. Some examples are given here allowing to study the relations between parameters and to establish some selection effects. Only A-clusters with measured redshifts are used here. All cosmologically dependent values are refered to a model with $H_o$ = 100 km s$^{-1}$ Mpc$^{-1}$ and $q_o$ = +1, with exception for the data of Schneider, Gunn and Hoessel (1983a,b), hereafter SGH, which are for $H_o$ = 60 and $q_o$ = 0.5.

We use the following parameters: (1) the redshift z is compiled from many sources, published and not yet published. We have gathered radial velocity for more than 600 A-clusters; (2) $m_{10}$ (Abell, 1958); (3) $m_1$ (Leir and van den Bergh, 1977 - hereafter LB); (4) $\Delta m = m_{10} - m_1$; (5) $L_x$ is the X-ray luminosity, erg s$^{-1}$, reduced to the interval 2 - 6 keV and is compiled from many sources; (6) S is the flux, Jy, at 1400 MHz (Owen et al., 1982); (7) and (8) - $M_{10}$ and $M_1$ are the absolute magnitudes for the tenth- and first-ranked galaxy, defined according to $M_{1,10} = m_{1,10} - 5 \log z - K(z) - 42.5201$, where the redshift dimming $K(z)$ is taken from Sandage (1973). Here Abell (1958) correction for the galactic absorption is used; (9) the richness group p (Abell, 1958); (10) P is the mean apparent population, gal/sq.deg., defined as P = $N / \pi r_A^2$, where the Abell radius is $r_A$ = 0.0286 $(1 + z)^2/z$ and N = 40, 65, 105,.. for p = 0, 1, 2,.., and (11) is Bautz-Morgan (BM) type from LB, codded as usually: I = 1, I - II = 2,.., III = 5.

Table 1 presents two correlation matrices. All correlation coefficients above the leading diagonal refer to A-clusters having $z < 0.1$. There are only 35 clusters with all known characteristics. The sample mean values and sample standard deviations for the characteristics are given on the right side of Table 1. The correlation coefficients below the leading. diagonal refer to A-clusters without restriction on redshift - there are 46 clusters and the sample means and standard deviations are given on the left side of Table 1.

All correlation coefficients in Tables 1 - 2 are multiplied by 100. Our Tables permit to calculate easily all linear regressions between all the characteristics. That is why we give together with correlation co- efficients and the number of clusters, the means and standard deviations.

Table 2 contains a correlation matrix for the SGH sample of 83 A-clus- ters. We added to SGH characteristics data from our file of Abell clus- ters of galaxies (numbers 1 - 8). The SGH characteristics are from GI, number 9, to the last number, 14. The denotations are generally accord- ing to SGH with the exception of $RM_1$, $RM_2$, and $RM_3$, which are the re- duced absolute magnitudes of the first, second and third rank galaxy.

It is not reasonable to compare data from our file with data of SGH, even though the cosmologically dependent values are defined for differ- ent models. But it is curious to see how the increasing of the accuracy in photometry improves the significance of all relations between global or specific cluster parameters for many extragalactic problems.

Correlation matrices given here, together with some other matrices, pub- lished elsewhere (Kalinkov et al., 1984a,b), present a complex statisti- cal structure of the optical, radio, and X-ray data for A-clusters. Many conclusions may be derived on the basis of this vast amount of informa- tion. A few examples are adduced below.

For 68 A-clusters we have the fundamental relation

$$L_x = 7.8 \times 10^{44} \ (z^2 S)^{0.34} \quad \text{or} \quad z^2 S = 8.2 \times 10^{-22} \ L_x^{0.42}.$$

For 88 clusters next relations are also significant:

$$\log L_x = 45.40 + (1.2 \pm 0.1) \log z; \ \log L_x = 35.57 - (0.40 \pm 0.08) M_{10};$$

$$\log L_x = 43.68 + (0.21 \pm 0.03) p; \ \log L_x = 42.82 + (0.54 \pm 0.04) \log P.$$

Adding other A-clusters with not measured, but calibrated redshifts (Kalinkov et al., 1984a) we obtain for 168 clusters:

$$\log L_x = 45.58 + (1.40 \pm 0.08) \log z; \ \log L_x = 37.57 - (0.31 \pm 0.07) M_{10};$$

$$\log L_x = 43.92 + (0.16 \pm 0.03) p; \ \log L_x = 42.75 + (0.62 \pm 0.04) \log P.$$

Table 1.  Correlation  Matrices  for  Abell  Clusters

| | (n=46) Mean St.dev | | 1 | 2 | 3 | 4 | 5 | 6 | 7 | 8 | 9 | 10 | 11 | (n=35) Mean St.dev. |
|---|---|---|---|---|---|---|---|---|---|---|---|---|---|---|
| $\log z$ | -1.16 0.26 | 1 | 100 | 86 | 85 | -22 | 69 | 3 | -37 | - 8 | 25 | 91 | 3 | -1.27 0.18 |
| $m_{10}$ | 15.80 1.14 | 2 | 94 | 100 | 82 | 4 | 55 | 2 | 16 | 9 | 8 | 72 | -10 | 15.33 0.86 |
| $m_1$ | 14.26 1.52 | 3 | 94 | 92 | 100 | -54 | 42 | - 6 | -15 | 46 | 11 | 72 | 11 | 13.60 1.03 |
| $\Delta m$ | 1.53 0.65 | 4 | -55 | -38 | -72 | 100 | 7 | 13 | 50 | -66 | - 6 | -20 | -34 | 1.73 0.59 |
| $\log L_x$ | 43.95 0.43 | 5 | 72 | 63 | 60 | -29 | 100 | 27 | -34 | -37 | 62 | 82 | -20 | 43.84 0.40 |
| $\log z^2 S$ | -2.60 0.46 | 6 | 43 | 40 | 38 | -19 | 37 | 100 | - 2 | -16 | 24 | 13 | -12 | -2.73 0.40 |
| $M_{10}$ | -21.02 0.51 | 7 | -60 | -28 | -47 | 62 | -52 | -28 | 100 | 32 | -34 | -44 | -25 | -20.88 0.47 |
| $M_1$ | -22.55 0.52 | 8 | 10 | 21 | 44 | -65 | -13 | - 4 | 20 | 100 | -22 | -16 | 15 | -22.61 0.54 |
| $p$ | 1.61 1.27 | 9 | 66 | 57 | 61 | -42 | 60 | 50 | -52 | 2 | 100 | 64 | -32 | 1.11 0.87 |
| $\log P$ | 2.08 0.65 | 10 | 96 | 88 | 89 | -54 | 75 | 49 | -62 | 7 | 85 | 100 | -12 | 1.78 0.41 |
| BM | 3.5 1.5 | 11 | 37 | 27 | 42 | -49 | 9 | 16 | -39 | 24 | 13 | 30 | 100 | 3.1 1.5 |

Table 2.  Correlation  Matrix  for  83  Abell  Clusters  (SGH)

| | Mean St.dev. | No | 1 | 2 | 3 | 4 | 5 | 6 | 7 | 8 | 9 | 10 | 11 | 12 | 13 | 14 |
|---|---|---|---|---|---|---|---|---|---|---|---|---|---|---|---|---|
| $\log z$ | -0.96 0.18 | 1 | 100 | 77 | 62 | - 6 | -83 | -66 | 66 | - 1 | 92 | 93 | 93 | -45 | -45 | -58 |
| $m_{10}$ | 16.95 0.52 | 2 | 77 | 100 | 54 | 23 | -38 | -50 | 67 | -23 | 67 | 68 | 68 | -42 | -23 | -33 |
| $m_1$ | 15.08 0.70 | 3 | 62 | 54 | 100 | -69 | -48 | 12 | 74 | 28 | 64 | 65 | 65 | -12 | -30 | -28 |
| $\Delta m$ | 1.87 0.60 | 4 | - 6 | 23 | -69 | 100 | 23 | -57 | -28 | -52 | -16 | -17 | -17 | -23 | 11 | 4 |
| $M_{10}$ | -20.93 0.67 | 5 | -83 | -38 | -48 | 23 | 100 | 67 | -46 | -13 | -80 | -80 | -81 | 30 | 39 | 49 |
| $M_1$ | -22.81 0.79 | 6 | -66 | -50 | 12 | -57 | 67 | 100 | -17 | 28 | -55 | -55 | -56 | 43 | 25 | 38 |
| $p$ | 1.68 1.33 | 7 | 66 | 67 | 74 | -28 | -46 | -17 | 100 | - 6 | 59 | 60 | 59 | -34 | -36 | -38 |
| BM | 3.3 1.4 | 8 | - 1 | -23 | 28 | -52 | -13 | 28 | - 6 | 100 | 22 | 20 | 19 | 52 | 17 | 10 |
| GI | 16.23 0.80 | 9 | 92 | 67 | 64 | -16 | -80 | -55 | 59 | 22 | 100 | 100 | 99 | - 6 | -33 | -49 |
| r | 16.05 0.92 | 10 | 93 | 68 | 65 | -17 | -80 | -55 | 60 | 20 | 100 | 100 | 100 | - 9 | -35 | -50 |
| g | 16.77 1.04 | 11 | 93 | 68 | 65 | -17 | -81 | -56 | 59 | 19 | 99 | 100 | 100 | -11 | -36 | -51 |
| $RM_1$ | 21.08 0.36 | 12 | -45 | -42 | -12 | -23 | 30 | 43 | -34 | 52 | - 6 | - 9 | -11 | 100 | 38 | 34 |
| $RM_2$ | 21.90 0.55 | 13 | -45 | -23 | -30 | 11 | 39 | 25 | -36 | 17 | -33 | -35 | -36 | 38 | 100 | 80 |
| $RM_3$ | 22.39 0.65 | 14 | -58 | -33 | -28 | 4 | 49 | 38 | -38 | 10 | -49 | -50 | -51 | 34 | 80 | 100 |

For the flux density at 1400 MHz we have

$\log S = -1.32 - (0.86 \pm 0.13) \log z$,    for 192 A-clusters.

Every significant relation with the distance should be regarded as a manifestation of the selection. For example only several correlation coefficients among the elements of the first row and the first column from Table 1 are non-significant. It may be seen that $\log z$ correlates very well with $m_{10}$, $m_1$, $\log L_x$, $M_{10}$, $p$, and $\log P$, but there is no correlation with $M_1$. The last is a clear demonstration of the fact, that the luminosity of the first-ranked galaxy is very little influenced by the selection.

Table 2 shows that GI, r, and g can be used more efficiently as regressors for calibrating the distance to A-clusters, instead of $m_{10}$, $m_1$ and so on. Since the SGH sample extends to a very large distance range, $M_{10}$ and $M_1$ unusually well correlate with $\log z$. But $RM_1$, $RM_2$, and $RM_3$ also correlate with $\log z$. It means that there is a selection effect which has to be removed from the SGH data.

It is impossible to mention here all interesting cases, which will be examined in a forthcoming paper.

## References

Abell, G. O.: 1958, Astrophys. J. Suppl. 3, 211
Kalinkov, M., Stavrev, K., Kuneva, I.: 1984a, Adv. Space Res. 3, 10-12, 227
Kalinkov, M., Stavrev, K., Kuneva, I.: 1984b, C. r. Acad. bulg. Sci. 37, No 10
Leir, A. A., van den Bergh, S.: 1977, Astrophys. J. Suppl. 34, 381 (LB)
Owen, F. N., White, R. A., Hilldrup, K. C., Hanisch, R. J.: 1982, Astron. J. 87, 1083
Sandage, A.: 1973, Astrophys. J. 183, 711
Schneider, D. P., Gunn, J. E., Hoessel, J. G.: 1983a, Astrophys. J. 264, 337 (SGH)
Schneider, D, P., Gunn, J. E., Hoessel, J. G.: 1983b, Astrophys. J. 268, 478 (SGH)

# A COMPILATION OF UBVRI PHOTOMETRY
# FOR GALAXIES IN THE ESO/UPPSALA CATALOGUE

Andris Lauberts and Elaine M. Sadler

European Southern Observatory, D-8046 Garching bei München, F.R.G.

ABSTRACT:    We have compiled a catalogue of all presently available
UBVRI aperture photometry for galaxies in the ESO/Uppsala catalogue.
It comprises a total of about 1000 galaxies south of -17° declination.

We have recently completed a compilation of UBVRI photoelectric
aperture photometry for southern galaxies in the ESO/Uppsala catalogue
(Lauberts 1982). The data were originally collected for the purpose of
calibrating PDS photometry of B and R Schmidt plates covering the
entire catalogue area (Lauberts and Valentijn 1983). This project will
eventually result in the publication of B magnitudes, B-R colours and
some additional parameters (e.g. position angle, axial ratio) for
about 16,000 galaxies south of -17°.

Our reason for publishing the data in this form is to provide a
convenient compilation of photometry for other observers, especially
those who wish to calibrate photographic and/or CCD photometry. The
data have, as far as possible, been checked for errors. The catalogue
is also useful in showing which galaxies still need further observations, and thus may help in coordinating the efforts of observers. It
includes all sources of photoelectric photometry known to us and
published before the end of 1983, as well as some unpublished data.

The data set is available both in printed form and on magnetic
tape. It will be updated regularly, and we would appreciate receiving
preprints/reprints of new photometric data. Part I of the catalogue
has now been published (Lauberts and Sadler 1984) and comprises a
complete listing of published UBVRI photometry. A limited number of
copies of part II (magnitude-aperture and colour-aperture plots for
each galaxy) are also available from the authors.

REFERENCES:

Lauberts, A. (1982) "The ESO/Uppsala Survey of the ESO(B) Atlas" (Munich: ESO).

Lauberts, A. and Valentijn, E.A. (1983) The Messenger, 34, 10.

Lauberts, A. and Sadler, E.M. (1984) "A Compilation of UBVRI Photometry for Galaxies in the ESO/Uppsala Catalogue" ESO Scientific Report No. 3.

# THE NUCLEAR RADIO SOURCES IN THE ELLIPTICAL GALAXIES NGC 3309
## AND NGC 3311 IN THE CLUSTER ABELL 1060

P.O. Lindblad[1], S. Jörsäter[2] and Aa. Sandqvist[1]

[1] Stockholm Observatory, Saltsjöbaden, Sweden

[2] European Southern Observatory, 8046 Garching b. München, W. Germany

## INTRODUCTION

The giant elliptical galaxies NGC 3309 and NGC 3311 are situated at the center of the Abell 1060 cluster. The galaxy NGC 3311 appears to be centred at the X-ray emission from the cluster (Forman 1984, private comm.) and it has a very extended outer envelope. The nuclear region is, however, substantially fainter than that of NGC 3309, and NGC 3311 contains an absorption feature in the centre.

We have obtained VLA 6 cm continuum observations, as well as CCD photometry in various Gunn and Johnson filters with the Danish 1.5 m telescope at ESO, La Silla. Isophote plots of the two galaxies in the infrared (Gunn Z) are shown in Figs. 1 and 2.

The radio data is shown in Figs. 3-4. A jet-like source is coincident with the optical nucleus of NGC 3309. The total flux of this source at 4885 GHz is 4.8 mJy. At the position of the optical nucleus of NGC 3311 there is radio emission which is marginally above the noise in our radio data, the flux at 4885 GHz is about 0.3 mJy.

## THE JET

The major axes of the elliptical infrared isophotes of NGC 3309 are directed in position angle 48°, compared with 26° for the jet.

The northeastern (NE) extension of the jet ends in a "spur" (A in Fig. 4). The southwestern (SW) extension lies along a straight line. About 4.5 arcsec (1 kpc projected distance) from the nucleus the jet narrows into a "nozzle" (B), before blossoming into a "lobe" (C) further out in the galaxy. The minimum equipartition energies of the three components are about equal and lie in the range $1-2 \cdot 10^{53}$ ergs. The energy density of the jet is of the order of $7 \cdot 10^{-11}$ erg cm$^{-3}$, and that of the lobe is $1 \cdot 10^{-11}$ erg cm$^{-3}$.

## DISCUSSION

The strongest compact radio source at 6 cm continuum in the A 1060 cluster is thus NGC 3309 and not the supposed core cD galaxy NGC 3311.

Mushotzky (1984) has reported a two-component X-ray spectrum of A 1060, one with a temperature of $2.3 \cdot 10^7$ K and the other $7.0 \cdot 10^7$ K. Jones and Forman (1984) have determined the central gas density in

A 1060 from observations made with Einstein. Their value is $4 \cdot 10^{-3}$ cm$^{-3}$ (assuming H = 75 km s$^{-1}$ Mpc$^{-1}$). They also report a considerable excess X-ray luminosity coming from the central region of the cluster, which is interpreted as produced by a cooling accretion flow of the cluster gas onto the central dominant galaxy (N 3311). The inflow of gas into the central core of the cluster should be able to power the nuclear activity of NGC 3309 giving rise to the observed radio source ejection, analogous to the mechanism suggested by Silk (1976). If we assume that the $2.3 \cdot 10^{7}$ K X-ray emission arises from cluster gas near and in the outer parts of NGC 3309 we find a pressure of $1.3 \cdot 10^{-11}$ erg cm$^{-3}$. This is almost identical to the energy density in the radio lobe of NGC 3309, and thus braking of the jet by cluster gas may have created an equilibrium situation at the lobe. It seems quite likely that the radio activity in NGC 3309 is caused by interaction with the cooling accretion flow of the cluster gas, implying that a substantial fraction of this accretion takes place onto NGC 3309.

REFERENCES

Jones, C., Forman, W.: 1984, Astrophys. J. <u>276</u>, 38.
Mushotzky, R.F.: 1984, Physica Scripta <u>17</u>, 157.
Silk, J.: 1976, Astrophys. J. <u>208</u>, 646.

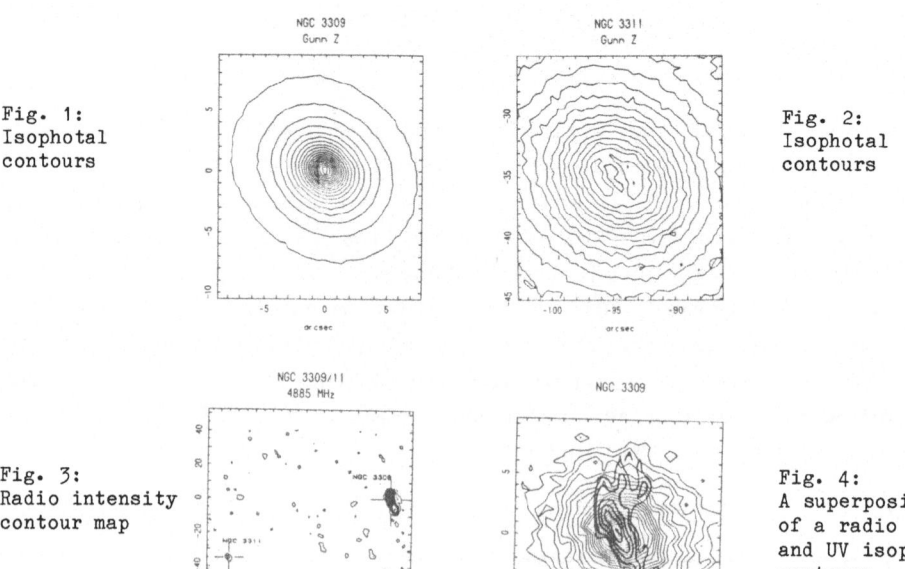

Fig. 1:
Isophotal
contours

Fig. 2:
Isophotal
contours

Fig. 3:
Radio intensity
contour map

Fig. 4:
A superposition
of a radio map
and UV isophotal
contours

# SURFACE PHOTOMETRY AND DETAILED STRUCTURE OF THE ELLIPTICAL
## GALAXY NGC 1199 AND ITS COMPACT COMPANIONS

J.-L. Nieto (1), F. Deschamps (1), A. Maury (2), Th. Roudier (3)


(1) Observatoires du Pic du Midi et de Toulouse
14, avenue Edouard Belin, F- 31400 Toulouse


(2) Telescope de Schmidt, CERGA
F- 06460 Saint-Vallier de Thiey


(3) Observatoires du Pic du Midi et de Toulouse
F- 65200 Bagnères de Bigorre

From deep photographic and spectrophotometric material, Arp (1978) has concluded that the SW compact companion of the elliptical galaxy NGC 1199 appears slightly in front of this galaxy on the line of sight in spite of its much higher redshift ($cz$ = 13,600 kms$^{-1}$). Arp's arguments rely essentially upon the fact that there is a circular area around the compact galaxy which is darker than the background light of the E galaxy.

To test these results, four deep IIaO plates in the blue light were obtained in November 1980 with the 1.5 Danish telescope at La Silla (ESO), under very good seeing conditions so that the full-width at half maximum of a stellar profile was between 1".45 and 1".85 for exposure times ranging from 150 to 240 minutes.

Results from a photometric analysis of the data were the following :

i) The light from NGC 1199 is confirmed to be absorbed near the SW companion by several tenths of magnitude. (Figure 1)

ii) Distance criteria applied to the high resdhift SW compact companion yield a distance to this object <u>much higher</u> than that given by its redshift. This could be explained by an absorption on the line of sight estimated to be between 1 and 2 magnitudes.

iii) NGC 1199 is a dust-lane elliptical galaxy with the dust lane along its major axis (Figure 2). Therefore it contains some strongly absorbing material as suggested by Benedict et al. (1978). These three results are completely <u>consistent with a cosmological scenario</u> for NGC 1199 and its SW companion : the light of the peculiar elliptical NGC 1199 is obscured by its strong absorption patches. These patches obscure also the background compact galaxy, producing a dark circular area around it ; this absorption makes this compact companion at an apparent distance much higher than that given by the magnitude-luminosity index relation.

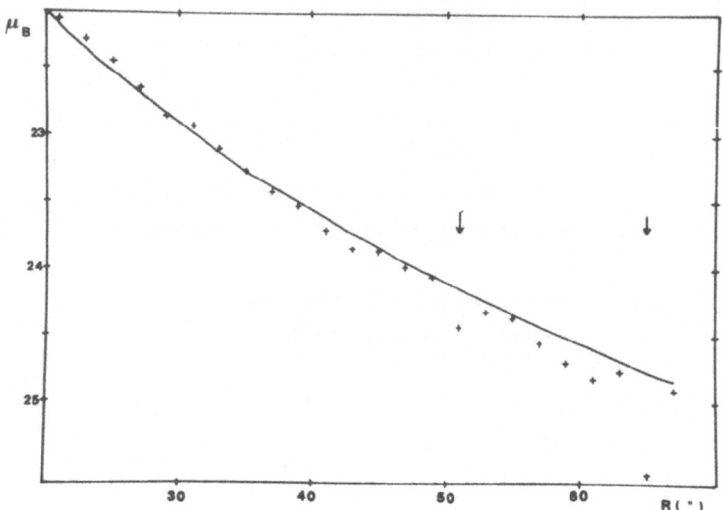

Figure 1 : The profile of the galaxy in a direction 10° away from the SW compact companion is plotted against the distance to the center (crosses). The solid line represents the equivalent profile of NGC 1199 whose scale has been corrected to take into account the geometry of the galaxy. The arrows show the location of the dark circular ring discussed by Arp where some absorption is detected .

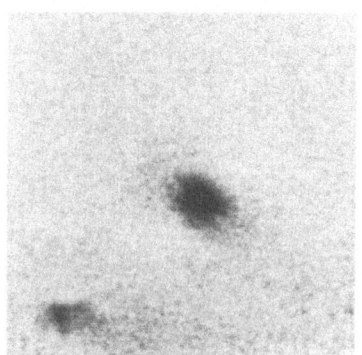

Figure 2 : A numerical unsharp masking treatment (Roudier, 1985) exhibits the central dust lane in NGC 1199. The scale is given by the distance between NGC 1199 and its SE companion being 38".

References :

Arp,H.C., 1978, Astrophys. J., 220, 441

Benedict,G.F., van Citters,G.W., Mc Graw J.T., Rybski, P.M., 1978, Bull. A.A.S., 10, 423

Roudier,Th., 1985, Thèse de 3ème cycle, Université Paul Sabatier, Toulouse

# THE VELOCITY FIELDS OF BARRED GALAXIES

D. Pfenniger
Geneva Observatory
CH-1290 Sauverny

Schwarzschild's method (1979) is used for constructing self-consistent 2-D models of a well pronounced and fast rotating SB whose 3-D dynamics was studied previously (Pfenniger, 1984). Particular care is taken to reduce the numerical errors.

For finding a self-consistent solution, a non-negative least-squares technique is used, which promises interesting applications to the interpretation of real SB for which, besides the mass distribution, only some radial velocities and the orientation are known.

By extremizing various quantities, like the total energy or angular momentum, the set of possible self-consistent solutions is explored.

The important results are:

1) Retrograde motion is in all cases required, between 10 and 30 percent of the mass is retrograde. Most retrograde orbits lie in the lens, just outside the bar, producing a very hot lens (Fig. 1).

2) Velocity dispersions are high but weakly anisotropic, consistent with the observations of Kormendy (1983) (Fig. 2).

3) SB may have fairly complicated flow patterns, eventually turbulent. Four types of flow have been found (Fig. 3, 4).

4) The proportion of semi-ergodic or stochastic orbits may be as large as 30 percent, but more probably below 10 percent of the mass. This feature is a key factor for the irreversible evolution of SB.

The original version of this chapter was revised.
An erratum to this chapter can be found at DOI 10.1007/BFb0030980

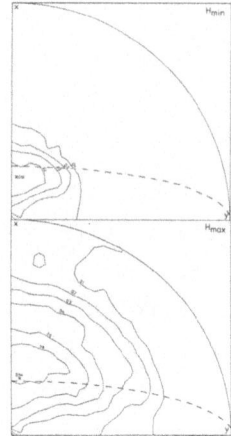

Fig. 1 : Relative local fraction
of mass of retrograde orbits for
the models with extreme values
of the total Hamiltonian. The
maximum always peaks along the
intermediate axis.

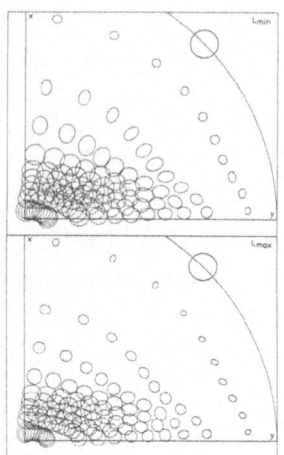

Fig. 2 : Velocity dispersion
fields of the models with extreme
values of the angular velocity
at corotation (radius of the
upper right circle).

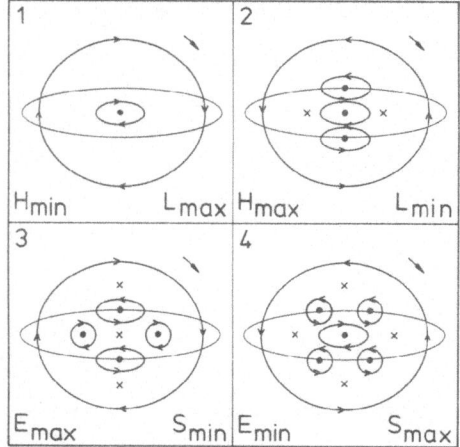

Fig. 3 : Schematic types of encountered flows in the rotating
frame of reference. The bar is
indicated by the longest ellipse.

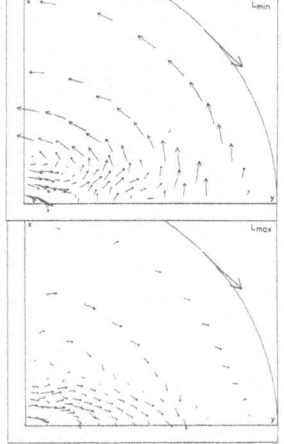

Fig. 4 : Velocity fields of the
models with extreme values of the
angular momentum, in the rotating
frame of reference, scaled with
the circular velocity at
corotation.

References
Kormendy, J. : 1983, Astrophys. J. 275, 529
Pfenniger, D. : 1984, Astron. Astrophys. 134, 373
Pfenniger, D. : 1984, Astron. Astrophys. (in press)
Schwarzschild, M. : 1979, Astrophys. J. 232, 236

# ABSORBING MATTER IN E AND SO GALAXIES

Patrick Poulain

Observatoire du Pic du Midi et de Toulouse

14, avenue Edouard Belin

31400 Toulouse, France

Elliptical galaxies have not been studied with the aim of determining their    mean internal absorption. One even thought that these objects were  dust-free  owing  to their weak gas content. But in recent years it  was   found   that  some  E  galaxies contain dust (Bertola et Galletta, 1978 ; Penston and Fosbury, 1978 ;   Kotanyi  and Ekers, 1979 ; Jorgensen and Norgaard-Nielsen, 1982).

If this absorbing matter is homogeneously distributed  throughout  the  galaxy,  it cannot be detected by image  analysis.  But  wide-band  photometry  in  a  suitable system gives means to determine a sufficiently accurate total colour excess  to  be able to obtain the intrinsic colour of the whole stellar population.

Therefore calculations were carried out, based on the UBVR measurements of  Sandage and Visvanathan (1978) and the JHK measurements of Persson et al. (1979). Using (V-R) and (J-K) indices which are practically independant  of  metal  content  I  have been able to calculate the colour excess and then infer the metallicity  rate  from the dereddened (U-V) and (B-V) indices.

With these results I have derived the empirical relationship between magnitude  and metallicity for Virgo and Coma  galaxies  (figure  1)  which  allows  to  determine absolute magnitudes from the distance modulus of these clusters. The  least  square method gives the solution.

$$V_{26} = - 8.10 \; \delta \; (U-V) + 14.71 \qquad\qquad (1)$$

Other relationships as dust density versus intrinsic  color  and  metallicity  rate versus equivalent width of the MgII line can be obtained also.

In addition, several galaxies with a large (V-R) excess turn out  to  bear  a  dust lane.

Nethertheless all these calculations rely upon measurements which are not homogeneous (various observers and various instruments) and this could explain the relative scattering. Therefore a project of obtaining UBVRI photometric observations of a large part of the Sandage and Visvanathan sample with the has been undertaken with the aim of improving the accuracy of the results.

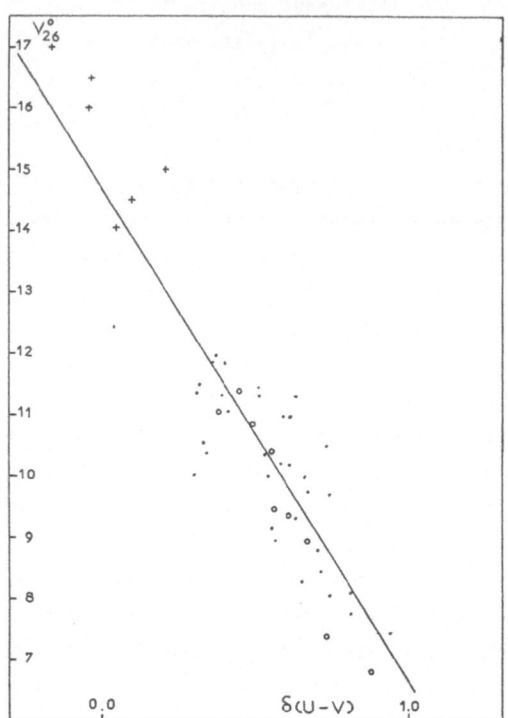

Fig. 1 : Absolute magnitude $V_{26}$ corrected from absorption versus metallicity rate. Dots are for Virgo galaxies, circles for Coma galaxies reduced to the distance of the Virgo cluster and crosses for Virgo dwarf galaxies.

References :

Bertola, F., and Galletta, G., 1978, Astrophys. J. **226**, L115

Jorgensen, H.E., and Norgaard-Nielsen, H.U., 1982, The Messenger **30**, 3.

Kotanyi, C.G., and Ekers, R.D., 1979, Astron. Astrophys. **73**, L1.

Penston, M.V., and Fosbury, R.A.E., 1978, Mon. Not. R. Astron. Soc. **183**, 479.

Persson, S.E., Frogel, J.A., and Aaronson, M., 1979, Astrophys. J. Supp. Series **39**, 61.

Sandage, A., and Visvanathan, N., 1978, Astrophys. J. **223**, 707.

# A PECULIARITY IN THE PHOTOMETRIC PROFILES
## OF SOME SO GALAXIES

F. Simien[1]   and   R. Michard[2]


[1]Observatoire de Lyon, F-69230 St Genis Laval
[2]Observatoire de Nice, B.P.139, F-06003 Nice Cedex

## I. THE DATA

We analyse the B-band surface photometry of four highly-inclined SO's, from a sample of 36 E/SO galaxies (Michard 1984).

## II. THE PHOTOMETRIC PROFILES

In the $r^{1/4}$ scale (Fig.1), the major-axis profiles show the upward convexity (bump above an $r^{1/4}$ distribution) usually interpreted as resulting from bulge and disk contributions (see, e.g., de Vaucouleurs 1977). On the other hand, the minor-axis profiles exhibit an upward concavity (dip below an $r^{1/4}$ distribution), a peculiarity which makes unsatisfactory a representation in terms of an $r^{1/4}$ bulge of constant flattening.

Fig.2 gives as an example a preliminary representation of the bulge of NGC 4638. The solid curves show a bulge model with constant axial ratio q=0.70, convolved with the Point Spread Function. Also shown is the minor axis of a similar bulge with q=0.53. A suitable bulge model, taking into account a realistic disk contribution, could be obtained only by relaxing one (an maybe both)of the following two conditions : 1) the bulge obeys the $r^{1/4}$ law, 2) q is constant. It seems that the latter condition implies a too high disk luminosity.

## III. CONCLUSION

A preliminary analysis suggests that : a) the flattening of many SO bulges is variable with r, and the amplitude of this variation depends on the disk/bulge luminosity ratio. b) the maximum bulge flattening is observed in the intermediate region, where the disk relative brightness is maximum.

We note that van der Kruit and Searle (1981) found evidence for a similar flattening effect in the outermost region of some spiral galaxy bulges. In a forthcoming paper, we will use a 2-D mapping of each galaxy in order to further study the characteristics of both bulge and disk. A confirmation of the present hypothesis would lead us to ask whether the observed effect is the reponse of the bulge to the potential of the disk (Freeman 1980). N-body simulations of galaxy formation should eventually settle the question (Barnes and White 1984); but a scenario which could account quantitatively for the flattening variation is yet to be investigated.

IV. REFERENCES

Barnes, J., and White, S.D.M. 1984, preprint.
de Vaucouleurs, G. 1977, in Evolution of Galaxies and Stellar
        Populations, ed. B.M. Tinsley and R.B. Larson, Yale Univ.
        Obs., New Haven.
Freeman, K.C. 1980, in Photometry Kinematics and Dynamics of Galaxies,
        ed. D.S. Evans, Univ. of Texas, Austin.
Michard, R. 1984, Astr. Ap. Suppl., in press.
van der Kruit, P.C., and Searle, L. 1981, Astr.Ap., 95, 105 and 116.

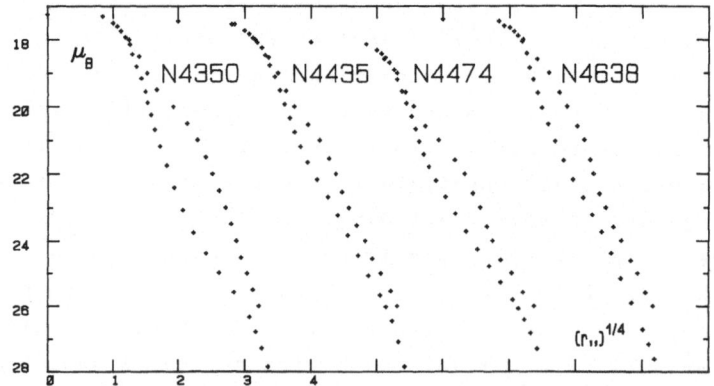

Fig.1   Major- and minor-axis profiles of four S0's

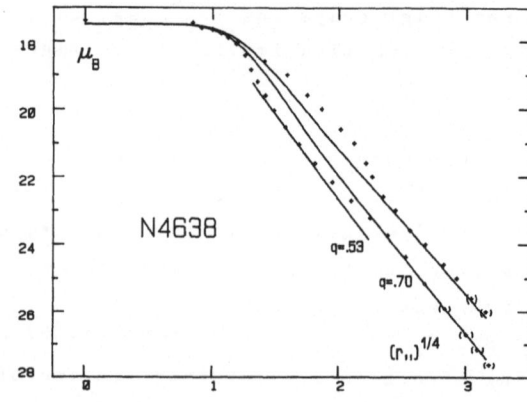

Fig.2   NGC 4638 and model profiles

# Erratum to: The Velocity Fields of Barred Galaxies

## D. Pfenniger

Erratum to: J-L. Nieto, *New Aspects of Galaxy Photometry,* DOI 10.1007/BFb0030977, © Springer-Verlag, 1985.

There was an error in the authors' name of this chapter, O. Pfenniger is incorrect. It was corrected to D. Pfenniger.

--------------------------------------------------------------------------------
The updated original version for this chapter can be found at
DOI 10.1007/BFb0030977

--------------------------------------------------------------------------------

# INDEX OF AUTHORS

# Lecture Notes in Physics

C. Hoffmeister, G. Richter, W. Wenzel

# Variable Stars

Translated from the German by S. Dunlop

1985. 170 figures, 64 tables. Approx. 350 pages
ISBN 3-540-13403-4

**Contents:** General Introduction. – Pulsating Variables. – Eruptive Variables. – Eclipsing Stars. – Supplement to the Classification. – The Discovery of Variable Stars. – The Significance of Variable Stars for Research on the Structure of the Galaxy and Stellar Evolution. – Oberservational Methods and Organizations. – Literature. – Subject Index.

M. Taube

# Evolution of Matter and Energy

## on a Cosmic and Planetary Scale

1985. 136 figures. Approx. 288 pages
ISBN 3-540-13399-2

**Contents:** Matter and Energy. The Interplay of Elementary Particles and Elementary Forces. – The Universe: How is it Observed Here and Now? Its Past and Possible Future. – The Origin and Nuclear Evolution of Matter. – Chemical Evolution and the Evolution of Life; The Cosmic Phenomena. – The Eternal Cycle of Matter on the Earth. – The Flow of Energy on the Earth. – The Biosphere: The Coupling of Matter and the Flow of Free Energy. – Is the Future Development of Mankind on this Planet Possible? – The Distant Future of Mankind; Terrestrial or Cosmic? – Bibliography. – Index.

Springer-Verlag
Berlin
Heidelberg
New York
Tokyo

# Selected Issues from
# Lecture Notes in Mathematics